Springer Series in Statistics

More information about this series at http://www.springer.com/series/692

Dominique Fourdrinier • William E. Strawderman
Martin T. Wells

Shrinkage Estimation

Springer

Dominique Fourdrinier
Mathématiques, BP 12
Université de Rouen
St-Étienne-du-Rouvray, France

William E. Strawderman
Department of Statistics
Rutgers University
Piscataway, NJ, USA

Martin T. Wells
Department of Statistical Science
Cornell University
Ithaca, NY, USA

ISSN 0172-7397 ISSN 2197-568X (electronic)
Springer Series in Statistics
ISBN 978-3-030-02184-9 ISBN 978-3-030-02185-6 (eBook)
https://doi.org/10.1007/978-3-030-02185-6

Library of Congress Control Number: 2018958646

This Springer imprint is published by the registered company Springer Nature Switzerland AG
The registered company address is: Gewerbestrasse 11, 6330 Cham, Switzerland

To our families

Loric, Cécile, and Armand

In memory of Susan
Rob, Myla, Bill, Jinny, Jim, Heather, Kay, Will, Tom, Evan, Emma, AJ, and Lily

Jill, Timothy, Elizabeth, and Peter

Preface

Starting in the 1930s, a mathematically rigorous approach to frequentist statistical inference, now called statistical decision theory, was introduced by Jerzy Neyman, Egon Pearson, and E.J.G. Pitman and formalized by Abraham Wald (Neyman and Pearson 1933, Pitman 1939, and Wald 1939, 1950). As far as estimation is concerned, statistical decision theory examines classes of functions of the data that can serve as possible estimators of the unknown parameter. These general estimators are compared through a risk function defined as the expected value over the sample space of the loss for every possible value of the parameters of interest. Alternatively, Bayesian decision theory examines estimators that are constructed by minimizing the expected loss, however now with respect to the parameter's posterior distribution. In general, statistical decision theory provides a rigorous foundation to formulate and solve decision problems under uncertainty.

In the case of the univariate normal distribution, the usual estimator of the population mean is the sample mean. The sample mean is the maximum likelihood estimator (MLE), the uniformly minimum variance unbiased estimator (UMVUE), the best invariant (or equivariant) and minimax estimator for nearly arbitrary symmetric loss, and is admissible for essentially arbitrary symmetric loss. Pitman (1939) suggested, on intuitive grounds, the use of best invariant procedures in certain problems of estimation concerning scale and location parameters. In the same year, Wald (1939) claimed admissibility for such best invariant estimators; unfortunately, as Peisakoff (1950) pointed out, there seemed to be a "lacuna" in Wald's proof. The possibility that some other estimator has a risk that is uniformly lower than that of the average existed. Hodges and Lehmann (1950) and Girshick and Savage (1951) first showed that no such estimator exists using the information inequality and then Blyth (1951) by using a limit of Bayes-type argument. That is, the usual sample mean is admissible, at least when estimating one unknown mean. In the bivariate normal case, the above properties also hold. Stein (1956) proved admissibility using an information inequality argument. In that same paper, Stein proved a result that astonished many and which motivates many of the ideas developed in this book. Stein's proof made it quite clear that admissibility should fail for any dimension greater than two.

In the p-variate normal case, Stein (1956) showed that estimators of the form $(1 - a/(b + \|X\|^2))X$ dominate the MLE/UMVUE, X, for a sufficiently small and b sufficiently large when $p \geq 3$. James and Stein (1961) sharpened the result and constructed an explicit class of dominating estimators, $(1 - a/\|X\|^2)X$ for $0 < a < 2(p - 2)$. Paradoxically the James-Stein estimator is itself inadmissible and can be dominated by another inadmissible estimate, its positive part. The James-Stein estimator can also be regarded as an "empirical Bayes rule," a term coined by Herbert Robbins. Ever since the pathbreaking results of Stein, the area of shrinkage estimation has flourished. The goal of this book is to provide a coherent framework for understanding this area. Our primary foci are on point and loss estimation for the mean vector for multivariate normal and spherically symmetric distributions. The coverage of topics reflects our personal perspective on shrinkage estimation, and we apologize for omissions. Nevertheless, we hope the material we present provides an adequate basis for the interested reader to pursue recent developments in shrinkage estimation. There are many open directions in the area with much more to be done.

Chapter 1 gives an overview of the statistical and decision theoretic terminology and results that will be used throughout the book. We assume that the reader is familiar with the basic statistical notions of parametric families of distributions, likelihood functions, maximum likelihood estimation, sufficiency, completeness, and unbiasedness. We review the results in Bayesian decision theory, minimaxity, admissibility, and invariance that will be used later in the book.

Chapter 2 is concerned with estimating the p-dimensional mean vector of a multivariate normal distribution under quadratic loss from a frequentist perspective. In Sect. 2.2, we give some intuition into why improvement over the MLE/UMVUE should be possible in higher dimensions and how much improvement might be expected. Section 2.3 is devoted to Stein's unbiased estimation of risk technique which provides the technical basis of many results in the area of multiparameter estimation. Section 2.4 is devoted to establishing improved procedures, such as the classical James-Stein estimator. In Sects. 2.5 and 2.6, we will provide a link between Stein's integration by parts lemma and Stokes' theorem and give insights into the Stein phenomenon in terms of nonlinear partial differential operators.

The Bayesian approach is well suited for the construction of possibly optimal estimators. The frequentist paradigm is complementary, as it is well suited for risk evaluations, but less well suited for estimator construction. In Chap. 3 we take a Bayesian view of shrinkage estimation. In Sect. 3.1 we derive a general sufficient condition for minimaxity of Bayes and generalized Bayes estimators in the known variance case; we also illustrate the theory with numerous examples. In Sect. 3.2 we extend the results of the previous section to the case when the variance is unknown. Section 3.3 considers the case of a known covariance matrix under and a general quadratic loss. The admissibility of Bayes estimators is discussed in Sect. 3.4. Interesting connections to maximum a posteriori (MAP) estimation, penalized likelihood methods, and shrinkage estimation are developed in Sect. 3.5. The fascinating connections to Stein estimation and estimation of a predictive density under Kullback-Leibler divergence are outlined in Sect. 3.6.

While Chaps. 2 and 3 consider estimation problems for the normal distribution setting, Chap. 4 introduces the general class of spherically symmetric distributions. Point estimation for this broad class is studied in subsequent chapters. In particular, Chap. 5 extends many of the results from Chaps. 2 and 3 to spherically symmetric distributions. Section 5.2 is devoted to a discussion of basic domination results for Baranchik-type estimators, while Sect. 5.3 examines more general estimators. Section 5.4 considers Bayes minimax estimation, and, finally, Sect. 5.5 discusses estimation with a concave loss.

In Chap. 6, we consider the general linear model with spherically symmetric error distributions when a residual vector is available. The inclusion of the residual term in estimates yields interesting and strong robustness properties. Section 6.1 gives the main results in this setting, and Sect. 6.2 discusses an interesting paradox concerning shrinkage estimators when the scale is known but when a residual vector is available. Section 6.3 extends some of the Bayes estimation results in Chap. 3 to spherically symmetric distributions when a residual vector is available. Section 6.4 develops a class of shrinkage estimators for a class of elliptically symmetric distributions. Section 6.5 studies improved estimation for concave loss when a residual vector is present.

Chapter 7 considers the problem of estimating a location vector which is constrained to lie in a convex subset of $\mathbb{R}^p$. Much of the chapter is devoted to one of two types of constraint sets, balls, and polyhedral cones. However, Sect. 7.2 is devoted to general convex constraint sets and more particularly to a striking result of Hartigan.

In Chap. 8 we switch gears away from location parameter estimation and focus on loss estimation and data-dependent evidence reports. In Sect. 8.2, we develop the quadratic loss estimation problem for a multivariate normal mean. Section 8.3 is devoted to the multivariate normal mean case where the variance is unknown. Extensions to spherically and elliptically symmetric distributions are given in Sects. 8.4. In Sect. 8.5 we use loss estimation ideas to develop a modern perspective on Akaike's information criterion (AIC), Mallows' C_p, and estimated degrees of freedom for model selection and propose generalizations to spherically and elliptically symmetric distributions. We conclude Chap. 8 by discussing confidence set assessment and the differential operators and dimension cut-off when estimating a loss in Sects. 8.6 and 8.7, respectively.

The text is intended for graduate students and researchers who want to learn about the theory underlying shrinkage estimation. The reader should have some exposure to graduate-level probability theory, mathematical statistics, and linear models. The necessary topics from analysis are developed in the Appendix.

This book project has a ridiculously lengthy history that likely dates back to summers during the late 1990s we spent together in Rouen. Given the duration of this project, there are many people who have substantially influenced the writing and rewriting of particular parts of this book. Their contributions are too diverse to specify, and their influence has been by means of their excellent contributions to the shrinkage estimation literature. In particular, Charles Stein had an immeasurable influence on our research; his work and behavior were an inspiration to many. We warmly thank our coauthors and colleagues Jim Berger, Jim Booth, Ann Brandwein, Anirban DasGupta, Persi Diaconis, Ed Green, Tatsuya Kubokawa, Éric Marchand, Yuzo Maruyama, Christian Robert, Andrew Ruhkin and Rob Strawderman, as well as our friends Sumanta Basu, Jacob Bien, Ed George, and Gene Hwang, and our late colleagues Larry Brown and George Casella, who are sorely missed. Also thanks to our outstanding students in courses at Cornell, Rouen and Rutgers, Ben Baer, Haim Bar, Didier Chételat, Irina Gaynanova, Daniel Gilbert, Fatiha Mezoued, Raj Narayanan, Galina Nogin, Ali Righi, Liz Schifano, and Stavros Zinonos. Thanks also to the National Science Foundation, National Institutes of Health, and Simons Foundation which have supported much of our research and writing.

Saint-Étienne-du-Rouvray, France — Dominique Fourdrinier
Piscataway, NJ, USA — William E. Strawderman
Ithaca, NY, USA — Martin T. Wells
November 2018

Contents

Chapter 1
Decision Theory Preliminaries

1.1 Introduction

In this chapter we give an overview of statistical and decision theoretic concepts and results that will be used throughout the book. We assume that the reader is familiar with the basic statistical notions of parametric families of distributions, likelihood functions, maximum likelihood estimation, sufficiency, completeness and unbiasedness at the level of, for example, Casella and Berger (2001), Shao (2003), or Bickel and Doksum (2001). In the following, we will discuss, often without proof, some results in Bayesian decision theory, minimaxity, admissibility, invariance, and general linear models.

1.2 The Multivariate Normal Distribution

For theoretical and practical reasons, the normal distribution plays a central role in statistics. The central limit theorem is one reason for its importance; given $X_1, \ldots, X_n$ independent and identically distributed (i.i.d.) random variables with mean μ and variances $\sigma^2 < \infty$, $(\bar{X}_n - \mu)/(\sigma/\sqrt{n})$ converges in distribution to the standard normal distribution, $\mathcal{N}(0, 1)$. Hence whatever the distribution of the X_i's, the distribution of the sample mean $\bar{X}_n$ can be approximated by a normal distribution with mean equal to μ and variance equal to σ^2/n. Essentially the same theorem has been used to provide theoretical justification for the empirical fact that many observed quantities tend to be approximately normally distributed.

In this section, we recall basic properties of the univariate and multivariate normal distributions. By definition, the univariate normal distribution $\mathcal{N}(\theta, \sigma^2)$ with mean $\theta \in \mathbb{R}$ and variance $\sigma^2 > 0$ has the density $(\sqrt{2\pi}\,\sigma)^{-1} \exp\{-(x - \theta)^2/(2\sigma^2)\}$ with respect to the Lebesgue measure in $\mathbb{R}^1$. For technical reasons, we also include the case where $\sigma^2 = 0$, which corresponds to the point mass at θ. In

D. Fourdrinier et al., *Shrinkage Estimation*, Springer Series in Statistics,
https://doi.org/10.1007/978-3-030-02185-6_1

this case, the distribution is singular and has no density with respect to the Lebesgue measure.

As the distribution of any random vector $X \in \mathbb{R}^n$ is characterized by the distribution of all linear functions of the form $a^{\mathrm{T}}X$ for $a \in \mathbb{R}^n$, the following multivariate extension is natural (see, e.g., Johnson and Kotz 1972).

Definition 1.1 A random vector $X \in \mathbb{R}^n$ has a normal distribution if, for all $a \in \mathbb{R}^n$, $a^{\mathrm{T}}X$ is distributed as an univariate normal distribution.

Note that the means and variances of the individual components exist by definition and so do the individual covariances, by the Cauchy-Schwarz inequality. Also, the characteristic function of a univariate standard normal $\mathcal{N}(0, 1)$ random variable X_j is given by $\varphi_{X_j}(t) = E[\exp\{i\, t\, X_j\}] = \exp\{-t^2/2\}$. Hence, if $X = (X_1, \ldots, X_n)$ where the X_j are i.i.d. standard normal, the characteristic function of X is equal to $\varphi_X(u) = E[\exp\{i\, u^{\mathrm{T}}X\}] = \exp\{-u^{\mathrm{T}}u/2\}$. Furthermore, if $Y = AX+\theta$ where A is a $p \times n$ matrix and θ is a $p \times 1$ vector, $\varphi_Y(v) = E[\exp\{i\, v^{\mathrm{T}}(\theta + AX)\}] = \exp\{i\, v^{\mathrm{T}}\theta\} \exp\{-v^{\mathrm{T}}\Sigma v/2\}$ where $\Sigma = A\, A^{\mathrm{T}}$ is the covariance matrix and θ, the mean vector of Y. This shows that the distribution of Y is determined by its mean vector θ and its covariance matrix Σ. Hence Definition 1.1 is not vacuous and the multivariate normal distribution exists for any mean vector θ and any positive semi-definite covariance matrix Σ (take $A = \Sigma^{1/2}$). We denote this distribution by $\mathcal{N}_p(\theta, \Sigma)$.

It follows from the form of the above characteristic function that, if X is distributed as $\mathcal{N}_n(\theta, \Sigma)$ and B is a $q \times n$ matrix and ν is a $q \times 1$ vector, then $Z = B\,X + \nu$ is distributed as $\mathcal{N}_q(\theta_Z, \Sigma_Z)$ where $\theta_Z = B\,\theta + \nu$ and $\Sigma_Z = B\,\Sigma\,B^{\mathrm{T}}$. Hence, in particular, all marginal distributions are normal. Specifically, decomposing

$$X = \begin{pmatrix} X_1 \\ X_2 \end{pmatrix} \sim \mathcal{N}_n\left(\begin{pmatrix} \theta_1 \\ \theta_2 \end{pmatrix}, \begin{pmatrix} \Sigma_{11} & \Sigma_{12} \\ \Sigma_{21} & \Sigma_{22} \end{pmatrix} \right)$$

with $\dim X_i = \dim \theta_i = n_i$ and where $\Sigma_{i,j}$ is $n_i \times n_j$ $(1 \le i, j \le 2)$, we have $X_i \sim \mathcal{N}_{n_i}(\theta_i, \Sigma_{ii})$. Note that X_1 is independent of X_2 if and only if $\Sigma_{12} = 0$.

We can find the conditional distribution of X_1 given X_2 as follows. Suppose there exists an $n_1 \times n_2$ matrix A such that $X_1 - A\,X_2$ is independent of X_2. Then the distribution of $X_1 - A\,X_2$ is normal with mean $\theta_1 - A\,\theta_2$ and covariance matrix $\Sigma_{11} - A\,\Sigma_{21}$. Hence the conditional distribution of X_1 given X_2 is normal with mean $\theta_1 + A\,(X_2 - \theta_2)$ and covariance matrix $\Sigma_{11} - A\,\Sigma_{21}$. However such an A is easy to find since $\mathrm{cov}(X_1 - A\,X_2, X_2) = \Sigma_{12} - A\,\Sigma_{22}$. If Σ_{22} is non-singular, $A = \Sigma_{12}\,\Sigma_{22}^{-1}$. If Σ_{22} is singular then $A = \Sigma_{12}\,\Sigma_{22}^{-}$, where Σ_{22}^{-} is a generalized inverse (see Muirhead 1982).

We now consider the existence of a density with respect to the Lebesgue measure on $\mathbb{R}^n$ for a random vector X distributed as $\mathcal{N}_n(\theta, \Sigma)$. Note that, when Σ is singular, there exists an $a \in \mathbb{R}^n$ such that $a \neq 0$ and $\Sigma a = 0$ and hence, for any such a,

$\mathrm{Var}(a^{\mathrm{T}}X) = a^{\mathrm{T}}\Sigma\, a = 0$. It follows that $a^{\mathrm{T}}X = a^{\mathrm{T}}\theta$ almost surely and hence that $X - \theta$ is almost surely in a proper subspace of $\mathbb{R}^n$; thus the distribution of X is singular and X has no density in $\mathbb{R}^n$.

If however Σ is nonsingular, then a density exists. To see this, let $\Sigma = A\, A^{\mathrm{T}}$ for some nonsingular $n \times n$ matrix and let $X = A\, Z + \theta$ where Z is a vector of i.i.d. standard normal random variables in $\mathbb{R}^1$, as in the comment after Definition 1.1. The standard change of variables formula gives the density of X as

$$\frac{1}{(2\pi)^n |\Sigma|^{1/2}} \exp\left\{-\frac{1}{2}(x-\theta)^{\mathrm{T}}\Sigma^{-1}(x-\theta)\right\}. \tag{1.1}$$

It is important to note that, when $\Sigma = \sigma^2 I_n$ and $\theta = 0$, the density (1.1) is a function of $\|x\|^2$. Consequently, for any orthogonal transformation h, the distribution of $Y = h(X)$ is the same as that of X. Many properties of the normal $\mathcal{N}_n(0, \sigma^2 I_n)$ follow from this invariance property and hold for other distributions similarly invariant. We formalize this in the following definition.

Definition 1.2 Let $\mathcal{O}$ be the group of orthogonal transformations on $\mathbb{R}^n$. A random vector $X \in \mathbb{R}^n$ (equivalently the distribution P of X) is orthogonally invariant if, for any $h \in \mathcal{O}$, the distribution of $Y = h(X)$ is the same as that of X.

In other words, for any bounded and continuous function f and for any $h \in \mathcal{O}$,

$$\int_{\mathbb{R}^n} f(h(x))\, dP(x) = \int_{\mathbb{R}^n} f(x)\, dP(x)\,. \tag{1.2}$$

It is worth noting that, if an orthogonally invariant random vector X has a density f with respect to the Lebesgue measure, it is of the form $f(x) = g(\|x\|^2)$ for a certain function g from $\mathbb{R}_+$ into $\mathbb{R}_+$ (see Theorem 4.2 in Chap. 4). In that case, we will denote $X \sim g(\|x\|^2)$. Also, if for some fixed $\theta \in \mathbb{R}^p$, $X - \theta$ is orthogonally invariant, we will write $X \sim g(\|x - \theta\|^2)$.

As a simple example of the use of this notion, note that, if X is orthogonally invariant and $P[X = 0] = 0$, then the unit vector, which lies on the unit sphere, $X/\|X\|$ is orthogonally invariant as well. We will see in Sect. 1.3 that there exists only one distribution orthogonally invariant on the unit sphere. It follows that, for any function φ from $\mathbb{R}^n$ into $\mathbb{R}^k$, the distribution of $\varphi\,(X/\|X\|)$ does not depend on the distribution of X as long as X is orthogonally invariant. One of the best known and most useful of such statistics is the Fisher (F-) statistic

$$\frac{\|\pi_1(X)\|^2/k_1}{\|\pi_2(X)\|^2/k_2}$$

where π_1 and π_2 are orthogonal projections from $\mathbb{R}^n$ onto orthogonal subspaces of dimension k_1 and k_2, respectively.

1.3 The Uniform Distribution on a Sphere

We already noticed the existence of an orthogonally invariant distribution on the unit sphere S of $\mathbb{R}^n$. A closely related alternative approach is through the uniform measure σ_R on the sphere S_R of radius R centered at 0 which can be defined for any Borel set Ω of S_R, as

$$\sigma_R(\Omega) = \frac{n}{R}\lambda(\{ru \in \mathbb{R}^n \mid 0 < r < R,\ u \in \Omega\}) \tag{1.3}$$

where λ is the Lebesgue measure on $\mathbb{R}^n$. Thus the measure of Ω is proportional to the Lebesgue measure of the cone spanned by Ω. The constant of proportionality n/R is standard and is chosen so that the total surface area of the sphere S_R agrees with the usual formulas relating $\sigma_R(S_R)$ to the volume of the ball B_R of radius R, that is, $\sigma_R(S_R) = n/R\ \lambda(B_R)$. For example, for $n = 2$, $\sigma_R(S_R) = 2/R\ \lambda(B_R) = 2\pi\, R$ or, for $n = 3$, $\sigma_R(S_R) = 3/R\ \lambda(B_R) = 4\pi\, R^2$. As a consequence $\sigma_R(S_R) = \sigma_1(S_1)\, R^{n-1}$.

The uniform distribution on S_R is naturally defined through σ_R.

Definition 1.3 The uniform distribution $\mathscr{U}_R$ on S_R is defined, for any Borel subset Ω of S_R, by

$$\mathscr{U}_R(\Omega) = \frac{\sigma_R(\Omega)}{\sigma_R(S_R)} = \frac{\sigma_R(\Omega)}{\sigma_1(S_1)\, R^{n-1}}. \tag{1.4}$$

The orthogonal invariance of $\mathscr{U}_R$ and σ_R follows immediately from the orthogonal invariance of the Lebesgue measure λ. The following lemma establishes a uniqueness property for $\mathscr{U}_R$.

Lemma 1.1 *The uniform distribution $\mathscr{U}_R$ on S_R is the unique orthogonally invariant distribution on S_R.*

Proof We follow the approach of Cellier and Fourdrinier (1990) which is adapted from the proof given by Philoche (1977) and relies on the uniqueness of the Haar measure on the group $\mathscr{O}$ of orthogonal transformations (as it is developed, for instance, by Nachbin 1965). More precisely, we use the fact that there exists a unique probability measure ν on $\mathscr{O}$ which is invariant under left and right translations, that is, which satisfies

$$\int_{\mathscr{O}} \phi(h^{-1} g)\, d\nu(g) = \int_{\mathscr{O}} \phi(g)\, d\nu(g)\,,$$

and

$$\int_{\mathscr{O}} \phi(g\, h^{-1})\, d\nu(g) = \int_{\mathscr{O}} \phi(g)\, d\nu(g)\,, \tag{1.5}$$

for any function ϕ defined on $\mathcal{O}$ and for any $h \in \mathcal{O}$. This measure is the so-called Haar measure on $\mathcal{O}$.

Clearly, it suffices to consider the case where $R = 1$, that is, for $S_1 = S$ and $\mathcal{U}_1 = \mathcal{U}$. Let $\mathcal{C}(S)$ be the set of real valued continuous functions on S. For any $f \in \mathcal{C}(S)$, for any $g \in \mathcal{O}$ and $x \in S$, define the functions $f_x(g)$ and $f_g(x)$ by

$$f_x(g) = f_g(x) = f(g^{-1}(x)) \,. \tag{1.6}$$

Let f be fixed in $\mathcal{C}(S)$. As the group $\mathcal{O}$ operates transitively on S, the integral $\int_{\mathcal{O}} f_x(g)\, d\nu(g)$ does not depend on $x \in S$. Indeed, for any $x \in S$ and any $y \in S$, there exists $h \in \mathcal{O}$ such that $x = h(y)$ so that, by (1.5) and (1.6),

$$\begin{aligned} \int_{\mathcal{O}} f_x(g)\, d\nu(g) &= \int_{\mathcal{O}} f_{h(y)}(g)\, d\nu(g) \\ &= \int_{\mathcal{O}} f_y(h^{-1} \circ g)\, d\nu(g) \\ &= \int_{\mathcal{O}} f_y(g)\, d\nu(g) \,. \end{aligned} \tag{1.7}$$

Similarly, for every orthogonally invariant distribution P on S, the integral $\int_{\mathcal{O}} f_g(x)\, dP(x)$ does not depend on $g \in \mathcal{O}$ since

$$\int_S f_g(x)\, dP(x) = \int_S f(g^{-1}(x))\, dP(x) = \int_S f(x)\, dP(x) \,, \tag{1.8}$$

according to (1.2). Then, by (1.7), (1.8) and Fubini's theorem,

$$\begin{aligned} \int_S f(x)\, dP(x) &= \int_{\mathcal{O}} \left(\int_S f(x)\, dP(x) \right) d\nu(g) \\ &= \int_{\mathcal{O}} \left(\int_S f_g(x)\, dP(x) \right) d\nu(g) \\ &= \int_S \left(\int_{\mathcal{O}} f_x(g)\, d\nu(g) \right) dP(x) \\ &= \int_{\mathcal{O}} f_x(g)\, d\nu(g) \,. \end{aligned}$$

Therefore, for any $f \in \mathcal{C}(S)$,

$$\int_S f(x)\, dP(x) = \int_S f(x)\, d\mathcal{U}(x) \,,$$

which implies that $P = \mathcal{U}$. □

The following result, mentioned in Sect. 1.2, is then immediate.

Lemma 1.2 *If* $X \in \mathbb{R}^n$ *is an orthogonally invariant random vector such that* $P[X = 0] = 0$ *then* $X/\|X\|$ *is distributed as* $\mathscr{U}_1$.

It is worth noting that σ_R (and hence $\mathscr{U}_R$) can be expressed through the usual parametrization in terms of n-dimensional spherical coordinates. Indeed let $V = (0, \pi)^{n-2} \times (0, 2\pi)$ and for $(t_1, \ldots, t_{n-1}) \in V$, $\varphi_R(t_1, \ldots, t_{n-1}) = (x_1, \ldots, x_n)$ with

$$\begin{aligned}
x_1 &= R \sin t_1 \sin t_2 \ldots \sin t_{n-2} \sin t_{n-1} \\
x_2 &= R \sin t_1 \sin t_2 \ldots \sin t_{n-2} \cos t_{n-1} \\
x_3 &= R \sin t_1 \sin t_2 \ldots \cos t_{n-2} \\
&\vdots \\
x_{n-1} &= R \sin t_1 \cos t_2 \\
x_n &= R \cos t_1 .
\end{aligned} \tag{1.9}$$

Note that φ_R maps V onto S_R, except for the set A of σ_R−measure 0, where $A = \{x = (x_1, \ldots, x_n) \in \mathbb{R}^n \mid x_1 = 0, x_2 \leq 0 \text{ and } \|x\| = R\}$.

Lemma 1.3 *For any Borel subset* Ω *of* S_R,

$$\sigma_R(\Omega) = R^{n-1} \int_{\varphi_R^{-1}(\Omega)} \sin^{n-2} t_1 \sin^{n-3} t_2 \ldots \sin t_{n-2}\, dt_1\, dt_2 \ldots dt_{n-1} . \tag{1.10}$$

Proof The usual n-dimensional spherical coordinates express x as $r\, \varphi_1(t_1, \ldots, t_{n-1})$ and the set on the right hand side of (1.3) can be written as $(0, R] \times \varphi_R^{-1}(\Omega)$. Hence, recalling that the Jacobian of the transformation in (1.9) is $r^{n-1} \sin^{n-2} t_1 \sin^{n-3} t_2 \ldots \sin t_{n-2}$ (see e.g. Muirhead 1982), we have

$$\begin{aligned}
\sigma_R(\Omega) &= \frac{n}{R} \lambda\left((0, R) \times \varphi_R^{-1}(\Omega)\right) \\
&= \frac{n}{R} \int_0^R r^{n-1} \int_{\varphi_R^{-1}(\Omega)} \sin^{n-2} t_1 \sin^{n-3} t_2 \ldots \sin t_{n-2}\, dt_1\, dt_2 \ldots dt_{n-1}\, dr \\
&= R^{n-1} \int_{\varphi_R^{-1}(\Omega)} \sin^{n-2} t_1 \sin^{n-3} t_2 \ldots \sin t_{n-2}\, dt_1\, dt_2 \ldots dt_{n-1} .
\end{aligned}$$

□

An immediate consequence is that, if X is distributed as $\mathscr{U}_R$, then the angles t_i are independent with density proportional to $\sin^{n-i-1} t_i$ on $(0, \pi)$ for $1 \leq i \leq n-2$ and

t_{n-1} is uniform on $(0, 2\pi)$. Note that $(\mathscr{U}_R)_{R>0}$ is a scale family of distributions in the sense that $\mathscr{U}_R(\Omega) = \mathscr{U}_1(\Omega/R)$ since, in (1.9) we have, $\varphi_R^{-1}(\Omega) = \varphi_1^{-1}(\Omega/R)$.

We will have occasion to use the following lemma which is just a re-expression in terms of σ_R of the usual formula for integration in n-dimensional spherical coordinates.

Lemma 1.4 *For any Lebesgue integrable function h, we have*

$$\int_{\mathbb{R}^n} h(x)\,dx = \int_0^\infty \int_{S_R} h(x)\,d\sigma_R(x)\,dR.$$

Proof Lemma 1.3 implies that

$$\int_{S_R} h(x)\; d\sigma_R(x)$$

$$= \int_V h(\varphi_R(t_1, \ldots, t_{n-1}))\; R^{n-1}\; \sin^{n-2} t_1 \ldots \sin t_{n-2}\; dt_1, \ldots, dt_{n-1} \qquad (1.11)$$

and the result follows. □

Corollary 1.1 *The area measure of the unit sphere in $\mathbb{R}^n$ is given by*

$$\sigma_1(S_1) = \frac{2\pi^{n/2}}{\Gamma(n/2)}.$$

Proof We apply Lemma 1.4 with $h(x) = (2\pi)^{-n/2} \exp\left\{-\|x\|^2/2\right\}$. Then

$$1 = \int_{\mathbb{R}^n} \frac{1}{(2\pi)^{n/2}} \exp\left\{-\frac{\|x\|^2}{2}\right\} dx = \int_0^\infty \frac{1}{(2\pi)^{n/2}} \exp\{-r^2/2\}\, \sigma_1(S_1)\, r^{n-1}\; dr$$

where we used the fact that $h(x)$ is a function of $\|x\|$ and that $\sigma_r(S_r) = \sigma_1(S_1)\, r^{n-1}$. Letting $t = r^2/2$ reduces the integral to a multiple of a gamma function. More precisely, we have

$$1 = \frac{\sigma_1(S_1)}{2\pi^{n/2}}\; \Gamma(n/2)$$

which is the desired result. □

It is worth noting that the normalizing constant $(2\pi)^{-n/2}$ of the normal density is usually obtained through Lemma 1.4 in dimension 2.

It is convenient to extend the notions of uniform measure and distribution on S_R to any sphere in $\mathbb{R}^n$.

Definition 1.4 For any $R > 0$ and for any $\theta \in \mathbb{R}^n$, let $S_{R,\theta} = \{x \in \mathbb{R}^n \mid \|x - \theta\| = R\}$ be the sphere of radius R and center θ. The uniform distribution $\mathcal{U}_{R,\theta}$ (respectively the uniform measure $\sigma_{R,\theta}$) on $S_{R,\theta}$ is the uniform distribution $\mathcal{U}_R$ (respectively the uniform measure σ_R) translated by θ, that is,

$$\mathcal{U}_{R,\theta}(\Omega) = \mathcal{U}_1\left(\frac{\Omega - \theta}{R}\right),$$

for any Borel set Ω of $S_{R,\theta}$. For completeness, we denote the point mass at θ as $\mathcal{U}_{0,\theta}$.

Note that the definition of $\mathcal{U}_{R,\theta}$ (and $\sigma_{R,\theta}$) can be extended to be a distribution (measure) on $\mathbb{R}^n$ by $\mathcal{U}_{R,\theta}(A) = \mathcal{U}_{R,\theta}(A \cap S_{R,\theta})$ for any Borel set A of $\mathbb{R}^n$.

Formula (1.10) is an example of what is sometimes called superficial (or natural) measure on an $n-1$ dimensional submanifold of $\mathbb{R}^n$. Briefly, let O be an open set in $\mathbb{R}^{n-1}$ and φ be a differentiable function mapping O into $\mathbb{R}^n$ with rank $n-1$. Let $g = \sqrt{\det(J^{\mathrm{T}}J)}$ where J is the $n \times (n-1)$ Jacobian matrix of φ. Then the superficial measure σ on $\varphi(O)$ is defined by

$$\sigma(\Omega) = \int_{\varphi^{-1}(\Omega)} g(t_1, \ldots, t_{n-1})\, dt_1 \ldots dt_{n-1}$$

for any Borel set Ω in $\varphi(O)$.

It is easy to check that, for the transformation given by (1.9), the function g is the integrand in the right hand side of (1.10). The superficial measure is connected in an essential way to Stokes' theorem which we will use extensively. There is more discussion in Sects. 2.5 and A.5. See also Stroock (1990).

1.4 Bayesian Decision Theory

In this section, we introduce loss functions, risk functions, and some results in Bayesian decision theory. Suppose $X \sim f_\theta(x)$ where $f_\theta(x)$ is a density with respect to a σ -finite measure μ on $\mathcal{X}$ a measurable subset of $\mathbb{R}^n$ ($\mathcal{X}$ is the sample space) and $\theta \in \Omega$ a measurable subset of $\mathbb{R}^p$ (Ω is the parameter space). We require that $f_\theta(x)$ be jointly measurable on $\mathcal{X} \times \Omega$.

In the problem of estimating a measurable function $g(\theta)$ from $\mathbb{R}^p$ into $g(\Omega) \subset \mathbb{R}^k$, an estimator is a measurable function $\delta(X)$ from $\mathbb{R}^n$ into $\mathcal{D} \subset \mathbb{R}^k$ ($\mathcal{D}$ is the decision space). Typically we would require $\mathcal{D} \subset g(\Omega)$ but, occasionally, it is more convenient to allow $\mathcal{D}$ to contain $g(\Omega)$.

The measure of closeness of an action $d \in \mathcal{D}$ to the "true value" of $g(\theta)$ is given by a (jointly measurable) loss function $L(\theta, d)$, where, for any $\theta \in \Omega$, $L(\theta, g(\theta)) = 0$ and, for any $d \in \mathcal{D}$, $L(\theta, d) \geq 0$. Hence there is no loss if the "correct decision"

$d = g(\theta)$ is made and is a nonnegative loss for whatever decision is made. A larger value of the loss corresponds to a worse decision.

A simple example for the case of $g(\Omega) \subset \mathbb{R}^1$ and $\mathscr{D} \subset \mathbb{R}^1$ is $L(\theta, d) = (d - g(\theta))^2$, the so called squared error loss. Another common choice is $L(\theta, d) = |d - g(\theta)|$ or, more generally, $L(\theta, d) = \rho(g(\theta), d)$ where $\rho(g(\theta), g(\theta)) = 0$ and $\rho(g(\theta), d)$ is monotone nondecreasing in d when $d \geq g(\theta)$, and monotone nonincreasing in d when $d \leq g(\theta)$, a so called bowl-shaped loss.

In higher dimensions, when $\mathscr{D} \subset \mathbb{R}^k$ and $\Omega \subset \mathbb{R}^k$, similar examples would be

$$L(\theta, d) = ||d - g(\theta)||^2 = \sum_{i=1}^{k} (d_i - g_i(\theta))^2$$

(the sum of squared errors loss or quadratic loss),

$$L(\theta, d) = \sum_{i=1}^{k} |d_i - g_i(\theta)|$$

(the sum of absolute errors loss) and

$$L(\theta, d) = (d - g(\theta))^{\mathrm{T}} Q (d - g(\theta)) ,$$

where Q is a positive semidefinite matrix (the weighted quadratic loss).

To help in the assessment of estimators (or, more generally, decision procedures), it is useful to introduce the risk function $\mathscr{R}(\theta, \delta) = E_\theta[L(\theta, \delta(X))]$. The risk function only depends on the estimator $\delta(\cdot)$ (and not just on its value, $\delta(x)$, at a particular observation, $X = x$) and, of course, on θ.

Frequentist decision theory is mainly concerned with the choice of estimators which, in some sense, make $\mathscr{R}(\theta, \delta)$ small. Bayesian decision theory, in particular, is largely focused on minimizing the average of $\mathscr{R}(\theta, \delta)$ with respect to some (positive) weight function (measure) π, referred to as the prior measure or prior distribution. It suffices for our purpose to suppose that the prior measure π is a finite measure on Ω and, without loss of generality, to assume it is a probability measure (i.e. $\pi(\Omega) = 1$).

Definition 1.5 (Bayes procedures) For any (measurable) function δ from $\mathscr{X}$ into $\mathscr{D}$ the Bayes risk of δ (with respect to π) is

$$\begin{aligned} r(\pi, \delta) &= \int_\Omega \mathscr{R}(\theta, \delta)\, d\pi(\theta) \\ &= \int_\Omega \left[\int_{\mathscr{X}} L(\theta, \delta(x))\, f_\theta(x)\, d\mu(x) \right] d\pi(\theta) . \end{aligned} \tag{1.12}$$

A (proper) Bayes procedure, $\delta_\pi(X)$, with respect to the (proper) prior π, is any estimator δ_π such that

$$r(\pi) = r(\pi, \delta_\pi) = \inf_\delta r(\pi, \delta) . \tag{1.13}$$

The quantity $r(\pi)$ is referred to as the Bayes risk of π or simply the Bayes risk.

In certain settings, it is not necessary to require that π be a finite measure but only to require that there exists a $\delta(X)$ such that (1.12) is finite. Note also that the joint measurability of $f_\theta(X)$, and also of $L(\theta, \delta(X))$, implies that the double integral in (1.12) makes sense.

It is helpful to define joint and marginal distributions as follows.

Definition 1.6

(1) The joint distribution of (X, θ) is

$$P[X \in A, \theta \in B] = \int_B \left[\int_A f_\theta(x) \, d\mu(x) \right] d\pi(\theta) . \tag{1.14}$$

(2) The marginal distribution of θ is the prior distribution $\pi(\cdot)$ since

$$P[\theta \in B] = \int_B \left[\int_{\mathscr{X}} f_\theta(x) \, d\mu(x) \right] d\pi(\theta) = \int_B d\pi(\theta) = \pi(B) . \tag{1.15}$$

(3) The marginal distribution of X is

$$\begin{aligned} M(A) &= P[X \in A] \\ &= \int_\Omega \left[\int_A f_\theta(x) \, d\mu(x) \right] d\pi(\theta) \\ &= \int_A \left[\int_\Omega f_\theta(x) \, d\pi(\theta) \right] d\mu(x) \quad \text{by Fubini's theorem} \\ &= \int_A m(x) \, d\mu(x) \end{aligned} \tag{1.16}$$

where

$$m(x) = \int_\Omega f_\theta(x) \, d\pi(\theta) .$$

Hence it follows that the marginal distribution of X is defined and is absolutely continuous with respect to μ, and has density m.

Definition 1.7 The posterior distribution of θ given x is defined such that (for $m(x) \neq 0$)

$$d\pi(\theta|x) = \frac{f_\theta(x)}{m(x)}\, d\pi(\theta)\,. \tag{1.17}$$

Note that the posterior distribution as defined in (1.17) is absolutely continuous with respect to the measure π, and hence, has density

$$\frac{f_\theta(x)}{m(x)}$$

with respect to π. It is well defined for all x such that $m(x) > 0$, and hence M-almost everywhere.

The above observations and (again) Fubini's theorem allow an immediate convenient re-expression of (1.12).

Lemma 1.5 *The Bayes risk in* (1.12) *may be expressed as*

$$\begin{aligned} r(\pi, \delta) &= \int_{\mathscr{X}} \left[\int_{\Omega} L(\theta, \delta(x))\, d\pi(\theta|x) \right] dM(x) \\ &= \int_{\mathscr{X}} \left[\int_{\Omega} L(\theta, \delta(x))\, d\pi(\theta|x) \right] m(x)\, d\mu(x). \end{aligned} \tag{1.18}$$

It follows that a Bayes estimate $\delta_\pi(x)$ may be calculated, for μ-almost every x, by minimizing the so-called posterior loss function or posterior expected loss of δ.

Lemma 1.6 *Suppose that there exists an estimator with finite Bayes risk and that, for M-almost every x, there exists a value $\delta_\pi(x)$ minimizing*

$$E[L(\theta, \delta(X))|x] = \int_{\Omega} L(\theta, \delta(x))\, \frac{f_\theta(x)}{m(x)}\, d\pi(\theta). \tag{1.19}$$

Then, provided it is a measurable function, $\delta_\pi(X)$ is a Bayes estimator and $E[L(\theta, \delta(X))|x]$ is said to be the posterior risk.

For details on the measurability aspects of Bayes estimators, see Brown and Purves (1973).

Corollary 1.2 *Under the assumptions of Lemma* 1.6,

(1) *if $L(\theta, d) = (d - g(\theta))^{\mathrm{T}} Q (d - g(\theta))$ where Q is positive (semi) definite, the Bayes estimator is given by $\delta_\pi(X) = E[g(\theta)|X]$ and*
(2) *if $L(\theta, d) = (d - g(\theta))^{\mathrm{T}} Q(\theta) (d - g(\theta))$ where $Q(\theta)$ is positive definite, the Bayes estimator is given by*

$$\delta_\pi(X) = (E[Q(\theta)|X])^{-1} E[Q(\theta)\, g(\theta)|X]\,.$$

Uniqueness of the Bayes estimator follows under the assumption of strict convexity of $L(\theta, d)$ in d, finiteness of the integrated risk of $\delta_\pi(X)$ and absolute continuity of μ with respect to the marginal distribution M of X (i.e. μ and M are mutually absolutely continuous).

It is often convenient to deal with prior measures π that are not finite. In such cases, there is typically no procedure $\delta(\cdot)$ for which (1.12) is finite. However, it is often the case that the posterior distribution given formally by (1.17) exists and is a finite measure that can be normalized to be a probability distribution. In such a case, an estimator $\delta_\pi(X)$ minimizing (1.19) is called a generalized Bayes (or formal Bayes) estimator.

Example 1.1 (Normal location families) Suppose $X \sim \mathcal{N}_p(\theta, \sigma^2 I_p)$ with σ^2 known and the prior measure π (not necessarily finite) satisfies

$$m(x) = \int_{\mathbb{R}^p} \left(\frac{1}{\sqrt{2\pi}\,\sigma}\right)^p \exp\left(-\frac{1}{2\sigma^2}||x-\theta||^2\right) d\pi(\theta) < \infty$$

for all $x \in \mathbb{R}^p$. Note that the marginal m is an analytic function since it can be expressed as

$$m(x) = \left(\frac{1}{\sqrt{2\pi}\,\sigma}\right)^p \exp\left(-\frac{1}{2\sigma^2}||x||^2\right) \int_{\mathbb{R}^p} \exp\left(-\frac{1}{2\sigma^2}||\theta||^2\right) \exp\left(-\frac{1}{\sigma^2}x^{\mathrm{T}}\theta\right) d\pi(\theta)\,,$$

which shows that it is proportional to the Laplace transform of a density with respect to π. Then, for a loss of the form $L(\theta, d) = (d - g(\theta))^{\mathrm{T}} Q (d - g(\theta))$, where Q is positive definite, the Bayes (or generalized Bayes) estimator and the posterior risk involve derivatives of m; more specifically, the gradient $\nabla m(x) = \left(\partial/\partial x_1 m(x), \cdots, \partial/\partial x_p m(x)\right)$ and the Laplacian $\Delta m(x) = \sum_{i=1}^p \partial^2/\partial x_i^2\, m(x)$.

Indeed we have

$$\begin{aligned}\delta_\pi(X) &= E[\theta|X] \\ &= X + \frac{\int_{\mathbb{R}^p}(\theta - X)\exp\left(-\frac{1}{2\sigma^2}||X-\theta||^2\right) d\pi(\theta)}{\int_{\mathbb{R}^p}\exp\left(-\frac{1}{2\sigma^2}||X-\theta||^2\right) d\pi(\theta)} \\ &= X + \sigma^2 \frac{\nabla m(X)}{m(X)}\,,\end{aligned} \tag{1.20}$$

where the interchange of integration and differentiation is justified by standard results for exponential families. See Brown (1986) and also Lemma A.4 in the Appendix. Expression (1.20) is due to Brown (1971) and is also useful in analyzing the risk properties of Bayes estimators. Similar expressions for spherically symmetric location families will be developed in Chaps. 5 and 6.

Consider now the posterior risk $E[\|\theta - \delta_\pi(X)\|^2 | x]$. According to (1.20), we have

$$E[\|\theta - \delta_\pi(X)\|^2 | x] = E\left[\left\|\theta - X - \sigma^2 \frac{\nabla m(X)}{m(X)}\right\|^2 \middle| x\right]$$

$$= E\left[\left\{\|\theta - X\|^2 + \sigma^4 \left\|\frac{\nabla m(X)}{m(X)}\right\|^2 - 2\sigma^2 (\theta - X)^{\mathrm{T}} \frac{\nabla m(X)}{m(X)}\right\} \middle| x\right].$$

Now

$$E\left[(\theta - X)^{\mathrm{T}} \frac{\nabla m(X)}{m(X)} \middle| x\right] = \sigma^2 \left\|\frac{\nabla m(x)}{m(x)}\right\|^2$$

since, by (1.20),

$$E\left[(\theta - X) \middle| x\right] = \sigma^2 \frac{\nabla m(x)}{m(x)}.$$

Hence

$$E[\|\theta - \delta_\pi(X)\|^2 | x] = E[\|\theta - X)\|^2 | x] - \sigma^4 \left\|\frac{\nabla m(X)}{m(X)}\right\|^2.$$

Also

$$E\left[\|\theta - X\|^2 \middle| x\right] = p\,\sigma^2 + \sigma^4 \frac{\Delta m(x)}{m(x)}$$

since, again by standard results for exponential families,

$$\frac{\Delta m(x)}{m(x)} = \frac{\Delta \int_{\mathbb{R}^p} \exp\left(-\frac{1}{2\sigma^2}||x - \theta||^2\right) d\pi(\theta)}{\int_{\mathbb{R}^p} \exp\left(-\frac{1}{2\sigma^2}||x - \theta||^2\right) d\pi(\theta)}$$

$$= \frac{\int_{\mathbb{R}^p} \Delta \exp\left(-\frac{1}{2\sigma^2}||x - \theta||^2\right) d\pi(\theta)}{\int_{\mathbb{R}^p} \exp\left(-\frac{1}{2\sigma^2}||x - \theta||^2\right) d\pi(\theta)}$$

$$= \frac{\int_{\mathbb{R}^p} \left(\frac{||x-\theta||^2}{\sigma^4} - \frac{p}{\sigma^2}\right) \exp\left(-\frac{1}{2\sigma^2}||x - \theta||^2\right) d\pi(\theta)}{\int_{\mathbb{R}^p} \exp\left(-\frac{1}{2\sigma^2}||x - \theta||^2\right) d\pi(\theta)}$$

$$= E\left[\frac{\|\theta - X\|^2}{\sigma^4} - \frac{p}{\sigma^2} \middle| x\right].$$

Therefore the posterior risk equals

$$E[\|\theta - \delta_\pi(X)\|^2 | \, x] = p\,\sigma^2 + \sigma^4 \left\{ \frac{\Delta m(x)}{m(x)} - \left\| \frac{\nabla m(x)}{m(x)} \right\|^2 \right\}. \tag{1.21}$$

Now suppose $\theta \sim \mathscr{N}_p(\nu, \tau^2 I_p)$ (i.e. π is a normal distribution with mean vector ν and covariance matrix $\tau^2 I_p$). Then the marginal $m(x)$ equals

$$\left(\frac{1}{\sqrt{2\pi}\,\sigma}\right)^p \left(\frac{1}{\sqrt{2\pi}\,\tau}\right)^p \int_{\mathbb{R}^p} \exp\left(-\frac{1}{2\sigma^2}||x-\theta||^2\right) \exp\left(-\frac{1}{2\tau^2}||\theta-\nu||^2\right) d\theta$$

$$= \left(\frac{1}{\sqrt{2\pi}\,\sqrt{\sigma^2+\tau^2}}\right)^p \exp\left(-\frac{1}{2\,(\sigma^2+\tau^2)}||x||^2\right)$$

since the convolution of $\mathscr{N}_p(0, \sigma^2 I_p)$ and $\mathscr{N}_p(\nu, \tau^2 I_p)$ is $\mathscr{N}_p(\nu, (\sigma^2+\tau^2)\, I_p)$. Hence the Bayes estimator is

$$\begin{aligned} \delta_\pi(X) &= X + \frac{\sigma^2\,(-(X-\nu))}{\sigma^2+\tau^2} \\ &= \frac{\tau^2}{\sigma^2+\tau^2}\, X + \frac{\sigma^2}{\sigma^2+\tau^2}\, \nu \\ &= \nu + \frac{\tau^2}{\sigma^2+\tau^2}\,(X-\nu) \qquad (1.22) \\ &= \nu + \left(1 - \frac{\sigma^2}{\sigma^2+\tau^2}\right)(X-\nu)\,. \end{aligned}$$

If the generalized prior distribution π is the Lebesgue measure ($d\pi(\theta) = d\theta$), then $m(X) \equiv 1$ and the generalized Bayes estimator is given by

$$\delta_\pi(X) = X + \sigma^2 \frac{\nabla 1}{1} = X\,.$$

It is often convenient, both theoretically and for computational reasons, to express (proper and generalized) prior distributions hierarchically, typically in two or three stages. The first stage of the hierarchy is often a conjugate prior, i.e. one such that the posterior distribution is in the same class as the prior distribution. In Example 1.1, the class of $\theta \sim \mathscr{N}_p(\nu, \tau^2 I_p)$ priors is a conjugate family since the posterior is given by

$$\theta | x \sim \mathscr{N}_p\left(\frac{\tau^2\, x + \sigma^2\, \nu}{\sigma^2+\tau^2}, \frac{\sigma^2\,\tau^2}{\sigma^2+\tau^2}\, I_p\right).$$

At the second stage, one could put a prior (or generalized prior) distribution on the first stage prior variance τ^2. A convenient way to do this in certain settings (see, for example, Chap. 4 where this device is used to produce improved shrinkage estimators for the normal model) is as follows.

Suppose the first stage prior variance τ^2 is expressed as $\tau^2 = \sigma^2 (1-\lambda)/\lambda$ for $0 < \lambda < 1$. Then $\sigma^2 + \tau^2 = \sigma^2/\lambda$ and

$$\frac{\tau^2 x + \sigma^2 \nu}{\sigma^2 + \tau^2} = (1-\lambda)\, x + \lambda\, \nu = \nu + (1-\lambda)\,(x-\nu)\,.$$

Hence a second stage prior $H(\lambda)$ with prior density $h(\lambda)$ for $0 < \lambda < 1$ (hierarchical, generalized, or proper) leads to the marginal density

$$m(x) = \int_0^1 \left(\frac{\lambda}{2\pi\sigma^2}\right)^{p/2} \exp\left(-\frac{\lambda}{2\sigma^2}||x-\nu||^2\right) h(\lambda)\, d\lambda$$

and the Bayes estimator

$$\begin{aligned}
\delta_\pi(X) &= X + \sigma^2 \frac{\nabla m(X)}{m(X)} \\
&= X - \frac{\int_0^1 \lambda^{p/2+1} \exp\left(-\frac{\lambda}{2\sigma^2}||X-\nu||^2\right) h(\lambda)\, d\lambda}{\int_0^1 \lambda^{p/2} \exp\left(-\frac{\lambda}{2\sigma^2}||X-\nu||^2\right) h(\lambda)\, d\lambda}(X-\nu) \\
&= \nu + E[(1-\lambda)|X]\,(X-\nu)\,.
\end{aligned}$$

Empirical Bayes estimators are closely related to hierarchical Bayes estimators. If the first stage prior $\pi(\theta|\tau)$ is viewed as specifying a class of priors indexed by a parameter τ, then the first stage marginal

$$m(x|\tau) = \int f_\theta(x)\, d\pi(\theta|\tau)$$

may be viewed as a likelihood depending on the data x and the parameter τ. One may choose to estimate the parameter τ in a classical frequentist way such as a maximum likelihood estimator (MLE) or perhaps a UMVU estimator, and then calculate a Bayes estimator by the first stage Bayesian model substituting the estimated λ. Such estimators are called empirical Bayes estimators.

For example, in the above normal model, the first stage marginal distribution (parametrized by τ^2 with ν fixed and known) is

$$X|\tau^2 \sim \mathcal{N}_p(\nu, (\sigma^2 + \tau^2) I_p)\,.$$

Since ν is fixed and known, $||X-\nu||^2$ is a complete sufficient statistic and the MLE of τ^2 is $\hat{\tau}^2 = \max(0, ||X-\nu||^2/p - \sigma^2)$, giving an empirical Bayes estimate of θ (based on (1.22))

$$\begin{aligned}\delta^{EB}(X) &= \nu + \frac{\hat{\tau}^2}{\sigma^2 + \hat{\tau}^2}(X-\nu) \\ &= \nu + \left(1 - \frac{p\sigma^2}{||X-\nu||^2}\right)_+ (X-\nu)\,.\end{aligned}$$

Alternatively, the UMVU estimator of $1/(\sigma^2+\tau^2)$ is $1/\widehat{(\sigma^2+\tau^2)} = (p-2)/||X-\nu||^2$, so a different empirical Bayes estimator based on (1.22) would be

$$\nu + \left(1 - \frac{(p-2)\,\sigma^2}{||X-\nu||^2}\right)(X-\nu).$$

The first of these is a version of the James-Stein positive-part estimator while the second is the classical James-Stein estimator. The risk properties of these estimators are examined in Chap. 2, when the distribution of X is normal, and in Chap. 5, when the distribution of X is spherically symmetric.

1.5 Minimaxity

In the development of Bayes estimators, the risk function was integrated with respect to a prior. Minimax estimation takes another approach and does not depend on a prior.

Definition 1.8 An estimator $\delta_0(X)$ is minimax if

$$\sup_{\theta\in\Omega} \mathscr{R}(\theta,\delta_0) = \inf_{\delta\in\mathscr{D}} \sup_{\theta\in\Omega} \mathscr{R}(\theta,\delta)\,,$$

where $\mathscr{D}$ is the class of all estimators.

It is occasionally useful to take $\mathscr{D}$ to be a subset of the class of all estimators (for example, all linear estimators) in which case δ_0 would be said to be minimax in $\mathscr{D}$.

We give two results which have proved useful for finding minimax estimators (see Lehmann and Casella (1998) for proofs).

Lemma 1.7 *If a proper prior π has an associated Bayes estimator $\delta_\pi(X)$ and if $\sup_{\theta\in\Omega}\mathscr{R}(\theta,\delta_\pi) = r(\pi,\delta_\pi)(= r(\pi))$, then $\delta_\pi(X)$ is minimax. The prior π is also least favorable in the sense that $r(\pi',\delta_\pi) \le r(\pi,\delta_\pi)$ for all prior distributions π'.*

One easy and useful corollary of this result is that a Bayes estimator with constant (finite) risk is minimax. The second result is more useful in the case where the parameter space is noncompact.

Lemma 1.8 *If $\delta_0(X)$ is an estimator such that $\sup_{\theta \in \Omega} \mathscr{R}(\theta, \delta_0) = r$ and if there exists a sequence of priors (π_n) such that $\lim_{n\to\infty} r(\pi_n, \delta_{\pi_n}) = r$ then $\delta_0(X)$ is minimax. The sequence of priors (π_n) is what is known as a least favorable sequence in the sense that, for any prior π, we have $r(\pi) \leq r$.*

This second result is useful for establishing minimaxity of the usual estimator X in the normal location problem.

Example 1.2 (Minimaxity of X for $X \sim \mathscr{N}_p(\theta, \sigma^2 I)$, σ^2 known) Let $X \sim \mathscr{N}_p(\theta, \sigma^2 I_p)$ with σ^2 known and loss equal to $L(\theta, d) = ||d - \theta||^2$. Suppose the sequence of priors, (π_n), on θ is $\mathscr{N}_p(0, n I_p)$. Then the posterior distribution is $\mathscr{N}_p(n/(n+\sigma^2)\, X, n\,\sigma^2/(n+\sigma^2)\, I_p)$ and the posterior risk is $n\,\sigma^2/(n+\sigma^2)\, p$ which is also the Bayes risk. Since $r(\pi_n) = [n\,\sigma^2/(n+\sigma^2)\, p] \to p\,\sigma^2 \equiv \mathscr{R}(\theta, X)$, it follows that X is minimax.

Example 1.3 (Minimaxity of X for $X \sim f(||X - \theta||^2)$) Similarly, if $X \sim f(||X - \theta||^2)$ where $E[||X - \theta||^2] = p\,\sigma^2 < \infty$, then the sequence of priors $\pi_n(\theta) = f^{*n}(\theta)$, where f^{*n} is the n-fold convolution of f with itself, leads to a proof that X is minimax. To see this, note that, if $U_1, \ldots, U_n$ are i.i.d. copies of $U_0 \sim f(||u||^2)$, then $\theta = \sum_{i=1}^n U_i \sim f^{*n}(||\theta||^2)$. Also $U_0 = X - \theta \sim f(||u_0||^2)$ and is independent of $\theta = \sum_{i=1}^n U_i$ and $X = (X - \theta) + \theta = \sum_{i=0}^n U_i$. It follows that the Bayes estimator corresponding to π_n may be represented as

$$\begin{aligned}
\delta_{\pi_n}(X) &= E[\theta | X] \\
&= E\left[\sum_{i=1}^n U_i \,\middle|\, \sum_{i=0}^n U_i\right] \\
&= n\, E\left[U_1 \,\middle|\, \sum_{i=0}^n U_i\right] \\
&= \frac{n}{n+1}\, E\left[\sum_{i=0}^n U_i \,\middle|\, \sum_{i=0}^n U_i\right] \\
&= \frac{n}{n+1}\, X\,.
\end{aligned}$$

The corresponding Bayes risk is

$$\begin{aligned}
E^{\theta}[E^{X|\theta}[||\delta_{\pi_n}(X) - \theta||^2]] &= E^{\theta}\left[E^{X|\theta}\left[\left|\left|\frac{n}{n+1} X - \theta\right|\right|^2\right]\right] \\
&= E^{\theta}\left[p\left(\frac{n}{n+1}\right)^2 \sigma^2 + \sum_{i=1}^p \left(\frac{1}{n+1}\theta_i\right)^2\right]
\end{aligned}$$

$$= E^{\theta}\left[p\left(\frac{n}{n+1}\right)^2\sigma^2 + \left(\frac{1}{n+1}\right)^2 ||\theta||^2\right]$$
$$= p\left(\frac{n}{n+1}\right)^2\sigma^2 + \left(\frac{1}{n+1}\right)^2 p\,n\,\sigma^2$$
$$\to p\,\sigma^2 = E[||X-\theta||^2]\,.$$

Hence X is minimax and (π_n) is a least favorable sequence.

Example 1.4 (Minimaxity of X in the unknown σ^2 case) In this example, we assume $(X, U) \sim \mathcal{N}_{p+k}((\theta, 0)^{\mathrm{T}}, \sigma^2 I)$ when $\dim X = \dim \theta = p$ and $\dim U = \dim 0 = k$. Suppose the loss is $||\delta - \theta||^2/\sigma^2$. We need the following easy result (see Lehmann and Casella 1998).

Lemma 1.9 *Suppose $\delta(X)$ is minimax in a problem for $X \sim f$ with $f \in \mathscr{F}_0$. Suppose $\mathscr{F}_0 \subset \mathscr{F}_1$ and $\sup_{f\in\mathscr{F}_0} \mathscr{R}(f, \delta) = \sup_{f\in\mathscr{F}_1} \mathscr{R}(f, \delta)$. Then $\delta(X)$ is minimax for $f \in \mathscr{F}_1$.*

The argument of Example 1.2 suffices to show that X is minimax for any fixed σ^2. Since the risk of X is constant and equal to p for the entire family for the scale invariant loss $||\delta - \theta||^2/\sigma^2$, it follows that X is minimax in the unknown scale case.

1.6 Admissibility

An admissible estimator is one which cannot be uniformly improved upon in terms of risk. An inadmissible estimator is one for which an improved estimator can be found. More formally we have the following definition.

Definition 1.9

(1) $\delta(X)$ is inadmissible if there exists an estimator $\delta'(X)$ for which $\mathscr{R}(\theta, \delta') \le \mathscr{R}(\theta, \delta)$ for all $\theta \in \Omega$, with strict inequality for some θ.
(2) $\delta(X)$ is admissible if it is not inadmissible.

The most direct method to prove that an estimator is inadmissible is to find a better one. Much of this book is concerned with exactly this process of finding and developing improved estimators, typically by combining information from all coordinates. Hence, in a certain sense, we are more concerned with inadmissibility issues than with admissibility.

Proving admissibility can often be difficult but there are a few basic techniques that can sometimes be applied with reasonable ease. The most basic is the following.

Lemma 1.10 *A unique (proper) Bayes estimator is admissible. (Here uniqueness is meant in the sense of probability 1 for all f_θ, $\theta \in \Omega$).*

Typically, a minimax estimator in a location parameter problem is not proper Bayes so Lemma 1.10 will be of little help in this setting.

A general sufficient condition for admissibility of generalized Bayes estimators has been given by Blyth (1951). To apply Blyth's method for a target generalized prior density $g(\theta)$, select an increasing sequence of proper prior densities approaching g, $g_1 \leq g_2 \leq \cdots \leq g$. Each g_n is not necessarily normalized, so it just satisfies $\int_\Omega g_n(\theta)d\theta < \infty$ for any fixed n. Let δ_g and δ_{gn} be the generalized Bayes estimator with respect to $g(\theta)$ and the proper Bayes estimator with respect to $g_n(\theta)$, respectively. The non-standardized Bayes risk difference between δ_g and δ_{g_n} with respect to $g_n(\theta)$ is given by

$$\Delta_n = \int_\Omega \left[\mathscr{R}(\theta, \delta_g) - \mathscr{R}(\theta, \delta_{g_n})\right] g_n(\theta)d\theta. \tag{1.23}$$

Blyth (1951) showed under certain conditions that if $\Delta_n \to 0$ as $n \to \infty$, δ_g is admissible. The following version of Blyth's methods is from Brown and Hwang (1982).

Theorem 1.1 *Suppose that the parameter space Ω is open, and the estimators with continous risk functions form a complete class. Suppose X has a density $f(\|x-\theta\|)$, where $f(\cdot)$ is strictly positive. Assume that there is an increasing sequence $(g_n)_{n\geq 1}$ of proper densities such that $\int_{\|\theta\|\leq 1} g_1(\theta)d\theta > c$ for some positive c and that $\Delta_n \to 0$ as $n \to \infty$. Then δ_g is admissible under quadratic loss $L(\theta, d) = \|d - \theta\|^2$.*

Proof Suppose that δ_g is inadmissible and let δ' be such that $\mathscr{R}(\theta, \delta') \leq \mathscr{R}(\theta, \delta_g)$ for all θ with strict inequality for some θ. Let $\delta'' = (\delta_g + \delta')/2$. Then, using Jensen's inequality,

$$\begin{aligned}
\mathscr{R}(\theta, \delta'') &= \int \|\delta''(x) - \theta\|^2 f(\|x - \theta\|)dx \\
&< \frac{1}{2}\left(\int_{\mathscr{X}} \|\delta_g(x) - \theta\|^2 f(\|x - \theta\|)dx + \int_{\mathscr{X}} \|\delta'(x) - \theta\|^2 f(\|x - \theta\|)dx\right) \\
&= \frac{1}{2}\left[\mathscr{R}(\theta, \delta') + \mathscr{R}(\theta, \delta_g)\right] \\
&\leq \mathscr{R}(\theta, \delta_g),
\end{aligned}$$

for any θ. Since $\mathscr{R}(\theta, \delta'')$ and $\mathscr{R}(\theta, \delta_g)$ are both continuous functions of θ, there exists an $\epsilon > 0$ such that $\mathscr{R}(\theta, \delta'') < \mathscr{R}(\theta, \delta_g) - \epsilon$ for $\|\theta\| \leq 1$. Then

$$\begin{aligned}
\Delta_n &= \int_\Omega \left[\mathscr{R}(\theta, \delta_g) - \mathscr{R}(\theta, \delta_{g_n})\right] g_n(\theta)d\theta \\
&\geq \int_\Omega [\mathscr{R}(\theta, \delta_g) - \mathscr{R}(\theta, \delta'')]g_n(\theta)d\theta \\
&\geq \int_{\|\theta\|\leq 1} [\mathscr{R}(\theta, \delta_g) - \mathscr{R}(\theta, \delta'')]g_1(\theta)d\theta
\end{aligned}$$

$$\geq \epsilon c$$
$$> 0, \tag{1.24}$$

which contradicts $\Delta_n \to 0$. □

A good choice of the sequence of proper priors approaching the target prior is critical for proving the admissibility of a generalized Bayes estimator. For example, when $p = 1$ under normality and spherical symmetry, Blyth (1951) showed that the most natural estimator X, which is generalized Bayes with respect to $g(\theta) = 1$, is admissible by using a sequence of conjugate priors.

However, even in 2 dimensions, a sequence of normal priors with covariance equal to multiples of the identity, fails to show admissibility of X and the technique completely fails in 3 and higher dimensions. An alternative sequence for 2 dimensions was found by James and Stein (1961) to demonstrate admissibility. Under spherically symmetry, James and Stein (1961) showed for $p = 2$ that $g_n(\theta) = h_n^2(\theta)$ works where

$$h_n(\theta) = \begin{cases} 1 & \|\theta\| \leq 1 \\ 1 - \frac{\log \|\theta\|}{\log n} & 1 \leq \|\theta\| \leq n/2 \\ \frac{\alpha(n, \|\theta\|)}{\|\theta\| \{\log \|\theta\|\}} & \|\theta\| > n/2 \end{cases}$$

and $\alpha(n, \|\theta\|)$ is chosen so that, for fixed θ, $\alpha(n, \|\theta\|) \|\theta\|^{-1} \{\log \|\theta\|\}^{-1} \to 1$ as $n \to \infty$ and $h_1 \leq h_2 \leq \cdots \leq 1$. On the other hand, Stein (1956) showed that when $p \geq 3$ the standard estimator X is inadmissible under normality (and more generally under the condition that the fourth moment exists). Brown (1966) showed the dimension cutoff of $p = 3$ for inadmissibility of the best equivariant estimator ($\delta(X) = X$ in the spherically symmetric case) was quite general.

Brown (1971) gave very general conditions on the generalized prior which gives admissibility under quadratic loss for normal families and which resolves most admissibility issues in the multivariate normal case (with σ^2 known). Here is a version of Brown's result.

Theorem 1.2 *Let $X \sim \mathcal{N}_p(\theta, \sigma^2 I)$. Suppose π is a generalized prior distribution and loss is $L(\theta, d) = ||d - \theta||^2$. Define, for $||x|| = r$,*

$$\bar{m}(r) = \int m(x)\, d\mathcal{U}_r(x)$$

and

$$\underline{m}(r) = \int (1/m(x))\, d\mathcal{U}_r(x)$$

where $\mathcal{U}_r$ is the uniform distribution on the sphere of radius r and $m(x)$ is the marginal distribution.

(1) *(Admissibility) If* $||\delta_\pi(x) - x||$ *is uniformly bounded and*

$$\int_c^\infty (r^{p-1}\,\bar{m}(r))^{-1}\,dr = \infty$$

for some $c > 0$*, then* $\delta_\pi(X)$ *is admissible.*

(2) *(Inadmissibility) If*

$$\int_c^\infty r^{1-p}\,\underline{m}(r)\,dr < \infty$$

for some $c > 0$*, then* $\delta_\pi(X)$ *is inadmissible.*

Although Brown (1971) is quite general, the proof of the deep result depends on complex analytical concepts, solutions of the exterior Dirichlet problem for a class of elliptic boundary value problems, and may be difficult to apply. Brown's result involves a continous time Markov process on the sample space. In contrast, Eaton et al. (1992) develops an approach to admissibility which involves a discrete time Markov chain on the parameter space.

In the setting of estimating the natural mean vector of an exponential family under a quadratic loss function, Brown and Hwang (1982) gave a sufficient condition for generalized Bayes estimators to be admissible when the generalized prior density $g(\theta)$ is differentiable. Here is a version of their result for the normal case.

Theorem 1.3 *Let* $X \sim \mathcal{N}_p(\theta, \sigma^2 I_p)$ *and* $L(\theta, d) = ||d - \theta||^2$*. Let* δ_g *be the generalized Bayes estimator which is given by*

$$\delta_g(X) = X + \sigma^2 \frac{\nabla m_g(X)}{m_g(X)} = X + \sigma^2 \nabla \log m_g(X)\,,$$

where m_g *is the marginal distribution under the prior density* g*. Assume that* g *satisfies*

(1) $\int_{\{\theta:\|\theta\|>1\}} \frac{g(\theta)}{\|\theta\|^2 \max\{\log\|\theta\|,\, \log 2\}^2} < \infty,$

(2) $\int_{\mathbb{R}^p} \frac{\|\nabla g(\theta)\|^2}{g(\theta)} d\theta < \infty$*, and*

(3) $\sup\{R(\theta, \delta_g) : \theta \in K\} < \infty$ *for all compact sets* K*.*

Then $\delta_g(X)$ *is admissible.*

Without loss of generality, we assume that $\sigma^2 = 1$. The proof uses the following notation and results:

(i) for a measurable function $\psi(\cdot)$ and for $x \in \mathbb{R}^p$,

$$m(\psi|x) = \left(\frac{1}{\sqrt{2\pi}}\right)^p \int_{\mathbb{R}^p} \psi(\theta)\, \exp\!\left(-\frac{1}{2}\|x - \theta\|^2\right) d\theta\,,$$

(ii) $\int_{\mathbb{R}^p} m(\psi|x)\,dx = \int_{\mathbb{R}^p} \psi(\theta)\,d\theta$, and
(iii) if $\psi(\cdot)$ is (weakly) differentiable, $\nabla_x m(\psi|x) = m(\nabla_\theta \psi|x)$.

Note that $m(\psi|x) = E_x[\psi(\theta)]$ where $\theta \sim \mathcal{N}_p(x, I_p)$. In an abuse of notation, we occasionally use $m(\theta\,\psi|x) = E_x[\theta\,\psi(\theta)]$.

Result (ii) follows from Fubini's theorem. Result (iii) follows from the fact that, for any $1 \le i \le p$, denoting $\theta_{-i} = (\theta_1, \ldots, \theta_{i-1}, \theta_{i+1}, \ldots, \theta_p)$ and $x_{-i} = (x_1, \ldots, x_{i-1}, x_{i+1}, \ldots, x_p)$,

$$\begin{aligned}
\frac{\partial}{\partial x_i} m(\psi|x) &= \int_{\mathbb{R}^p} \psi(\theta)\,(\theta_i - x_i)\left(\frac{1}{\sqrt{2\pi}}\right)^p \exp\!\left(-\frac{1}{2}\|x-\theta\|^2\right) d\theta \\
&= \int_{\mathbb{R}^{p-1}} \int_{\mathbb{R}} \psi(\theta)\,(\theta_i - x_i)\,\frac{1}{\sqrt{2\pi}} \exp\!\left(-\frac{1}{2}(x_i-\theta_i)^2\right) d\theta_i \\
&\qquad\qquad \left(\frac{1}{\sqrt{2\pi}}\right)^{p-1} \exp\!\left(-\frac{1}{2}\|x_{-i}-\theta_{-i}\|^2\right) d\theta_{-i} \\
&= \int_{\mathbb{R}^{p-1}} \int_{\mathbb{R}} \frac{\partial}{\partial \theta_i}\psi(\theta)\,\frac{1}{\sqrt{2\pi}} \exp\!\left(-\frac{1}{2}(x_i-\theta_i)^2\right) d\theta_i \\
&\qquad\qquad \left(\frac{1}{\sqrt{2\pi}}\right)^{p-1} \exp\!\left(-\frac{1}{2}\|x_{-i}-\theta_{-i}\|^2\right) d\theta_{-i} \\
&= m\left(\frac{\partial}{\partial \theta_i}\psi \middle| x\right).
\end{aligned}$$

The first equality follows from the standard interchange of differentiation and integration for exponential families. The third equality follows from Stein's lemma (Theorem 2.1) in one dimension applied to the inner integral.

In the following, it should be clear from the context whether the symbol ∇ refers to ∇_x or ∇_θ and we omit the subscript in most of the proof.

Proof The key insight of the proof lies in the decomposition of Δ_n given by (1.23) using the triangle and Cauchy-Schwarz inequalities. Take the sequence of priors $g_n(\theta) = h_n^2(\theta)\,g(\theta)$ where

$$h_n(\theta) = \begin{cases} 1 & \|\theta\| \le 1 \\ 1 - \frac{\log\|\theta\|}{\log n} & 1 \le \|\theta\| \le n \\ 0 & \|\theta\| > n \end{cases}$$

for $n = 1, 2, 3, \ldots$. The Bayes risk difference between δ_g and δ_{g_n}

$$\Delta_n = \int_{\mathbb{R}^p} \left[\mathcal{R}(\theta, \delta_g) - \mathcal{R}(\theta, \delta_{g_n})\right] g_n(\theta) d\theta$$

can be expressed, thanks to (i) and Fubini's theorem, as

$$\begin{aligned}
\Delta_n &= \int_{\mathbb{R}^p} m\left(\left[\|\delta_g(x)-\theta\|^2 - \|\delta_{g_n}(x)-\theta\|^2\right] g_n \big| x\right) dx \\
&= \int_{\mathbb{R}^p} m\left(\left[\|\delta_g(x)\|^2 - \|\delta_{g_n}(x)\|^2 - 2\,(\delta_g(x)-\delta_{g_n}(x))^{\mathrm{T}}\theta\right] g_n \big| x\right) dx \\
&= \int_{\mathbb{R}^p} \left[\left\{\|\delta_g(x)\|^2 - \|\delta_{g_n}(x)\|^2\right\} m(g_n \big| x) - 2\,(\delta_g(x)-\delta_{g_n}(x))^{\mathrm{T}} m(\theta g_n \big| x)\right] dx \\
&= \int_{\mathbb{R}^p} \left\{\|\delta_g(x)\|^2 - \|\delta_{g_n}(x)\|^2 - 2\,(\delta_g(x)-\delta_{g_n}(x))^{\mathrm{T}}\delta_{g_n}(x)\right\} m(g_n \big| x) dx \\
&= \int_{\mathbb{R}^p} \|\delta_g(x)-\delta_{g_n}(x)\|^2\, m(g_n|x)\, dx
\end{aligned}$$

since $m(\theta\, g_n|x) = \delta_{g_n}(x)\, m(g_n \big| x)$. Note we have also used the factorization

$$m(h(x)\,\psi \big| x) = h(x)\, m(\psi \big| x)\,. \tag{1.25}$$

Then, as

$$\delta_g(x) = x + \nabla \log m(g|x)\,,\ \delta_{g_n}(x) = x + \nabla \log m(g_n|x) \text{ and } g_n = g\, h_n^2\,,$$

we have

$$\begin{aligned}
\Delta_n &= \int_{\mathbb{R}^p} \left\| \frac{\nabla m(g|x)}{m(g|x)} - \frac{\nabla m(g\, h_n^2|x)}{m(g\, h_n^2|x)} \right\|^2 m(g\, h_n^2|x)\, dx \\
&\le 2 \int_{\mathbb{R}^p} \left\| \frac{m(g\, \nabla h_n^2|x)}{m(g\, h_n^2|x)} \right\|^2 m(g\, h_n^2|x)\, dx \\
&\quad +2 \int_{\mathbb{R}^p} \left\| \frac{m(\nabla g|x)}{m(g|x)} - \frac{m(h_n^2\, \nabla g|x)}{m(g\, h_n^2|x)} \right\|^2 m(g\, h_n^2|x)\, dx \\
&\equiv 2\,(A_n + B_n)\,,
\end{aligned}$$

since

$$\nabla g_n = \nabla(g\, h_n^2) = h_n^2\, \nabla g + g\, \nabla h_n^2 \quad \text{and} \quad \|a+b\|^2 \le 2\,(\|a\|^2 + \|b\|^2)\,.$$

To show $A_n \to 0$ note that

$$\begin{aligned}\|m(g\,\nabla h_n^2\,|x)\|^2 &= 4\,\|m(g\,h_n\nabla h_n\,|x)\|^2\\ &\le 4\,\{m(g\,h_n\,\|\nabla h_n\|\,|x)\}^2 \qquad \text{(by Jensen's inequality)}\\ &\le 4\,\{[m^{1/2}(g\,h_n^2\,|x)]\,[m^{1/2}(g\,\|\nabla h_n\|^2\,|x)]\}^2 (\text{ by Cauchy-Schwarz})\\ &= 4\,[m(g\,h_n^2\,|x)]\,[m(g\,\|\nabla h_n\|^2\,|x)]\,.\end{aligned}$$

Hence

$$\begin{aligned}A_n &= 4\int_{\mathbb{R}^p}\left\|\frac{m(g\,h_n\nabla h_n|x)}{m(g\,h_n^2|x)}\right\|^2 m(g\,h_n^2|x)\,dx\\ &\le 4\int_{\mathbb{R}^p} m(g\,\|\nabla h_n\|^2\,|x)\,dx\\ &= 4\int_{\mathbb{R}^p}\|\nabla h_n(\theta)\|^2\,g(\theta)\,d\theta\,,\end{aligned}$$

where equality follows by property (ii). Calculating the gradient term gives

$$\begin{aligned}\|\nabla h_n(\theta)\|^2 &= \frac{1}{\|\theta\|^2\log^2(n)}\,\mathbb{1}_{[1\le\|\theta\|\le n]}\\ &\le \frac{1}{\|\theta\|^2\max\{\log\|\theta\|,\log 2\}^2}\,\mathbb{1}_{[1\le\|\theta\|]}\,. \end{aligned} \tag{1.26}$$

Since $\|\nabla h_n(\theta)\|^2 \to 0$ for all θ, Condition (1) and (1.26) imply, by the dominated convergence theorem, $A_n \to 0$ as $n \to \infty$.

Next, note that, since $g_n \to g$, the integrand $b_n(x)$ of B_n tends to zero for all $x \in \mathbb{R}^p$ and, using the factorization property in (1.25), can be expressed as

$$\begin{aligned}b_n(x) &= \frac{\left\|m\left(\frac{m(\nabla g|x)}{m(g|x)}\,h_n^2\,g - h_n^2\,\nabla g\middle|x\right)\right\|^2}{m(h_n^2\,g|x)}\\ &= \frac{\left\|m\left(g_n\left\{\frac{m(\nabla g|x)}{m(g|x)} - \frac{\nabla g}{g}\right\}\middle|x\right)\right\|^2}{m(h_n^2\,g|x)}\,.\end{aligned}$$

By Jensen and Cauchy-Schwarz inequality applied as above, it follows that

$$b_n(x) \le m\left(\left\|g_n\left\{\frac{m(\nabla g|x)}{m(g|x)} - \frac{\nabla g}{g}\right\}\right\|^2\middle|x\right).$$

Now, as $0 \le h_n \le 1$, we have

$$b_n(x) \le m\left(\left\| g\left\{\frac{m(\nabla g|x)}{m(g|x)} - \frac{\nabla g}{g}\right\}\right\|^2 \middle| x\right)$$
$$= m\left(\frac{\|\nabla g\|^2}{g}\middle| x\right) - \frac{\|m(\nabla g|x)\|^2}{m(g|x)},$$

where equality follows by expanding and using the factorization property in (1.25). Hence

$$b_n(x) \le m\left(\frac{\|\nabla g\|^2}{g}\middle| x\right)$$

and it follows that

$$B_n \le \int_{\mathbb{R}^p} m\left(\frac{\|\nabla g\|^2}{g}\middle| x\right) dx = \int_{\mathbb{R}^p} \frac{\|\nabla g\|^2}{g}\, d\theta < \infty$$

by condition (2). Therefore $B_n \to 0$ as $n \to \infty$ by the dominated convergence theorem.

Finally, $A_n + B_n \to 0$ as $n \to \infty$ and δ_g is admissible by Blyth's method (Theorem 1.1) . □

As an application of Theorem 1.3 it is a easy to show that the estimator X is admissible for $p = 1$ and $p = 2$. Indeed, if $g = 1$, then $\delta_g(X) = X$ since $\nabla g = 0$ and the conditions of Theorem 1.3 are trivial to verify. In the general case with $p \ge 3$, $\delta_g(X) = X$ is inadmissible. In this case, Condition (2) holds but Condition (1) fails.

Consider the class of priors with $g(\theta) \le \|\theta\|^{2-p-\epsilon}$ for some $\epsilon > 0$ and with $\|\nabla g(\theta)/g(\theta)\|^2 = O(\|\theta\|^{-1})$. In this case, the conditions of Theorem 1.3 are easy to check. Hence $\delta_g(X)$ is admissible.

In the case where $g(\theta) \le \|\theta\|^{2-p}$,

$$\left\|\frac{\nabla g(\theta)}{g(\theta)}\right\|^2 = O(\|\theta\|^{-1}), \quad \text{and} \quad \left\|\frac{\partial^2 g(\theta)}{\partial\theta_i \partial\theta_j}\right\| = O(\|\theta\|^{-2}),$$

it can be shown, using an extension of Lemma 3.4.1 of Brown (1971) that the conditions of Theorem 1.3 are satisfied. Hence $\delta_g(X)$ is admissible. If $g(\theta) \sim \|\theta\|^r$ as $|\theta\| \to \infty$ and is smooth, Brown (1971, 1979) show that $\delta_g(X) - X \sim r\, X/\, \|X\|^2$ as $\|X\| \to \infty$. The case $r = 2 - p$ gives a form $X \sim (1 - (p-2)/\, \|X\|^2)\, X$ that motivates the James-Stein estimator.

Maruyama (2009) extends Brown and Hwang's decomposition method to the spherically symmetric case and gives a condition for admissibility as strong as that of Brown (1971) under normality. A minimization problem that corresponds to the A_n term in Brown and Hwang is formulated with clever use of an assumption that the target prior is regularly varying. Maruyama's construction also works well for proving that the corresponding B_n approaches 0 as $n \to \infty$ in the spherically symmetric case. As a result, Maruyama gives a strong sufficient condition for the admissibility of generalized Bayes estimators by using an adaptive sequence of proper priors. Maruyama and Takemura (2008) deals with the same problem as Maruyama (2009) and give a sufficient condition for admissibility without the assumption that the target prior is regularly varying.

1.7 Invariance

Large classes of problems are invariant under a variety of groups of transformations on the sample space $\mathscr{X}$, associated groups of transformations acting on the parameter space Ω, and the action space $\mathscr{D}$. In such cases, it seems natural to search for (optimal) procedures that behave in a manner consistent with the group structure. There is also a (generalized) Bayes connection, in that optimal procedures, when they exist, can be viewed as generalized Bayes estimators with respect to right invariant Haar measure which may be considered a natural "objective" prior. Almost all the problems considered in this book are invariant under the location or location-scale group (when σ^2 is unknown).

We give a brief discussion of some of the general theory (for more details, see e.g. Lehmann and Casella 1998 and Schervish 1997). Suppose $X \in \mathscr{X} \sim P_\theta$ with $\theta \in \Omega$ is an identifiable family and G is a group of one-to-one and onto transformations on $\mathscr{X}$. Suppose also that, for all $\theta \in \Omega$ and $g \in G$, if $X \sim P_\theta$, then there exists a $\theta' \in \Omega$ such that $gX \sim P_{\theta'}$. In this case, we may associate a transformation $\bar{g}$ on Ω defined as $\bar{g}\theta = \theta'$. It can be shown that $\bar{g}$ is one-to-one and onto. The collection $\bar{G} = \{\bar{g}|g \in G\}$ also forms a group of one-to-one transformations acting on Ω. Under these conditions, the statistical model is said to be invariant under G.

As an example, suppose the distributions of X form a location parameter family in $\mathbb{R}^p$ with density $f(x-\theta)$. The location group G in $\mathbb{R}^p$ consists of transformations of the form $g_a : x \mapsto x + a$ where $a \in \mathbb{R}^p$. If $X \sim f(x-\theta)$, then $\tilde{X} = g_a(X) \sim f(\tilde{x} - (\theta + a))$ so that $\bar{g}_a : \theta \mapsto \theta + a$. In this case, $\bar{G}$ and G essentially coincide although they act on different (but equivalent) spaces.

If the statistical problem is to estimate $h(\theta)$, under the loss function $L(\theta, d)$, the problem is said to be invariant if there is a g^* acting on the action space $\mathscr{D}$ corresponding to each $g \in G$ such that $L(\theta, d) = L(\bar{g}\theta, g^*d)$ for every $\theta \in \Omega$, $d \in \mathscr{D}$ and $g \in G$. In the above location problem, if $h(\theta) = \theta$ and $L(\theta, d) = \rho(||d-\theta||^2)$, the transformation g_a^* corresponding to g_a is $g_a^* : d \mapsto d + a$, so that essentially $G = \bar{G} = G^*$.

An estimator, δ, is said to be equivariant if $\delta(g\,X) = g^*\delta(X)$ for all $g \in G$ and $X \in \mathscr{X}$. In the above location problem, this implies that $\delta(X + a) = \delta(X) + a$. Choosing $a = -X$, this implies $\delta(X) = X + \delta(0)$.

The following is a key property of equivariant estimators in invariant problems.

Lemma 1.11 *If the problem is invariant and δ is equivariant, then the risk of δ is constant on orbits of $\bar{G}$, i.e. $\mathscr{R}(\bar{g}\,\theta, \delta) = \mathscr{R}(\theta, \delta)$ for all $\theta \in \Omega$ and $\bar{g} \in \bar{G}$. If the group $\bar{G}$ acting on Ω is transitive (i.e. for all $\theta_1, \theta_2 \in \Omega$, there exists $\bar{g} \in \bar{G}$ such that $\theta_2 = \bar{g}\theta_1$) then it follows that the risk of an equivariant estimator is constant on Ω.*

Proof The lemma immediately follows from the equalities

$$\begin{aligned}
\mathscr{R}(\bar{g}\,\theta, \delta) &= E_{\bar{g}\theta}[L(\bar{g}\,\theta, \delta(X))] \\
&= E_{\theta}[L(\bar{g}\,\theta, \delta(g\,X))] && \text{since } g\,X \sim P_{\bar{g}\theta} \\
&= E_{\theta}[L(\bar{g}\,\theta, g^*\delta(X))] && \text{since } \delta \text{ is equivariant} \\
&= \mathscr{R}(\theta, \delta) && \text{since the problem is invariant.}
\end{aligned}$$

□

This constancy of risk holds in location problems because the group $\bar{G}$ is transitive: for any $\theta_1, \theta_2 \in \mathbb{R}^p$, $\theta_2 = \theta_1 + (\theta_2 - \theta_1)$ so that $\bar{g}_{\theta_2 - \theta_1}\theta_1 = \theta_2$.

The risk constancy of equivariant estimators gives hope of finding a best one, or minimum risk equivariant (MRE) estimator, since all that is required is the existence of an estimator that attains the infimum among the set of constant risks. The following lemma settles the issue for the location problem with quadratic loss; the proofs of these results are given in Lehmann and Casella (1998).

Lemma 1.12

(1) *For the multivariate location problem with loss $L(\theta, d) = ||d - \theta||^2$, the MRE $\delta_0(X)$ exists and is unique provided $E_0[||X||^2] < \infty$.*
(2) *$\delta_0(X) = X - E_0[X]$*
(3) *$\delta_0(X) = \int_{\mathbb{R}^p} \theta\, f(X-\theta)\, d\theta \,\big/ \int_{\mathbb{R}^p} f(X - \theta)\, d\theta$, i.e. δ_0 is the generalized Bayes estimator with respect to the Lebesgue measure on $\mathbb{R}^p$ (this is known as the Pitman estimator).*
(4) *The MRE coincides with the UMVUE of θ provided the UMVUE exists and is equivariant.*

Things are somewhat simpler in the spherically symmetric case (see the comment after Definition 1.2).

Lemma 1.13 *If $X \sim f(||x - \theta||^2)$ and $L(\theta, d) = ||d - \theta||^2$ then*

(1) *$\delta_0(X) = X$ is MRE;*
(2) *the MRE is also UMVUE provided the family of distributions is complete.*

See also Sect. 4.5.4. For a general location-scale family

$$(X, U) \sim \frac{1}{\sigma^{p+k}} f\left(\frac{||x-\theta||^2 + ||u||^2}{\sigma^2}\right),$$

the results for estimation of the parameter θ under loss $L((\theta, \sigma^2), d) = ||d - \theta||^2/\sigma^2$ are quite similar. In particular, the family is invariant under the group of transformations $g_{a,b,P}(x, u) = (a + b\,x, b\,P\,u)$, where $a \in \mathbb{R}^p$, $b > 0$, P orthogonal, is such that $\bar{g}_{a,b,P}(\theta, \sigma^2) = (a + b\,\theta, b^2\,\sigma^2)$, and thus $\overline{G}$ is transitive. Similarly, $g^*_{a,b,P}\,d = a + b\,d$ and $\delta(X, U)$ is equivariant if $\delta(a + b\,X, b\,P\,U) = a + b\,\delta(X, U)$.

Choosing P such that $P\,u = (||u||, 0, \ldots, 0)^{\mathrm{T}}$, $b = 1/||u||$, and $a = -x/||u||$, implies

$$\begin{aligned}\delta(X, U) &= \left[\frac{X}{||U||} + \delta(0, (1, 0, \ldots, 0)^{\mathrm{T}})\right] \Big/ \left[\frac{1}{||U||}\right] \\ &= X + c\,||U||\end{aligned}$$

where $c = \delta(0, (1, 0, \ldots, 0)^{\mathrm{T}}) \in \mathbb{R}^p$ is arbitrary.

Lemma 1.14 *Suppose (X, U) has the density function*

$$\frac{1}{\sigma^{p+k}} f\left(\frac{||x-\theta||^2 + ||u||^2}{\sigma^2}\right)$$

and the invariant loss is

$$L((\theta, \sigma^2), d) = \frac{||d-\theta||^2}{\sigma^2}.$$

Then

(1) *$\delta_0(X) = X$ is MRE and unbiased.*
(2) *The MRE is also the UMVUE provided the family of distributions is complete.*
(3) *$\delta_0(X)$ is generalized Bayes with respect to the right invariant prior on $(\theta, \sigma^2) \in \mathbb{R}^p \times \mathbb{R}^+$, that is,*

$$\delta_0(X) = \frac{\int\int \frac{\theta}{\sigma^2} \frac{1}{\sigma^{p+k}} f\left(\frac{||x-\theta||^2 + ||u||^2}{\sigma^2}\right) \frac{1}{\sigma^2}\, d\theta\, d\sigma^2}{\int\int \frac{1}{\sigma^2} \frac{1}{\sigma^{p+k}} f\left(\frac{||x-\theta||^2 + ||u||^2}{\sigma^2}\right) \frac{1}{\sigma^2}\, d\theta\, d\sigma^2}.$$

The minimaxity of the MRE of the location parameter in the location and location-scale families follows also from the so-called Hunt-Stein theorem since the location and location-scale groups are amenable. See Kiefer (1957), Robert (1994), Lehmann and Casella (1998), Bondar and Milnes (1981), and Eaton (1989) for details.

Chapter 2
Estimation of a Normal Mean Vector I

2.1 Introduction

This chapter is concerned with estimating the p-dimensional mean vector of a multivariate normal distribution under quadratic loss. Most of the chapter will be concerned with the case of a known covariance matrix of the form $\Sigma = \sigma^2 I_p$ and "usual quadratic loss," $L(\theta, \delta) = \|\delta - \theta\|^2 = (\delta - \theta)^{\mathrm{T}}(\delta - \theta)$. Generalizations to known general covariance matrix Σ, and to general quadratic loss, $L(\theta, \delta) = (\delta - \theta)^t Q(\delta - \theta)$, where Q is a $p \times p$ symmetric non-negative definite matrix will also be considered. Let $X \sim \mathcal{N}_p(\theta, \sigma^2 I_p)$ where σ^2 is assumed known and it is desired to estimate the unknown vector $\theta \in \mathbb{R}^p$. The "usual" estimator of θ is $\delta_0(X) = X$, in the sense that it is the maximum likelihood estimator (MLE), the uniformly minimum variance unbiased estimator (UMVUE), the least squares estimator (LSE), and under a wide variety of loss functions it is the minimum risk equivariant estimator (MRE), and is minimax. The estimator $\delta_0(X)$ is also admissible under a wide class of invariant loss functions if $p = 1$ or 2. However, Stein (1956) showed that X is inadmissible if $p \geq 3$ for the loss $L(\theta, \delta) = \|\delta - \theta\|^2$. This result was surprising at the time and has led to a large number of developments in multi-parameter estimation. One important aspect of this "Stein phenomenon" (also known as the Stein paradox at one time, see Efron and Morris 1977) is that it illustrates the difference between estimating one component at a time and simultaneously estimating the whole mean vector. Indeed, if we wish to estimate any particular component, θ_i, of the vector θ, then the estimator $\delta_{0i}(X) = X_i$ remains admissible whatever the value of p (see for example Lehmann and Casella (1998), Lemma 5.2.12). James and Stein (1961) showed that the estimator $\delta_a^{JS}(X) = (1 - a\,\sigma^2/\|X\|^2)X$ dominates $\delta_0(X)$ for $p \geq 3$ provided $0 < a < 2\,(p-2)$. They also showed that the risk of $\delta_{p-2}^{JS}(X) = \left(1 - (p-2)\,\sigma^2/\|X\|^2\right) X$ at $\theta = 0$ is equal to $2\,\sigma^2$ for all $p \geq 3$ indicating that substantial gain in risk over the usual estimator is possible for large p, since the risk of $\delta_0(X)$ is equal to the constant $p\,\sigma^2$.

D. Fourdrinier et al., *Shrinkage Estimation*, Springer Series in Statistics,
https://doi.org/10.1007/978-3-030-02185-6_2

In Sect. 2.2, we will give some intuition into why improvement over $\delta_0(X)$ should be possible in higher dimensions and how much improvement might be expected. Section 2.3 is devoted to Stein's unbiased estimation of risk technique which provides the technical basis of many results in the area of multi-parameter estimation. Section 2.4 is devoted to establishing improved procedures such as the James-Stein estimator. In Sect. 2.5, we will provide a link between Stein's lemma and Stokes' theorem while, in Sect. 2.6, we will give some insight into Stein's phenomenon in terms of nonlinear partial differential operators.

2.2 Some Intuition into Stein Estimation

2.2.1 Best Linear Estimators

Suppose X is a p-dimensional random vector such that $E[X] = \theta$ and $Cov(X) = \sigma^2 I_p$ where θ is unknown and σ^2 is known. We do not require at this point that X have a multivariate normal distribution. Consider estimators of θ of the form $\delta_a(X) = (1-a)\,X$ under quadratic loss $L(\theta, \delta) = \|\delta - \theta\|^2 = \sum_{i=1}^p (\delta_i - \theta_i)^2$. The risk of $\delta_a(X)$ is given by

$$
\begin{aligned}
R(\theta, \delta_a) &= E\left[\sum_{i=1}^p ((1-a)\,X_i - \theta_i)^2\right] \\
&= \sum_{i=1}^p Var\,((1-a)\,X_i) + \sum_{i=1}^p (E\,[(1-a)\,X_i - \theta_i])^2 \\
&= (1-a)^2\,p\,\sigma^2 + a^2 \sum_{i=1}^p \theta_i^2 \\
&= (1-a)^2\,p\,\sigma^2 + a^2\,\|\theta\|^2 .
\end{aligned}
$$

The optimal choice of a, a_{opt}, which minimizes $R(\theta, \delta_a)$ is obtained by differentiating $R(\theta, \delta_a)$ with respect to a and equating the result to 0, that is,

$$
\begin{aligned}
\frac{\partial}{\partial a} R(\theta, \delta_a) &= -2\,(1-a)\,p\sigma^2 + 2\,a\,\|\theta\|^2 \\
&= 0
\end{aligned}
$$

and solving for a gives

$$
a_{opt} = \frac{p\,\sigma^2}{p\,\sigma^2 + \|\theta\|^2} .
$$

We see that a_{opt} depends on the unknown, θ but since

$$E\|X\|^2 = p\sigma^2 + \|\theta\|^2,$$

we may estimate a_{opt} as

$$\hat{a}_{opt} = \frac{p\sigma^2}{\|X\|^2},$$

and hence approximate the best linear "estimator"

$$\delta_{a_{opt}}(X) = \left(1 - \frac{p\sigma^2}{p\sigma^2 + \|\theta\|^2}\right) X$$

by

$$\hat{\delta}_{a_{opt}}(X) = \left(1 - \frac{p\sigma^2}{\|X\|^2}\right) X.$$

This is in fact a James-Stein type estimator

$$\hat{\delta}_{a_{opt}}(X) = \delta_p^{JS}(X),$$

which is close to the optimal James-Stein estimator (as we will see in Sect. 2.4 $\delta_{p-2}^{JS}(X)$ is optimal if X is normal). Hence, the James-Stein estimator can be viewed as an approximation to the best linear "estimator" that adapts to the value of $\|\theta\|^2$.

It is worth noting that $a_{opt} = p\sigma^2/(p\sigma^2+\|\theta\|^2)$ can typically be better estimated for large values of p since $E\|X\|^2/p = \sigma^2 + \|\theta\|^2/p$ and (if we assume X_i are symmetric about θ_i and that the $X_i - \theta_i$ are independent)

$$Var\left(\frac{\|X\|^2}{p}\right) = \frac{Var(X_1 - \theta_1)^2}{p} + \frac{4\,\|\theta\|^2\,\sigma^2}{p^2}$$

which tends to 0 uniformly as $p \to \infty$ provided $\|\theta\|^2/p$ is bounded. This helps to explain why there is a dimension effect and that it is easier to find dominating estimators for large p.

It is also interesting to note that normality plays no role in the above discussion indicating that we can expect James-Stein type estimators to improve on $\delta_0(X)$ in a fairly general location vector setting. This will be discussed further in Chaps. 5 and 6 for spherically symmetric distributions.

Note also, since the estimators are generally shrinking X toward 0, we expect the largest gains in risk to occur at $\theta = 0$. In particular the risk of $\delta_{a_{opt}}(X)$ at the true value of θ is given by

$$R(\theta, \delta_{a_{opt}}) = \frac{p\,\sigma^2\,\|\theta\|^2}{p\,\sigma^2 + \|\theta\|^2} = R(\theta, X)\,\frac{\|\theta\|^2}{p\,\sigma^2 + \|\theta\|^2}.$$

Hence, when $\|\theta\|^2$ is large, there is very little savings in risk, but when $\|\theta\|^2$ is close to 0, the improvement is substantial.

We will see later in Sect. 2.4 that this is also true for James-Stein-type estimators in the sense that there is very little savings in risk for large $\|\theta\|^2$ but substantial savings for small $\|\theta\|^2$ and especially so for large p.

2.2.2 *Some Geometrical Insight*

The argument here closely follows the discussion presented by Brandwein and Strawderman (1991a). We again suppose $E[X] = \theta \in \mathbb{R}^p$ and $Cov(X) = \sigma^2 I_p$ with σ^2 known. Since $E[\|X\|^2] = \|\theta\|^2 + p\,\sigma^2$, it seems that X is" too long" as an estimator of θ and that perhaps the projection of θ onto X or something close to it would be a better estimator than X. Again, the projection of θ onto X will depend on θ and so will not be a valid estimator, but perhaps we can find a reasonable approximation. Since the projection of θ on X has the form $(1-a)\,X$ we are trying to approximate the constant a. Note $E(\theta - X)^{\mathrm{T}}\theta = 0$, and hence we expect θ and $X - \theta$ to be nearly orthogonal which implies that we expect $0 < a < 1$.

In what follows, we assume that θ and $X-\theta$ are exactly orthogonal. The situation is shown in Fig. 2.1.

From the two right triangles in Fig. 2.1 we note

$$\|(1-a)\,X\|^2 + \|Y\|^2 = \|\theta\|^2 \quad \text{and} \quad \|a\,X\|^2 + \|Y\|^2 = \|X-\theta\|^2.$$

Since

$$E\|X\|^2 = \|\theta\|^2 + p\,\sigma^2 \quad \text{and} \quad E\|X-\theta\|^2 = p\,\sigma^2,$$

reasonable approximations are

$$\|\theta\|^2 \cong \|X\|^2 - p\sigma^2 \quad \text{and} \quad \|X-\theta\|^2 \cong p\,\sigma^2.$$

Hence we have as approximations

$$\|(1-a)\,X\|^2 + \|Y\|^2 \cong \|X\|^2 - p\,\sigma^2 \quad \text{and} \quad \|a\,X\|^2 + \|Y\|^2 \cong p\,\sigma^2.$$

Subtracting to eliminate $\|Y\|^2$, that is,

$$\|(1-a)\,X\|^2 - \|a\,X\|^2 = (1-2\,a)\|X\|^2 \cong \|X\|^2 - 2\,p\,\sigma^2,$$

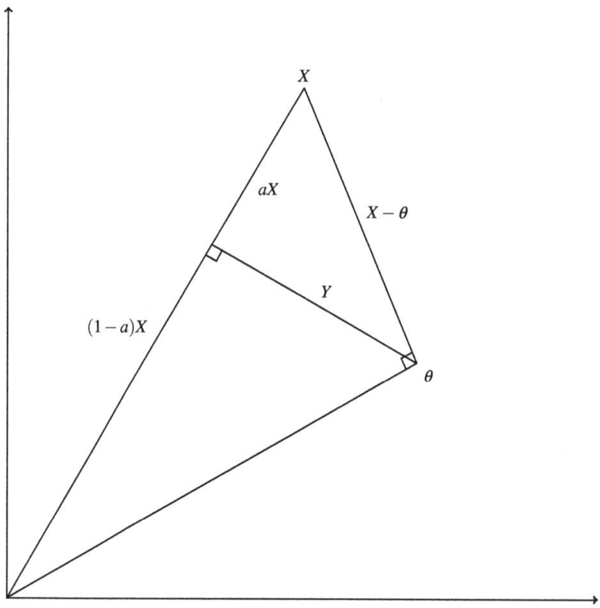

Fig. 2.1 Observation vector X in p dimensions with mean θ orthogonal to $X - \theta$

we obtain $a \cong p\,\sigma^2/\|X\|^2$. Hence, we may approximate the projection of θ on X as

$$(1-a)\,X \cong \left(1 - \frac{p\,\sigma^2}{\|X\|^2}\right) X = \delta_p^{JS}(X)\,, \tag{2.1}$$

remarkably the same James-Stein estimator suggested in Sect. 2.2.1. Once again, note that normality plays no role in the discussion. Stein (1962) gave a similar geometric arguement to contruct confidence sets for θ, centred at (2.1), as the orthogonal projection of θ on X. For more on the geometrical explanation of the inadmissibility of X as a point estimator see Brown and Zhao (2012).

2.2.3 *The James-Stein Estimator as an Empirical Bayes Estimator*

Assume in this subsection that $X \sim \mathcal{N}_p(\theta, \sigma^2 I_p)$ with (σ^2 known) and that the prior distribution on θ is $\mathcal{N}_p(0, \tau^2 I_p)$. As indicated in Sect. 1.4, the Bayes estimator of θ for quadratic loss is the posterior mean of θ given by $\delta(X) = E[\theta \mid X] = (1 - \sigma^2/(\tau^2 + \sigma^2))X$.

If we now assume that τ^2 is unknown we can derive an empirical Bayes estimator as follows; the marginal distribution of X is $\mathcal{N}_p(0, (\sigma^2 + \tau^2)\, I_p)$ and hence $\|X\|^2$, which is distributed as $(\sigma^2 + \tau^2)$ times a chi-square with p degrees of freedom, is a complete sufficient statistic for $\sigma^2+\tau^2$. It follows that $(p-2)/\|X\|^2$ is the UMVUE of $1/(\sigma^2 + \tau^2)$ and that $\delta^{JS}_{p-2}(X) = (1 - (p-2)\,\sigma^2/\|X\|^2)\,X$ can be viewed as an empirical Bayes estimator of θ.

Here we have explicitly used the assumption of normality but a somewhat analogous argument will be given in Sect. 5.1 for a general multivariate location family.

2.3 Improved Estimators via Stein's Lemma

In this section, we restrict our attention to the case where $X \sim \mathcal{N}_p(\theta, \sigma^2 I_p)$ with σ^2 known and where the loss function is $L(\theta, \delta) = \|\delta - \theta\|^2$. We will be concerned with developing expressions for the risk function of a general estimator of the form $\delta(X) = X + \sigma^2 g(X)$ for some function g from $\mathbb{R}^p$ into $\mathbb{R}^p$. This development is due to Stein (1973, 1981).

Through

$$\begin{aligned} L(\theta, \delta) &= \|X + \sigma^2 g(X) - \theta\|^2 \\ &= \|X - \theta\|^2 + \sigma^4 \|g(X)\|^2 + 2\,\sigma^2 (X - \theta)^{\mathrm{T}} g(X)\,, \end{aligned} \tag{2.2}$$

we will see that the risk of δ is finite if and only if $E_\theta[\|g(X)\|^2] < \infty$. Indeed, considering the expectation of the cross product term in (2.2), we have

$$E_\theta[|(X-\theta)^{\mathrm{T}} g(X)|] \le \big(E_\theta[\|(X-\theta)\|^2]\big)^{1/2} \big(E_\theta[\|g(X)\|^2]\big)^{1/2},$$

by the Cauchy-Schwarz inequality. Therefore, as $E_\theta[\|(X-\theta)\|^2] < \infty$, it suffices that $E_\theta[\|g(X)\|^2] < \infty$ to have $E_\theta[\|X + g(X) - \theta\|^2] < \infty$, that is, $R(\theta, X + g(X)) < \infty$.

Conversely, assume that $R(\theta, X + g(X)) < \infty$. As

$$\begin{aligned} \|g(X)\|^2 &= \|X + g(X) - \theta - (X - \theta)\|^2 \\ &= \|X + g(X) - \theta\|^2 + \|X - \theta\|^2 - 2\,(X-\theta)^{\mathrm{T}}(X + g(X) - \theta) \end{aligned}$$

then applying the above argument gives $E_\theta[\|g(X)\|^2] < \infty$ since, by assumption, $E_\theta[\|X + g(X) - \theta\|^2] < \infty$, $E_\theta[\|(X-\theta)\|^2] < \infty$ and hence using again the Cauchy-Schwarz inequality

$$E_\theta[|(X-\theta)^{\mathrm{T}}(X+g(X)-\theta)|] \le \big(E_\theta[\|(X-\theta)\|^2]\big)^{1/2} \big(E_\theta[\|X+g(X)-\theta\|^2]\big)^{1/2}.$$

Under this fineteness condition the risk function of δ is given by

$$R(\theta, \delta) = p\,\sigma^2 + \sigma^4 E_\theta[\|g(X)\|^2] + 2\,\sigma^2 E_\theta[(X-\theta)^{\mathrm{T}} g(X)]\,.$$

Stein's lemma in (2.7) below allows an alternative expression for the last expectation, that is, $E_\theta[(X-\theta)^{\mathrm{T}} g(X)] = \sigma^2 E_\theta[\mathrm{div} g(X)]$ where $\mathrm{div} g(X) = \sum_{i=1}^{p} \frac{\partial}{\partial X_i} g_i(X)$ under suitable conditions on g. The great advantage that Stein's lemma gives is that the risk function can be expressed as the expected value of a function of X only (and not θ), that is,

$$R(\theta,\delta) = E_\theta[p\,\sigma^2 + \sigma^4\,\|g(X)\|^2 + 2\,\sigma^4\,\mathrm{div} g(X)], \tag{2.3}$$

and hence the expression

$$p\,\sigma^2 + \sigma^4 \left[\|g(X)\|^2 + 2\,\mathrm{div} g(X)\right]$$

can be interpreted as an unbiased estimate of the risk of δ (see Corollary 2.1 (3)). Actually, as X is a complete sufficient statistic, this unbiased estimator is the uniformly minimum variance unbiased estimator of the risk. To see that $E_\theta[(X-\theta)^{\mathrm{T}} g(X)] = \sigma^2 E_\theta[\mathrm{div} g(X)]$ is quite easy if g is sufficiently smooth. Suppose first that $p = 1$ and g is absolutely continuous. We show in Sect. A.5 in the Appendix that $\lim_{x\to\pm\infty} g(x)\exp\{-(x-\theta)^2/2\sigma^2\} = 0$ as soon as $E_\theta[|g'(X)|] < \infty$ (see also Hoffmann 1992 where g is assumed to be continuously differentiable). Then a simple integration by parts gives

$$\begin{aligned}
E_\theta[(X-\theta)g(X)] &= \frac{1}{(2\pi\sigma^2)^{1/2}} \int_{-\infty}^{\infty} (x-\theta)g(x)\exp\{-(x-\theta)^2/2\sigma^2\}\,dx \\
&= \frac{1}{(2\pi\sigma^2)^{1/2}} \int_{-\infty}^{\infty} \sigma^2 g(x)\left(\frac{-d}{dx}\exp\{-(x-\theta)^2/2\sigma^2\}\right)dx \\
&= \frac{\sigma^2}{(2\pi\sigma^2)^{1/2}} \int_{-\infty}^{\infty} g'(x)\exp\{-(x-\theta)^2/2\sigma^2\}\,dx \\
&= \sigma^2 E_\theta[g'(X)]\,.
\end{aligned}$$

In higher dimensions, let $g = (g_1,\ldots,g_p)$ be a function from $\mathbb{R}^p$ into $\mathbb{R}^p$. Also, for any $x = (x_1,\ldots,x_p) \in \mathbb{R}^p$ and for fixed $i = 1,\ldots,p$, set $x_{-i} = (x_1,\ldots,x_{i-1},x_{i+1},\ldots,x_p)$ and, with a slight abuse of notation, $x = (x_i, x_{-i})$. Then, using the independence of X_i and X_{-i}, we have

$$\begin{aligned}
E_\theta[(X_i-\theta_i)\,g_i(X)] &= E_\theta\big[E_\theta[(X_i-\theta_i)\,g_i(X_i,X_{-i})|\,X_{-i}]\big] \\
&= E_\theta\big[E_\theta[\sigma^2\,\partial_i g_i(X_i,X_{-i})|\,X_{-i}]\big] \\
&= \sigma^2 E_\theta[\partial_i g_i(X)]\,.
\end{aligned}$$

Now, summing on i gives $E_\theta[(X-\theta)^{\mathrm{T}} g(X)] = \sigma^2\,E_\theta[\mathrm{div} g(X)]$.

However, we wish to include estimators such as the James-Stein estimators

$$\delta_a^{JS}(X) = \left(1 - \frac{a\,\sigma^2}{\|X\|^2}\right) X \tag{2.4}$$

where the coordinate functions of $g(X) = (a\,\sigma^2/\|X\|^2)\,X$ are not smooth, since g explodes at 0. For this reason, Stein considered a weaker regularity condition for his identity to hold, that he called almost differentiability. In his proof, he essentially required that $g(x) = (g_1(x), g_2(x), \ldots, g_p(x))$ be such that, for each $i = 1, \ldots, p$, the coordinate $g_i(x)$ is absolutely continuous in x_i for almost every x_{-i}. Formally, he stated: "A function h from $\mathbb{R}^p$ into $\mathbb{R}^p$ is said to be almost differentiable if there exists a function $\nabla h = (\nabla_1 h, \ldots, \nabla_p h)$ from $\mathbb{R}^p$ into $\mathbb{R}^p$ such that, for all $z \in \mathbb{R}^p$,

$$h(x + z) - h(x) = \int_0^1 z^{\mathrm{T}}\,\nabla h(x + t\,z)\,dt\,, \tag{2.5}$$

for almost all $x \in \mathbb{R}^p$. A function $g = (g_1, \ldots, g_p)$ from $\mathbb{R}^p$ into $\mathbb{R}^p$ is said to be almost differentiable if all its coordinate functions g_i's are" (see Sect. A.1 in the Appendix for a detailed discussion).

We will establish Stein's identity under the weaker notion of weak differentiability which is of more common use in analysis and also in statistics (see e.g. Johnstone 1988). To this end, recall that the space of functions h from $\mathbb{R}^p$ into $\mathbb{R}$ such that h is locally integrable is defined by

$$L^1_{loc}(\mathbb{R}^p) = \left\{h : \mathbb{R}^p \to \mathbb{R} \mid \int_K |h(x)|\,dx < \infty \quad \forall K \subset \mathbb{R}^p \text{ with } K \text{ compact}\right\}.$$

Definition 2.1 A locally integrable function h from $\mathbb{R}^p$ into $\mathbb{R}$ is said to be weakly differentiable if there exist p locally integrable functions $\partial_1 h, \ldots \partial_p h$ such that, for any $i = 1, \ldots, p$,

$$\int_{\mathbb{R}^p} h(x)\,\frac{\partial \varphi}{\partial x_i}(x)\,dx = -\int_{\mathbb{R}^p} \partial_i h(x)\,\varphi(x)\,dx \tag{2.6}$$

for any infinitely differentiable function φ with compact support from $\mathbb{R}^p$ into $\mathbb{R}$.

Note that weak differentiability is a global, not local, property. The functions $\partial_i h$ in Definition 2.1 are denoted, as the usual derivatives, by $\partial/\partial x_i$. The vector $\partial h = (\partial_1 h, \ldots, \partial_p h) = (\partial h/\partial x_1, \ldots, \partial h/\partial x_p)$ denotes the weak gradient of h and the scalar $\mathrm{div} g = \sum_{i=1}^p \partial_i g_i$ denotes the weak divergence of g. The following proposition establishes a link between weak differentiability and those aspects of almost differentiability that Stein used (and we will use) in the proof of Stein's lemma.

Proposition 2.1 (Ziemer 1989) *Let h be a locally integrable function from $\mathbb{R}^p$ into $\mathbb{R}$. Then h is weakly differentiable if and only if there exists a representative*

h_0 which is equal to h almost everywhere such that, for any $i = 1, \ldots, p$, the function $h_0(x_i, x_{-i})$ is absolutely continuous in x_i for almost all values of x_{-i} and whose (classical) partial derivatives belong to $L^1_{loc}(\mathbb{R}^p)$. Also the classical partial derivatives of h_0 coincide with the weak partial derivatives of h almost everywhere.

Proposition 2.1 is essentially Theorem 2.1.4. of Ziemer (1989) who deals with functions h in $L^1(\Omega)$ where Ω is an open set of $\mathbb{R}^p$ (and, more generally, in $L^q(\Omega)$ with $q \geq 1$). However, his proof relies only on local integrability of h and its partial derivatives. So, the apparently stronger statement of Proposition 2.1 follows from his arguments. See also Theorem 8.27 of Bressan (2012).

As indicated in Proposition 2.1, the key feature of weak differentiability is the local integrability of the function and of all its partial derivatives. For the functions h of interest to us, the representative h_0 is the function itself so that the weak differentiability follows from the local integrability of h and its derivatives and its absolute continuity along almost all lines parallel to the axes. In particular, as the weak partial derivative is unique up to pointwise almost everywhere equivalence, the weak partial derivative of a continuously differentiable function coincides with the usual derivative (see e. g. Hunter 2014, Chap. 3).

As an example, consider the shrinkage factor $h(x) = x/\|x\|^2$ of the James-Stein estimator in (2.4). In Sect. A.2 in the Appendix, we show that h is weakly differentiable if and only if $p \geq 3$ and that $\operatorname{div} h(x) = (p-2)/\|x\|^2$. We also show that h is not almost differentiable in the sense of Stein given above. This last fact is due to the requirement that h be absolutely continuous in all directions while weak differentiability, in contrast, only requires absolute continuity in directions parallel to the axes. Again we note that Stein only used absolute continuity in the coordinates directions.

We give now a precise statement of Stein's lemma for weakly differentiable functions along the lines of Stein (1981). Note that we will see, in Sect. 2.5, that it is closely related to Stokes' theorem, which will provide an alternative proof.

Theorem 2.1 (Stein's lemma) *Let $X \sim \mathcal{N}_p(\theta, \sigma^2 I_p)$ and let g be a weakly differentiable function from $\mathbb{R}^p$ into $\mathbb{R}^p$. Then*

$$E_\theta[(X-\theta)^{\mathrm{T}} g(X)] = \sigma^2 \, E_\theta[\operatorname{div} g(X)]\,, \tag{2.7}$$

provided, for any $i = 1, \ldots, p$, either

$$E_\theta[|(X_i - \theta_i) g_i(X)|] < \infty \quad or \quad E_\theta[|\partial_i g_i(X)|] < \infty\,. \tag{2.8}$$

Formula (2.7) is often referred to as Stein's identity.

Proof Let $x = (x_1, \ldots, x_p) \in \mathbb{R}^p$ and set

$$\varphi(x) = \frac{\|x-\theta\|^2}{2\sigma^2} \quad \text{and} \quad \phi(x) = \frac{1}{(2\pi\sigma^2)^{p/2}} \exp(-\varphi(x))\,.$$

Equality (2.7) is equivalent to

$$E_\theta[\nabla\varphi(X)^{\mathrm{T}} g(X)] = \sigma^2\, E_\theta[\mathrm{div} g(X)]\,. \tag{2.9}$$

For fixed $i = 1, \ldots, p$, set $x_{-i} = (x_1, \ldots, x_{i-1}, x_{i+1}, \ldots, x_p)$ and, with a slight abuse of notation, set $x = (x_i, x_{-i})$. Note that

$$\frac{\partial \phi(x)}{\partial x_i} = -\frac{\partial \varphi(x)}{\partial x_i}\, \phi(x)$$

so that $\phi(x)$ can be written as

$$\phi(x) = \int_{-\infty}^{x_i} -\frac{\partial \varphi(\tilde{x}_i, x_{-i})}{\partial \tilde{x}_i}\, \phi(\tilde{x}_i, x_{-i})\, d\tilde{x}_i = \int_{x_i}^{\infty} \frac{\partial \varphi(\tilde{x}_i, x_{-i})}{\partial \tilde{x}_i}\, \phi(\tilde{x}_i, x_{-i})\, d\tilde{x}_i\,, \tag{2.10}$$

noticing that, by assumption, $\lim_{|x_i| \to \infty} \varphi(x_1, \ldots, x_p) = \infty$ implies

$$\lim_{|x_i| \to \infty} \phi(x_i, x_{-i}) = \frac{1}{(2\pi\,\sigma^2)^{p/2}} \lim_{|x_i| \to \infty} \exp\big(-\,\varphi(x_i, x_{-i})\big) = 0\,. \tag{2.11}$$

Fixing $i \in \{1, \ldots, p\}$ and assuming first $E_\theta[|\partial_i g_i(X)|] < \infty$, we can write using (2.10), for almost every x_{-i},

$$\begin{aligned}
&\int_{-\infty}^{\infty} \frac{\partial g_i(x_i, x_{-i})}{\partial x_i}\, \phi(x_i, x_{-i})\, dx_i \\
&= \int_{-\infty}^{0} \frac{\partial g_i(x_i, x_{-i})}{\partial x_i} \int_{-\infty}^{x_i} -\frac{\partial \varphi(\tilde{x}_i, x_{-i})}{\partial \tilde{x}_i}\, \phi(\tilde{x}_i, x_{-i})\, d\tilde{x}_i\, dx_i \\
&\quad + \int_{0}^{\infty} \frac{\partial g_i(x_i, x_{-i})}{\partial x_i} \int_{x_i}^{\infty} \frac{\partial \varphi(\tilde{x}_i, x_{-i})}{\partial \tilde{x}_i}\, \phi(\tilde{x}_i, x_{-i})\, d\tilde{x}_i\, dx_i \\
&= \int_{-\infty}^{0} -\frac{\partial \varphi(\tilde{x}_i, x_{-i})}{\partial \tilde{x}_i}\, \phi(\tilde{x}_i, x_{-i}) \int_{\tilde{x}_i}^{0} \frac{\partial g_i(x_i, x_{-i})}{\partial x_i}\, dx_i\, d\tilde{x}_i \\
&\quad + \int_{0}^{\infty} \frac{\partial \varphi(\tilde{x}_i, x_{-i})}{\partial \tilde{x}_i}\, \phi(\tilde{x}_i, x_{-i}) \int_{0}^{\tilde{x}_i} \frac{\partial g_i(x_i, x_{-i})}{\partial x_i}\, dx_i\, d\tilde{x}_i\,.
\end{aligned} \tag{2.12}$$

Now, according to Proposition 2.1, as g is weakly differentiable, we may assume without loss of generality that, for each $i = 1, \ldots, p$, the function $g_i(x_i, x_{-i})$ is absolutely continuous in x_i for almost all values of x_{-i} so that

$$-\int_{\tilde{x}_i}^{0} \frac{\partial g_i(x_i, x_{-i})}{\partial x_i}\, dx_i = \int_{0}^{\tilde{x}_i} \frac{\partial g_i(x_i, x_{-i})}{\partial x_i}\, dx_i = g_i(\tilde{x}_i, x_{-i}) - g_i(0, x_{-i})\,.$$

Then (2.12) becomes

$$\begin{aligned}
&\int_{-\infty}^{\infty} \frac{\partial g_i(x_i, x_{-i})}{\partial x_i}\, \phi(x_i, x_{-i})\, dx_i \\
=&\int_{-\infty}^{\infty} \frac{\partial \varphi(x_i, x_{-i})}{\partial x_i}\, \phi(x_i, x_{-i})\, [g_i(x_i, x_{-i}) - g_i(0, x_{-i})]\, dx_i \\
=&\int_{-\infty}^{\infty} \frac{\partial \varphi(x_i, x_{-i})}{\partial x_i}\, \phi(x_i, x_{-i})\, g_i(x_i, x_{-i})\, dx_i\,,
\end{aligned}$$

since, using again (2.11),

$$-\int_{-\infty}^{\infty} \frac{\partial \varphi(x_i, x_{-i})}{\partial x_i}\, \phi(x_i, x_{-i})\, dx_i = \int_{-\infty}^{\infty} \frac{\partial \phi(x_i, x_{-i})}{\partial x_i}\, dx_i = 0\,.$$

Finally, integrating with respect to x_{-i} gives

$$\begin{aligned}
E_\theta\!\left[\frac{\partial g_i(X)}{\partial x_i}\right] &= \int_{\mathbb{R}^p} \frac{\partial g_i(x_i, x_{-i})}{\partial x_i}\, \phi(x_i, x_{-i})\, dx_i\, dx_{-i} \\
&= \int_{\mathbb{R}^p} \frac{\partial \varphi(x_i, x_{-i})}{\partial x_i}\, \phi(x_i, x_{-i})\, g_i(x_i, x_{-i})\, dx_i\, dx_{-i} \\
&= E_\theta\!\left[\frac{\partial \varphi(X)}{\partial x_i}\, g_i(X)\right]
\end{aligned}$$

and hence, summing on i gives (2.9), which is the desired result.

To show (2.7) assuming $E_\theta[|(X_i - \theta_i)^{\mathrm{T}} g_i(X)|] < \infty$ for $i \in \{1, \ldots, p\}$, it suffices to essentially reverse the steps in the above argument. □

The following corollary is immediate from Stein's lemma and the above discussion. Recall that $L(\theta, d) = \|d - \theta\|^2$, $R(\theta, \delta) = E_\theta[L(\theta, \delta(X))] = E_\theta[\|\delta(X) - \theta\|^2]$, and $E_\theta[\|g(X)\|^2] < \infty$ implies that for any $i = 1, \ldots, p$, $E_\theta[|(X_i - \theta_i)\, g_i(X)|] < \infty$.

Corollary 2.1 *Let $g(X)$ be a weakly differentiable function from $\mathbb{R}^p$ into $\mathbb{R}^p$ such that $E_\theta[\|g(X)\|^2] < \infty$. Then*

(1) $R(\theta, X + \sigma^2 g(X)) = E_\theta[p\sigma^2 + \sigma^4 (\|g(X)\|^2 + 2\,\mathrm{div} g(X))]$;
(2) $\delta(X) = X + \sigma^2 g(X)$ *is minimax as soon as* $\|g(X)\|^2 + 2\,div\, g(X) \le 0$ *a.e. and dominates X provided there is strict inequality on a set of positive measure; and*
(3) $p\sigma^2 + \sigma^4 (\|g(X)\|^2 + 2\,\mathrm{div} g(X))$ *is an unbiased estimator (in fact the UMVUE) of* $R(\theta, X + \sigma^2 g(X))$.

We note once again that $\delta(X)$ is minimax since it dominates (or ties) the minimax estimator X. In the next few sections we apply the above corollary to show domination of the James-Stein estimators and several others over the usual estimator in three and higher dimensions.

2.4 James-Stein Estimators and Other Improved Estimators

In this section, we apply the integration by parts results of Sect. 2.3 to obtain several classes of estimators that dominate the classical minimax estimator $\delta_0(X)$ in dimension 3 and higher. The estimators of James and Stein, Baranchik, and certain estimators shrinking toward subspaces are the main application of this section. Bayes (generalized, proper, and pseudo) are considered in Chap. 3. Throughout this section, except for Theorem 2.4, let $X \sim \mathcal{N}_p(\theta, \sigma^2 I_p)$ and loss be $L(\theta, \delta) = \|\delta - \theta\|^2$. According to Corollary 2.1 it suffices to find weakly differentiable functions g from $\mathbb{R}^p$ into $\mathbb{R}^p$ such that $E_\theta[\|g(X)\|^2] < \infty$ and $\|g(X)\|^2 + 2 \text{ div } g(X) \le 0$ (with strict inequality on a set of positive measure) in order to show that $\delta(X) = X + \sigma^2 g(X)$ dominates X.

2.4.1 James-Stein Estimators

The class of James-Stein estimators is given by

$$\delta_a^{JS}(X) = \left(1 - \frac{a\,\sigma^2}{\|X\|^2}\right) X. \tag{2.13}$$

The basic properties of $\delta_a^{JS}(X)$ are given in the following result.

Theorem 2.2 *Under the above model*

(1) *The risk of $\delta_a^{JS}(X)$ is given by*

$$R(\theta, \delta_a^{JS}) = p\,\sigma^2 + \sigma^4\,(a^2 - 2\,a\,(p-2))E_\theta\left[\frac{1}{\|X\|^2}\right] \tag{2.14}$$

for $p \ge 3$.

(2) *$\delta_a^{JS}(X)$ dominates $\delta_0(X) = X$ for $0 < a < 2\,(p-2)$ and is minimax for $0 \le a \le 2\,(p-2)$ for all $p \ge 3$.*

(3) *The uniformly optimal choice of a is $a = p - 2$ for $p \ge 3$.*

(4) *The risk at $\theta = 0$ for the optimal James-Stein estimator $\delta_{p-2}^{JS}(X)$ is $2\,\sigma^2$ for all $p \ge 3$.*

Proof Observe that $\delta_a^{JS}(X) = X + \sigma^2 g(X)$ where $g(X) = -a/\|X\|^2\,X$. As noted in Sect. 2.3, $g(X)$ is weakly differentiable if $p \ge 3$. Also $E_\theta[\|g(X)\|^2] = a^2\,E_\theta[1/\|X\|^2]$ is finite if $p \ge 3$ since $\|X\|^2/\sigma^2$ has a non-central χ^2 distribution with p degrees of freedom and non-centrality parameter $\lambda = \|\theta\|^2/2\sigma^2$. Indeed by the usual Poisson representation of a non-central χ^2, we have $\|X\|^2/\sigma^2 \mid K \sim \chi^2_{p+2K}$ where $K \sim$ Poisson ($\lambda = \|\theta\|^2/2\sigma^2$) and hence,

$$E_\theta\left[\frac{\sigma^2}{\|X\|^2}\right] = E_\lambda\left[E\left[\frac{1}{\chi^2_{p+2K}}\Big|K\right]\right] = E_\lambda\left[\frac{1}{p+2K-2}\right] \leq \frac{1}{p-2} < \infty \tag{2.15}$$

if $p > 2$.

Also, according to (A.18), for any $x \neq 0$,

$$\mathrm{div}\left(\frac{x}{\|x\|^2}\right) = \frac{p-2}{\|x\|^2}. \tag{2.16}$$

Hence,

$$\|g(x)\|^2 + 2\,\mathrm{div} g(x) = (a^2 - 2\,a\,(p-2))\,\frac{1}{\|x\|^2}$$

and by Corollary 2.1, for $p \geq 3$,

$$R(\theta, \delta_a^{JS}) = p\,\sigma^2 + \sigma^4\,(a^2 - 2\,a\,(p-2))\,E_\theta\!\left(\frac{1}{\|X\|^2}\right).$$

This proves (1).

Part (2) follows since $a^2 - 2\,a\,(p-2) < 0$ for $0 < a < 2\,(p-2)$ and hence for such $a > 0$,

$$R(\theta, \delta_a^{JS}) < p\,\sigma^2 = R(\theta, \delta_0). \tag{2.17}$$

The minimaxity claim for $0 \leq a \leq 2\,(p-2)$ follows by replacing $<$ by $\leq$ in (2.17). It is interesting to note that $R\big(\theta, \delta^{JS}_{2(p-2)}\big) \equiv R(\theta, \delta_0) \equiv p\,\sigma^2$ and, more generally, $R(\theta, \delta_{2(p-2)-a}) \equiv R\big(\theta, \delta_a^{JS}\big)$.

Part (3) follows by noting that, for all θ, the risk of $R\big(\theta, \delta_a^{JS}\big)$ is minimized by choosing $a = p-2$ since this value minimizes the quadratic $a^2 - 2\,a\,(p-2)$.

To prove part (4) note that $\|X\|^2/\sigma^2$ has a central chi-square distribution with p degrees of freedom when $\theta = 0$. Hence, $E_0\left[\sigma^2/\|X\|^2\right] = E\Big[1/\chi^2_p\Big] = (p-2)^{-1}$ and therefore, provided $p \geq 3$,

$$\begin{aligned} R(0, \delta^{JS}_{p-2}) &= p\,\sigma^2 + \big((p-2)^2 - 2\,(p-2)^2\big)\,\frac{\sigma^2}{p-2} \\ &= p\,\sigma^2 - (p-2)\,\sigma^2 \\ &= 2\,\sigma^2\,. \end{aligned}$$

□

Hence we have that $\delta^{JS}_{p-2} = \left(1-(p-2)\,\sigma^2/\|X\|^2\right) X$ is the uniformly best estimator in the class of James-Stein estimators. This is the estimator that is typically referred to as the James-Stein estimator. Also note that at $\theta = 0$ the risk is $2\,\sigma^2$ regardless of p and so, large savings in risk are possible in a neighborhood of $\theta = 0$ for large p.

In Theorem 2.2, the fact that $p \geq 3$ is crucial (which is coherent with the admissibility of X for $p = 1$ and $p = 2$). Actually, a crucial part of the proof uses Stein's identity, which fails to hold if $p = 1, 2$ with $h(x) = x/\|x\|^2$. Indeed, when $p = 1$, $h(x) = 1/x$ and $\text{div}(x) = -1/x^2$ so that $E_0[X^{\mathrm{T}}h(X)] = 1$ and $E_0[\text{div}h(X)] = -\infty$. When $p = 2$, we also have $E_0[X^{\mathrm{T}}h(X)] = 1$ while $E_0[\text{div}h(X)] = 0$ since, for any $x \neq 0$, $\text{div}h(x) = 0$. It is interesting to note that, while the divergence of h exists and is 0 almost everywhere, h is not weakly differentiable since its partial derivatives are not locally integrable as shown in Sect. A.1 in the Appendix.

We may use (2.15) to give upper and lower bounds for the risk of δ^{JS}_a based on the following lemma.

Lemma 2.1 *Let $K \sim$ Poisson(λ). Then, for $b \geq 1$, we have*

$$\frac{1}{b+\lambda} \leq E_\lambda\left[\frac{1}{b+K}\right] \leq \frac{\frac{1-e^{-\lambda}}{\lambda}}{(b-1)\frac{1-e^{-\lambda}}{\lambda}+1} \leq \frac{1}{b-1+\lambda}.$$

Proof The first inequality follows directly from Jensen's inequality and the fact that $E_\lambda(K) = \lambda$. The second inequality follows since (also by Jensen's inequality)

$$\begin{aligned} E_\lambda\left[\frac{1}{b+K}\right] &= E_\lambda\left[\frac{\frac{1}{K+1}}{\frac{b-1}{K+1}+1}\right] \\ &\leq \frac{E_\lambda\left[\frac{1}{K+1}\right]}{(b-1)E_\lambda\left[\frac{1}{K+1}\right]+1} \\ &= \frac{\frac{1-e^{-\lambda}}{\lambda}}{(b-1)\frac{1-e^{-\lambda}}{\lambda}+1} \end{aligned}$$

and $E_\lambda\left[(K+1)^{-1}\right] = (1-\exp(-\lambda))/\lambda$.

Now, since $y/[(b-1)y+1]$ is increasing in y and $(1-\exp(-\lambda))/\lambda < \lambda^{-1}$, we have

$$\frac{\frac{1-e^{-\lambda}}{\lambda}}{(b-1)\,\frac{1-e^{-\lambda}}{\lambda}+1} \leq \frac{\frac{1}{\lambda}}{\frac{b-1}{\lambda}+1} = \frac{1}{b-1+\lambda}.$$

Hence the third inequality follows. □

The following bounds on the risk of δ_a^{JS} follow directly from (2.14), (2.15) and Lemma 2.1.

Corollary 2.2 (Hwang and Casella 1982) *For $p \geq 4$ and $0 \leq a \leq 2\,(p-2)$, we have*

$$p\,\sigma^2 + \frac{(a^2 - 2\,a\,(p-2))\,\sigma^2}{p - 2 + \|\theta\|^2/\sigma^2} \leq R(\theta, \delta_a^{JS}) \leq p\,\sigma^2 + \frac{(a^2 - 2\,a\,(p-2))\,\sigma^2}{p - 4 + \|\theta\|^2/\sigma^2}.$$

We note in passing that the upper bound may be improved at the cost of added complexity by using the second inequality in Lemma 2.1. The improved upper bound has the advantage that it is exact at $\theta = 0$. The lower bound is also valid for $p = 3$ and is also exact at $\theta = 0$.

2.4.2 *Positive-Part and Baranchik-Type Estimators*

James-Stein estimators are such that, when $\|X\|^2 < a\,\sigma^2$, the multiplier of X becomes negative and, furthermore, $\lim_{\|X\| \to 0} \|\delta_a^{JS}(X)\| = \infty$. It follows that, for any $K > 0$, there exits $\eta > 0$ such that $\|X\| < \eta$ implies $\|\delta_a^{JS}(X)\| > K$. Hence an observation that would lead to almost certain acceptance of $H_0 : \theta = 0$ gives rise to an estimate very far from 0. Furthermore the estimator is not monotone in the sense that a larger value of X for a particular coordinate may give a smaller estimate of the mean of that coordinate. For example, if $X = (X_0, 0, \ldots, 0)$ and $-\sqrt{a\,\sigma^2} < X_0 < 0$, then $\left(1 - a\,\sigma^2/\|X\|^2\right) X_0 > 0$ while, if $0 < X_0 < \sqrt{a\,\sigma^2}$, then $\left(1 - a\,\sigma^2/\|X\|^2\right) X_0 < 0$.

This behavior is undesirable. One possible remedy is to modify the James-Stein estimator to its positive-part, namely

$$\delta_a^{JS+}(X) = \left(1 - \frac{a\,\sigma^2}{\|X\|^2}\right)_+ X \tag{2.18}$$

where $t_+ = \max(t, 0)$. The positive past estimate is a particular example of a Baranchik-type estimator of the form

$$\delta_{a,r}^B(X) = \left(1 - \frac{a\,\sigma^2\,r(\|X\|^2)}{\|X\|^2}\right) X \tag{2.19}$$

where, typically $r(\cdot)$ is continuous and nondecreasing. The $r(\cdot)$ function for δ_a^{JS+} is given by

$$r(\|X\|^2) = \begin{cases} \frac{\|X\|^2}{a\,\sigma^2} & \text{if} \quad 0 < \|X\|^2 < a\,\sigma^2 \\ 1 & \text{if} \quad \|X\|^2 \geq a\,\sigma^2. \end{cases}$$

We show in this section that, under certain conditions, the Baranchik-type estimators improve on X and that the positive-part James-Stein estimator improves on the James-Stein estimator as well.

We first give conditions under which a Baranchik-type estimator improves on X.

Theorem 2.3 *The estimator given by* (2.19) *with* $r(\cdot)$ *absolutely continuous, is minimax for* $p \geq 3$ *provided*

(1) $0 < a \leq 2\,(p-2)$;
(2) $0 \leq r(\cdot) \leq 1$; *and*
(3) $r(\cdot)$ *is nondecreasing.*

Furthermore, it dominates X provided that both inequalities are strict in (1) *or in* (2) *on a set of positive measure, or if* $r'(\cdot)$ *is strictly positive on a set of positive measure.*

Proof Here $\delta^B_{a,r}(X) = X + \sigma^2\, g(X)$ where $g(X) = (-a\, r(\|X\|^2)/\|X\|^2)\; X$. As noted in Sect. A.2 of the Appendix, $g(\cdot)$ is weakly differentiable and

$$\begin{aligned} \operatorname{div} g(X) &= -a \left\{ r(\|X\|^2) \operatorname{div}\left(\tfrac{X}{\|X\|^2}\right) + \tfrac{X^{\mathrm{T}}}{\|X\|^2} \nabla r(\|X\|^2) \right\} \\ &= -a \left\{ r(\|X\|^2) \tfrac{p-2}{\|X\|^2} + 2\, r'(\|X\|^2) \right\}. \end{aligned}$$

Hence,

$$\begin{aligned} &\|g(X)\|^2 + 2\operatorname{div} g(X) \\ &= \frac{a^2\, r^2(\|X\|^2)}{\|X\|^2} - \frac{2\,a\,(p-2) r(\|X\|^2)}{\|X\|^2} - 4\, a\, r'(\|X\|^2) \\ &\leq \frac{r(\|X\|^2)}{\|X\|^2} (a^2 - 2\,a\,(p-2)) - 4\,a\,r'(\|X\|^2) \\ &\leq 0\,. \end{aligned} \tag{2.20}$$

The first inequality being satisfied by Conditions (2) while the last inequality uses all of Conditions (1), (2), and (3). Hence, minimaxity follows from Corollary 2.1. Under the additional conditions, it is easy to see that the above inequalities become strict on a set of positive measure so that domination over X is guaranteed. □

As an example of a dominating Baranchik-type estimator consider

$$\delta(X) = \left(1 - \frac{a\,\sigma^2}{b + \|X\|^2}\right) X$$

for $0 < a \leq 2\,(p-2)$ and $b > 0$. Here $r(\|X\|^2) = \|X\|^2/(\|X\|^2 + b)$ and is strictly increasing.

The theorem also shows that the positive-part James-Stein estimator dominates X for $0 < a \le 2(p-2)$. In fact, as previously noted, the positive-part James-Stein estimator even improves on the James-Stein estimator itself. This reflects the more general phenomenon that a positive-part estimator will typically dominate the non-positive-part version if the underlying density is symmetric and unimodal. Here is a general result along these lines.

Theorem 2.4 *Suppose X has a density $f(x-\theta)$ in $\mathbb{R}^p$ such that the function f is symmetric and unimodal in each coordinate separately for each fixed value of the other coordinates. Then, for any finite risk estimator of θ of the form*

$$\delta(X) = \left(1 - B\left(X_1^2, X_2^2, \ldots, X_p^2\right)\right) X,$$

the positive-part estimator

$$\delta_+(X) = \left(1 - B\left(X_1^2, X_2^2, \ldots, X_p^2\right)\right)_+ X$$

dominates $\delta(X)$ under any loss of the form $L(\theta, \delta) = \sum_{i=1}^p a_i(\delta_i - \theta_i)^2$ ($a_i > 0$ for all i) provided $P_\theta[B(X_1^2, X_2^R, \ldots, X_p^2) > 1] > 0$.

Proof Note that the two estimators differ only on the set where $B(\cdot) > 1$. Hence the ith term in $R(\theta, \delta) - R(\theta, \delta_+)$ is

$$\begin{aligned} &a_i\, E_\theta\left[\left\{(1 - B(X_1^2, \ldots, X_p^2))^2\, X_i^2 - 2\,\theta_i X_i\,(1 - B(X_1^2, \ldots, X_p^2)\right\} I_{B>1}(X)\right] \\ &\quad > -2\,\theta_i\, a_i\, E_\theta\left[X_i(1 - B(X_1^2, \ldots, X_p^2) I_{B>1}(X)\right]. \end{aligned}$$

Therefore it suffices to show that, for any nonnegative function $H(X_1^2, \ldots, X_p^2)$, $\theta_i E_\theta[X_i H(X_1^2, \ldots, X_p^2)] \ge 0$. This follows by symmetry if whenever $\theta_i \ge 0$, then $E_\theta[X_i \mid X_i^2 = t_i^2, X_j = t_j\ j \ne i] \ge 0$ for all i ($1 \le i \le p$) and all $(t_1, \ldots, t_p)$. However this expression is proportional to

$$\begin{aligned} &\mid t_i \mid \left[f\left((t_1-\theta_1)^2, (t_2-\theta_2)^2, \ldots, (\mid t_i \mid -\theta_i)^2, \ldots, (t_p-\theta_p)^2\right)\right. \\ &\left. -f\left((t_1-\theta_1)^2, (t_2-\theta_2)^2, \ldots, (-\mid t_i \mid -\theta_i)^2, \ldots, (t_p-\theta_p)^2\right)\right] \ge 0 \end{aligned}$$

since, for $\theta_i \ge 0$, $(\mid t_i \mid -\theta_i)^2 \le (-\mid t_i \mid -\theta_i)^2$ and since $f(X_1^2, X_2^2, \ldots, X_p^2)$ is nonincreasing in each argument. Hence the theorem follows. □

For the remainder of this current section we return to the assumption that $X \sim \mathcal{N}_p(\theta, \sigma^2 I_p)$.

The positive-part James-Stein estimators are inadmissible because of a lack of smoothness which precludes them from being generalized Bayes. The Baranchik class however contains "smooth" estimators which are generalized (and even proper) Bayes and admissible. Baranchik-type estimators will play an important role in Chap. 3.

We close this subsection with a generalization of the Baranchik result in Theorem 2.3. It is apparent from the proof of the theorem that it is only necessary that the second expression in (2.20) be nonpositive (and negative on a set of positive measure) in order for $\delta(X)$ to dominate X. In particular it is not necessary that $r(\cdot)$ be nondecreasing. The following result (see Efron and Morris 1976 and Fourdrinier and Ouassou 2000) gives a necessary and sufficient condition for the unbiased estimator of risk difference, $R(\theta, \delta) - R(\theta, X)$, for $\delta(X) = \left(1 - a\, r(\|X\|^2)/\|X\|^2\right) X$, to be nonpositive. The proof is by direct calculation.

Lemma 2.2 *Let* $g(X) = -a\left(r(\|X\|^2)/\|X\|^2\right) X$ *where* $r(y)$ *is an absolutely continuous function from* $\mathbb{R}^+$ *into* $\mathbb{R}$. *Then on the set where* $r(y) \neq 0$,

$$\|g(x)\|^2 + 2\,\mathrm{div} g(x) = a\left\{\frac{a\, r^2(y)}{y} - \frac{2(p-2) r(y)}{y} - 4\, r'(y)\right\}$$

$$= -4\, a^2 r^2(y) y^{\frac{p-2}{2}} \frac{d}{dy}\left[y^{-\frac{p-2}{2}}\left(\frac{1}{2(p-2)} - \frac{1}{a\, r(y)}\right)\right] \; a.e.,$$

where $y = \|x\|^2$.

The following corollary broadens the class of minimax estimators of Baranchik's form.

Corollary 2.3 *Suppose* $\delta(X) = \left(1 - a\, r(\|X\|^2)/\|X\|^2\right) X$ *with*

$$a\, r(y) = \left[\frac{1}{2\,(p-2)} + y^{(p-2)/2}\, H(y)\right]^{-1}$$

where $H(y)$ *is absolutely continuous, nonnegative and nonincreasing. Then* $\delta(X)$ *is minimax provided* $E_\theta\left[r^2(\|X\|^2)/\|X\|^2\right] < \infty$. *If in addition* $H(y)$ *is strictly monotone on a set of positive measure where* $r(y) \neq 0$, *then* $\delta(X)$ *dominates* X.

Proof The result follows from Corollaries 2.1 and 2.2 by noting that

$$H(y) = y^{-(p-2)/2}\left(\frac{1}{2(p-2)} - \frac{1}{a\, r(y)}\right).$$

□

An application of Corollary 2.3 gives a useful class of dominating estimators due to Alam (1973).

Corollary 2.4 *Let* $\delta(X) = \left(1 - a\, f(\|X\|^2)/(\|X\|^2)^{\tau+1}\right) X$ *where* $f(y)$ *is nondecreasing and absolutely continuous and where* $0 \leq a\, f(y)/y^\tau < 2\,(p - 2 - 2\,\tau)$ *for some* $\tau \geq 0$. *Then* $\delta(X)$ *is minimax and dominates* X *if* $0 < a\, f(y)/y^\tau$ *on a set of positive measure.*

Proof The proof follows from Corollary 2.3 by letting

$$a\,r(y) = \frac{a\,f(y)}{y^{\tau}} \quad \text{and} \quad H(y) = -y^{-(p-2)/2}\left(\frac{1}{2\,(p-2)} - \frac{y^{\tau}}{a\,f(y)}\right).$$

Clearly r is bounded so that $E_\theta\left[r^2(\|X\|^2)/\|X\|^2\right] < \infty$ and $H(y) \geq 0$. Also

$$\begin{aligned} H'(y) &= \frac{p-2}{2}y^{-p/2}\left(\frac{1}{2\,(p-2)} - \frac{y^{\tau}}{a\,f(y)}\right) \\ &\quad - y^{-(p-2)/2}\left(\frac{-\tau\,y^{\tau-1}}{a\,f(y)} + \frac{y^{\tau}\,f^{\tau}(y)}{a\,f^2(y)}\right) \\ &\leq y^{-\frac{p}{2}}\left[\frac{1}{4} - y^2\frac{p-2-2\tau}{2a\,f(y)}\right] \\ &\leq 0 \end{aligned}$$

since $f'(y) \geq 0$ and $0 < a\,f(y)/y^{\tau} < 2\,(p-2-2\,\tau)$. □

A simple example of a minimax Baranchik-type estimator with a nonmonotone $r(\cdot)$ is given by $r(y) = y^{1-\tau}/(1+y)$ for $0 < \tau < 1$ and $0 < a < 2(p-2-2\,\tau)$. To see this, apply Corollary 2.4 with $f(y) = y/(1+y)$ and note that $f(y)$ is increasing and $0 \leq f(y)/y^{\tau} = r(y) \leq 1$. Note also that $r'(y) = y^{-\tau}[(1-\tau) - \tau\,y]/(1+y)^2$, hence $r(y)$ is increasing for $0 < y < (1-\tau)/\tau^{-1}$ and decreasing for $y > (1-\tau)/\tau^{-1}$.

We will use the above corollaries in Chap. 3 to establish minimaxity of certain Bayes and generalized Bayes estimators.

2.4.3 Unknown Variance

In the development above, it was tacitly assumed that the covariance matrix was known and equal to a multiple of the identity matrix $\sigma^2\,I_p$. Typically, this covariance is unknown and should be estimated. The next result extends Stein's identity (2.7) to the case where it is of the form $\sigma^2\,I_p$ with σ^2 unknown.

Lemma 2.3 *Let $X \sim \mathcal{N}_p(\theta, \sigma^2 I_p)$ and let S be a nonnegative random variable independent of X such that $S \sim \sigma^2\chi^2_k$. Denoting by E_{θ,σ^2} the expectation with respect to the joint distribution of (X, S), we have the following two results, provided the corresponding expectations exist:*

(1) *if $g(x, s)$ is a function from $\mathbb{R}^p \times \mathbb{R}_+$ into $\mathbb{R}^p$ such that, for any $s \in \mathbb{R}_+$, $g(\cdot, s)$ is weakly differentiable, then*

$$E_{\theta,\sigma^2}\left[\frac{1}{\sigma^2}(X-\theta)^{\mathrm{T}}g(X, S)\right] = E_{\theta,\sigma^2}[\mathrm{div}_X g(X, S)]$$

where $\mathrm{div}_x g(x, s)$ is the divergence of $g(x, s)$ with respect to x;

(2) *if* $h(x, s)$ *is a function from* $\mathbb{R}^p \times \mathbb{R}_+$ *into* $\mathbb{R}$ *such that, for any* $x \in \mathbb{R}^p$, $h(x, \|u\|^2)$ *is weakly differentiable as a function of* u, *then*

$$E_{\theta,\sigma^2}\left[\frac{1}{\sigma^2}\,h(X,S)\right] = E_{\theta,\sigma^2}\left[2\,\frac{\partial}{\partial S}h(X,S) + (k-2)\,S^{-1}\,h(X,S)\right].$$

Proof Part (1) is Stein's lemma, from Theorem 2.1. Part (2) can be seen as a particular case of Lemma 1(ii) (established for elliptically symmetric distributions) of Fourdrinier et al. (2003), although we will present a direct proof. Part (2) also follows from well known identities for chi-square distributions.

The joint distribution of (X, S) can be viewed as resulting, in the setting of the canonical form of the general linear model, from the distribution of $(X, U) \sim \mathcal{N}((\theta, 0), \sigma^2 I_{p+k})$ with $S = ||U||^2$. Then we can write

$$\begin{aligned} E_{\theta,\sigma^2}\left[\frac{1}{\sigma^2}\,h(X,S)\right] &= E_{\theta,\sigma^2}\left[\frac{1}{\sigma^2}\,U^{\mathrm{T}}\,\frac{U}{||U||^2}\,h(X,||U||^2)\right] \\ &= E_{\theta,\sigma^2}\left[\mathrm{div}_U\left(\frac{U}{||U||^2}\,h(X,||U||^2)\right)\right] \end{aligned}$$

according to Part (1). Hence, expanding the divergence term, we have

$$\begin{aligned} E_{\theta,\sigma^2}\left[\frac{1}{\sigma^2}\,h(X,S)\right] &= E_{\theta,\sigma^2}\left[\frac{k-2}{||U||^2}\,h(X,||U||^2) + \frac{U^{\mathrm{T}}}{||U||^2}\,\partial_U h(X,||U||^2)\right] \\ &= E_{\theta,\sigma^2}\left[\frac{k-2}{S}\,h(X,S) + 2\,\frac{\partial}{\partial S}h(X,S)\right] \end{aligned}$$

since

$$\partial_U h(X,||U||^2) = 2\,\frac{\partial}{\partial S}h(X,S)\Big|_{S=||U||^2}\,U\,.$$

□

The following theorem provides an estimate of risk in the setting of an unknown variance when the loss is given by

$$\frac{||\delta-\theta||^2}{\sigma^2}\,. \tag{2.21}$$

Theorem 2.5 *Let* $X \sim \mathcal{N}_p(\theta, \sigma^2 I_p)$ *where* θ *and* σ^2 *are unknown and* $p \geq 3$ *and let* S *be a nonnegative random variable independent of* X *such that* $S \sim \sigma^2\chi^2_k$. *Consider an estimator of* θ *of the form* $\varphi(X, S) = X + S\,g(X, S)$ *with* $E_{\theta,\sigma^2}[S^2\,||g(X, S)||^2] < \infty$, *where* E_{θ,σ^2} *denotes the expectation with respect to the joint distribution of* (X, S).

Then an unbiased estimator of the risk under loss (2.21) *is*

$$\delta_0(X,S) = p + S\left\{(k+2)\,||g(X,S)||^2 + 2\,\mathrm{div}_X g(X,S) + 2\,S\frac{\partial}{\partial S}||g(X,S)||^2\right\}. \tag{2.22}$$

Proof According to the expression of $\varphi(X, S)$, its risk $R(\theta, \varphi)$ is the expectation of

$$\frac{1}{\sigma^2}\,||X-\theta||^2 + 2\,\frac{S}{\sigma^2}\,(X-\theta)^{\mathrm{T}} g(X,S) + \frac{S^2}{\sigma^2}\,||g(X,S)||^2. \tag{2.23}$$

Clearly,

$$E_{\theta,\sigma^2}\left[\frac{1}{\sigma^2}\,||X-\theta||^2\right] = p$$

and Lemma 2.3 (1) and (2) express, respectively, that

$$E_{\theta,\sigma^2}\left[\frac{1}{\sigma^2}(X-\theta)^{\mathrm{T}} g(X,S)\right] = E_{\theta,\sigma^2}[\mathrm{div}_X g(X,S)].$$

With $h(x, s) = s^2\,||g(x, s)||^2$ we have

$$E_{\theta,\sigma^2}\left[\frac{S^2}{\sigma^2}\,||g(X,S)||^2\right] = E_{\theta,\sigma^2}\left[S\left\{(k+2)\,||g(X,S)||^2 + 2\,S\,\frac{\partial}{\partial S}||g(X,S)||^2\right\}\right].$$

Therefore $R(\theta,\varphi) = E_{\theta,\sigma^2}[\delta_0(X,S)]$ with $\delta_0(X,S)$ given in (2.22), which means that $\delta_0(X, S)$ is an unbiased estimator of the risk $||\varphi(X,S)-\theta||^2/\sigma^2$. □

Corollary 2.5 *Under condition of Theorem* 2.5, *if, for any* $(x, s) \in \mathbb{R}^p \times \mathbb{R}_+$,

(i) $\partial/\partial s\,\|g(x,s)\|^2 \le 0$ *and*
(ii) $(k+2)\,||g(x,s)||^2 + 2\,\mathrm{div}_x g(x,s) + 2 \le 0$,

then $\varphi(X, S)$ *is minimax. It dominates* X *if either inequality is strict on a set of positive measure.*

In the following corollary, we consider an extension of the Baranchik form in Theorem 2.3.

Corollary 2.6 *Let*

$$\delta(X,S) = \left(1 - \frac{a\,S\,r(\|X\|^2/S)}{\|X\|^2}\right) X$$

If r *is nondecreasing and if* $0 < a\,r(\|X\|^2/S) < 2\,(p-2)/(k+2)$, *then* $\delta(X, S)$ *dominates* X *and is minimax.*

Proof Straightforward calculations show that the term in curly brackets in (2.22) equals

$$a\,\frac{r(\|X\|^2/S)}{\|X\|^2}\,((k+2)\,a\,r(\|X\|^2/S)-2\,(p-2))$$
$$-4\,a\,\frac{r'(\|X\|^2/S)}{S}(1+a\,r(\|X\|^2/S)). \qquad (2.24)$$

Therefore, if $0 < a\,r(\|X\|^2/S) < 2\,(p-2)/(k+2)$, then $\delta(X,S)$ dominates X and is minimax. □

Note that, in the case $r \equiv 1$, the bound on the constant a is $2\,(p-2)/(k+2)$. This is the estimator developed by James and Stein (1961) using direct methods.

2.4.4 Estimators That Shrink Toward a Subspace

We saw in Sect. 2.4.1, when σ^2 is known, that the James-Stein estimator shrinks toward $\theta = 0$ and that substantial risk savings are possible if θ is in a neighborhood of 0. If we feel that θ is close to some other value, say θ_0, a simple adaptation of the James-Stein estimator that shrinks toward θ_0 may be desirable. Such an estimator is given by

$$\delta^{JS}_{a,\theta_0}(X) = \theta_0 + \left(1 - \frac{a\,\sigma^2}{\|X-\theta_0\|^2}\right)(X-\theta_0). \qquad (2.25)$$

It is immediate that $R(\theta, \delta^{JS}_{a,\theta_0}(X)) = R(\theta-\theta_0, \delta^{JS}_a)$ since

$$\begin{aligned} R(\theta, \delta^{JS}_{a,\theta_0}) &= E_\theta\,\|\theta_0 + \left(1 + \frac{a\,\sigma^2}{\|X-\theta_0\|^2}\right)(X-\theta_0) - \theta\|^2 \\ &= E_{\theta-\theta_0}\|\left(1 + \frac{a\,\sigma^2}{\|X\|^2}\right)X - (\theta-\theta_0)\|^2 \\ &= R(\theta-\theta_0, \delta^{JS}_a(X)). \end{aligned}$$

Hence, for $p \geq 3$, δ^{JS}_{a,θ_0} dominates X and is minimax for $0 < a < 2\,(p-2)$, and $a = p-2$ is the optimal choice of a. Furthermore the risk of $\delta^{JS}_{a,\theta_0}(X)$ at $\theta = \theta_0$ is $2\,\sigma^2$ and so large gains in risk are possible in a neighborhood of θ_0. The same argument establishes the fact that, for any estimator, $\delta(X)$, we have $R(\theta, \theta_0 + \delta(X-\theta_0)) = R(\theta-\theta_0, \delta(X))$. Hence any of the minimax estimators of Sects. 2.4.1 and 2.4.2 may be modified in this way and minimaxity will be preserved.

More generally, we may feel that θ is close to some subspace V of dimension $s < p$. In this case, we may wish to shrink X toward the subspace V. One way to do this is to consider the class of estimators given by

$$P_V X + \left(1 - \frac{a\,\sigma^2 r(\|X - P_V X\|^2)}{\|X - P_V X\|^2}\right)(X - P_V X) \tag{2.26}$$

where $P_V X$ is the projection of X onto V.

A standard canonical representation is helpful. Suppose V is an s-dimensional linear subspace of $\mathbb{R}^p$ and $V^\perp$ is the $p - s$ dimensional orthogonal complement of V. Let $P = (P_1 \;\; P_2)$ be an orthogonal matrix such that the s columns of the $p \times s$ matrix P_1 span V and the $p - s$ columns of the $p \times (p - s)$ matrix P_2 span $V^\perp$.

For any vector $z \in \mathbb{R}^p$, let

$$W = \begin{pmatrix} W_1 \\ W_2 \end{pmatrix} = P^{\mathrm{T}} z$$

where W_1 is $s \times 1$ and W_2 is $(p - s) \times 1$. Then $P_V z = P_1 W_1$ and $\|P_V z\|^2 = \|P_1 W_1\|^2 = \|W_1\|^2$. Also $P_{V^\perp} z = P_2 W_2$ and $\|P_{V^\perp} z\|^2 = \|P_2 W_2\|^2 = \|W_2\|^2$. Further, if $X \sim \mathcal{N}_p(\theta, \sigma^2 I)$, then

$$P^{\mathrm{T}} X = \begin{pmatrix} Y_1 \\ Y_2 \end{pmatrix} \sim \mathcal{N}_p\left(\begin{pmatrix} \nu_1 \\ \nu_2 \end{pmatrix}, \sigma^2 \begin{pmatrix} I_s & 0 \\ 0 & I_{p-s} \end{pmatrix}\right)$$

where $P_1 \nu_1 = P_V \theta$ and $P_2 \nu_2 = P_{V^\perp} \theta$ so that

$$\|P_V X\|^2 = \|Y_1\|^2, \quad \|P_{V^\perp} X\|^2 = \|Y_2\|^2$$

and

$$\|P_V (X - \theta)\|^2 = \|Y_1 - \nu_1\|^2, \quad \|P_{V^\perp}(X - \theta)\|^2 = \|Y_2 - \nu_2\|^2.$$

The following result gives risk properties of the estimator (2.26).

Theorem 2.6 *Let V be a subspace of dimension $s \geq 0$. Then, for the estimator (2.26), we have*

$$R(\theta, \delta) = s\,\sigma^2 + E_{\nu_2}\left[\left\|\left(1 - \frac{a\,\sigma^2 r(\|Y_2\|^2)}{\|Y_2\|^2}\right) Y_2 - \nu_2\right\|^2\right]$$

where Y_2 and ν_2 are as above. Further, if $p - s \geq 3$ and a and $r(y)$ satisfy the assumptions of Theorem 2.3 (or Corollary 2.3 or Corollary 2.4) with $p - s$ in place of p, then $\delta(X)$ is minimax and dominates X if the additional conditions are satisfied.

Proof The proof involves showing that the risk decomposes into the sum of two components. The first component is essentially the risk of the usual estimator in a space of dimension s (i.e. of V) and the second represents the risk of a Baranchik-type estimator in a space of dimension $p - s$. The risk is

$$\begin{aligned} R(\theta, \delta) &= E_\theta\left[\left\|P_V X + \left(1 - \frac{a\,\sigma^2 r(\|X - P_V X\|^2)}{\|X - P_V X\|^2}\right)(X - P_V X) - \theta\right\|^2\right] \\ &= E_\theta\left[\left\|(P_V X - P_V \theta) + \left(1 - \frac{a\,\sigma^2 r(\|X - P_V X\|^2)}{\|X - P_V X\|^2}\right)(X - P_V X) - (\theta - P_V \theta)\right\|^2\right] \\ &= E_\theta[\|P_V(X - \theta)\|^2] \\ &\qquad + E_\theta\left[\left\|\left(1 - \frac{a\,\sigma^2 r(\|X - P_V X\|^2)}{\|X - P_V X\|^2}\right)(X - P_V X) - (\theta - P_V\theta)\right\|^2\right] \\ &= E_{\nu_1}[\|Y_1 - \nu_1\|^2] + E_{\nu_2}\left[\left\|\left(1 - \frac{a\,\sigma^2 r(\|Y_2\|^2)}{\|Y_2\|^2}\right) Y_2 - \nu_2\right\|^2\right] \\ &= s\,\sigma^2 + E_{\nu_2}[\| \left(1 - \frac{a\,\sigma^2 r(\|Y_2\|^2)}{\|Y_2\|^2}\right) Y_2 - \nu_2\|^2]\,. \end{aligned}$$

This gives the first part of the theorem. The second part follows since $Y_2 \sim \mathcal{N}_{p-s}(\nu_2, \sigma^2 I_{p-s})$, with $p - s \geq 3$. □

For example, if we choose $r(y) \equiv 1$ the risk of the resulting James-Stein type estimator

$$P_V X + \left(1 - \frac{a\,\sigma^2}{\|X - P_V X\|^2}\right)(X - P_V X)$$

is

$$p\,\sigma^2 + \sigma^4\,(a^2 - 2\,a\,(p - s - 2))\,E_\theta\left[\frac{1}{\|X - P_V X\|^2}\right].$$

This estimator is minimax if $0 \leq a \leq 2\,(p - s - 2)$ and dominates X if $0 < a < 2\,(p - s - 2)$ provided $p - s \geq 3$. The uniformly best choice of a is $p - s - 2$. If in fact $\theta \in V$, the risk of the corresponding optimal estimator is $(s + 2)\,\sigma^2$, since in this case $\nu_2 = P_{V^\perp}\theta = 0$ and $E_\theta\left[\sigma^2\|X - P_V X\|^{-2}\right] = E_0\left[\sigma^2\|Y_2\|^{-2}\right] E\left[1/\chi^2_{p-s}\right] = (p - s - 2)^{-1}$. If $\theta \notin V$, then $\nu_2 \neq 0$ and $\|Y_2\|^2$ has a non-central chi-square distribution with $p - s$ degrees of freedom and non-centrality parameter $\|\nu_2\|^2/2\,\sigma^2$.

One of the first instances of an estimator shrinking toward a subspace is due to Lindley (1962). He suggested that while we might not have a good idea as to the value of the vector θ, one may feel that the components are approximately equal. This suggests shrinking all the coordinates to the overall coordinate mean $\bar{X} = p^{-1}\sum_{i=1}^{p} X_i$ which amounts to shrinking toward the subspace V of dimension one generated by the vector $\mathbf{1} = (1, \ldots, 1)^{\mathrm{T}}$. The resulting optimal James-Stein type estimator is

$$\delta(X) = \bar{X}\,\mathbf{1} + \left(1 - \frac{(p-3)\,\sigma^2}{\|X - \bar{X}\mathbf{1}\|^2}\right)(X - \bar{X}\,\mathbf{1}).$$

Here, the risk is equal to $3\,\sigma^2$ if in fact all coordinates of θ are equal. If the dimension of the subspace V is also at least 3 we could consider applying a shrinkage estimator to $P_V X$ as well.

In the case where σ^2 is unknown, it follows from the results of Sect. 2.4.3 that replacing σ^2 in (2.26) by $S/(k+2)$ results in an estimator that dominates X under squared error loss and is minimax under scaled squared error loss (provided $r(\cdot)$ satisfies the conditions of Theorem 2.6).

It may sometimes pay to break up the whole space into a direct sum of several subspaces and apply shrinkage estimators separately to the different subspaces.

Occasionally it is helpful to shrink toward another estimator. For example, Green and Strawderman (1991) combined two estimators, one of which is unbiased, remarkably by shrinking the unbiased estimator toward the biased estimator to obtain a Stein-type improvement over the unbiased estimator.

The estimators discussed in this section shrink toward some "vague" prior information that θ is in or near the specified set. Consequently it shrinks toward the set but does not restrict the estimator to lie in the set. In Chap. 7 we will consider estimators that are restricted to lie in a particular set. We will see in Chap. 7 that, although vague and restricted constraints seem conceptually similar, it turns out that the analyses of risk functions in these two settings are quite distinct.

2.5 A Link Between Stein's Lemma and Stokes' Theorem

That a relationship exists between Stein's lemma and Stokes' theorem (the divergence theorem) is not surprising. Indeed, Stein's lemma expresses that, if X has a normal distribution with mean θ and covariance matrix proportional to the identity matrix, the expectation of the inner product of $X - \theta$ and a suitable function g is proportional to the expectation of the divergence of g. On the other hand, when the sets of integration are spheres $S_{r,\theta}$ and balls $B_{r,\theta}$ of radius $r \geq 0$ centered at θ, Stokes' theorem states that the integral of the inner product of g and the unit outward vector at $x \in S_{r,\theta}$, which is $(x - \theta)/\|x - \theta\|$, with respect to the uniform measure equals the integral of the divergence of g on $B_{r,\theta}$ with respect to the Lebesgue measure.

Typically, Stokes' theorem is considered for a more general open set Ω in $\mathbb{R}^p$ with boundary $\partial\Omega$ which could be less smooth than a sphere, and where the function g is often smooth. For example, Stroock (1990) considers a bounded open set Ω in $\mathbb{R}^p$ for which there exists a function φ from $\mathbb{R}^p$ into $\mathbb{R}$ having continuous third order partial derivatives with the properties that $\Omega = \{x \in \mathbb{R}^p \mid \varphi(x) < 0\}$ and the gradient $\partial\varphi$ of φ vanishes at no point where φ itself vanishes. Further he requires that g has continuous first order partial derivatives in a neighborhood of the closure $\bar{\Omega}$ of Ω. For such an open set, its boundary is $\partial\Omega = \{x \in \mathbb{R}^p \mid \varphi(x) = 0\}$. Then, Stroock states that

$$\int_{\partial\Omega} n^{\mathrm{T}}(x)\, g(x) d\sigma(x) = \int_{\Omega} \operatorname{div} g(x)\, dx \tag{2.27}$$

where $n(x)$ is the outer normal (the unit outward vector) to $\partial\Omega$ at $x \in \partial\Omega$ and σ is the surface measure (the uniform measure) on $\partial\Omega$. He provides an elegant proof of Stokes' theorem in (2.27) through a rigorous construction of the outer normal and the surface measure. It is beyond the scope of this book to reproduce Stroock's proof, especially as the link we wish to make with Stein's identity only needs to deal with Ω being a ball and with $\partial\Omega$ being a sphere. Note that Stroock's conditions are satisfied for a ball of radius $r \geq 0$ centered at $\theta \in \mathbb{R}^p$ with the function $\varphi(x) = \|x - \theta\| - r$. In that context, Stokes' theorem expresses that

$$\int_{S_{r,\theta}} \left(\frac{x - \theta}{\|x - \theta\|} \right)^{\mathrm{T}} g(x)\, d\sigma_{r,\theta}(x) = \int_{B_{r,\theta}} \operatorname{div} g(x)\, dx \tag{2.28}$$

where $\sigma_{r,\theta}$ is the uniform measure on $S_{r,\theta}$.

In the following, we will show that Stein's identity for continuously differentiable functions can be derived in a straightforward way from this ball-sphere version of Stokes' theorem. Furthermore, and perhaps more interestingly, we will see that the converse is also true: Stein's identity (for which we have an independent proof in Sect. 2.3) implies directly the classical ball-sphere version of Stokes' theorem.

Proposition 2.2 *Let $X \sim \mathcal{N}_p(\theta, \sigma^2 I_p)$ and let g be a continuously differentiable function from $\mathbb{R}^p$ into $\mathbb{R}^p$ such that either*

$$E_\theta[|(X - \theta)^{\mathrm{T}} g(X)|] < \infty \quad \text{or} \quad E_\theta[|\operatorname{div} g(X)|] < \infty\,. \tag{2.29}$$

Then Stein's identity in (2.7) *holds, that is,*

$$E_\theta[(X - \theta)^{\mathrm{T}} g(X)] = \sigma^2\, E_\theta[\operatorname{div} g(X)]\,. \tag{2.30}$$

Proof Integrating through uniform measures on spheres (see Lemma 1.4), we have

$$\begin{aligned} E_{\theta,\sigma^2}[(X - \theta)^{\mathrm{T}} g(X)] &= \int_{\mathbb{R}^p} (x - \theta)^{\mathrm{T}} g(x)\, \frac{1}{(2\pi\,\sigma^2)^{p/2}} \exp\left(-\frac{\|x - \theta\|^2}{2\sigma^2} \right) dx \\ &= \int_0^\infty \int_{S_{r,\theta}} \left(\frac{x - \theta}{\|x - \theta\|} \right)^{\mathrm{T}} g(x)\, d\sigma_{r,\theta}(x)\, \psi_{\sigma^2}(r)\, dr \end{aligned} \tag{2.31}$$

where

$$\psi_{\sigma^2}(r) = \frac{1}{(2\pi\,\sigma^2)^{p/2}}\, r\, \exp\left(-\frac{r^2}{2\sigma^2}\right) \tag{2.32}$$

and $\sigma_{r,\theta}$ is the uniform measure on $S_{r,\theta}$. Then applying Stokes' theorem in (2.28) to the inner most integral in (2.31) gives

$$E_{\theta,\sigma^2}[(X-\theta)^{\mathrm{T}} g(X)] = \int_0^\infty \int_{B_{r,\theta}} \mathrm{div} g(x)\, dx\, \psi_{\sigma^2}(r)\, dr\,. \tag{2.33}$$

Now, applying Fubini's theorem to the right-hand side of (2.33), we have

$$\begin{aligned}
\int_0^\infty \int_{B_{r,\theta}} \mathrm{div} g(x) dx\, \psi_{\sigma^2}(r)\, dr &= \int_{\mathbb{R}^p} \mathrm{div} g(x) \int_{\|x-\theta\|}^\infty \psi_{\sigma^2}(r)\, dr\, dx \\
&= \int_{\mathbb{R}^p} \mathrm{div} g(x) \frac{1}{(2\pi\,\sigma^2)^{p/2}} \left[-\sigma^2 \exp\left(-\frac{r^2}{2\sigma^2}\right)\right]_{\|x-\theta\|}^\infty dx \\
&= \sigma^2 \int_{\mathbb{R}^p} \mathrm{div} g(x) \frac{1}{(2\pi\,\sigma^2)^{p/2}} \exp\left(-\frac{\|x-\theta\|^2}{2\sigma^2}\right) dx \\
&= \sigma^2\, E_{\theta,\sigma^2}[\mathrm{div} g(X)]
\end{aligned} \tag{2.34}$$

since, according to (2.32),

$$\frac{\partial}{\partial r}\left\{\frac{1}{(2\pi\,\sigma^2)^{p/2}}\left[-\sigma^2 \exp\left(-\frac{r^2}{2\sigma^2}\right)\right]\right\} = \psi_{\sigma^2}(r)\,.$$

Therefore combining (2.33) and (2.34) we have that

$$E_{\theta,\sigma^2}[(X-\theta)^{\mathrm{T}} g(X)] = \sigma^2\, E_{\theta,\sigma^2}[\mathrm{div} g(X)]\,,$$

which is Stein's identity in (2.39).

To show Stein's identity in (2.7) assuming $E_\theta[|\mathrm{div} g(X)|] < \infty$, it suffices to essentially reverse the steps in the above development. □

Note that using Stokes' theorem in the proof of Proposition 2.2 allows the weaker condition (2.29) instead of Condition (2.8) used in Theorem 2.1.

Kavian (1993) showed that (2.27) and (2.28) continue to hold for weakly differentiable functions g, provided that g behaves properly in a neighborhood of the boundary. See also Lepelletier (2004). However, Stokes' theorem may fail if g is not sufficiently smooth in a neighborhood of the boundary. For example, it is clear that a weakly differentiable function may be redefined on the boundary of the ball $B_{r,\theta}$ without affecting either its weak differentiability or the integral of the right-hand side of (2.28). But, by properly defining g on $S_{r,\theta}$, the integral over $S_{r,\theta}$ on the left-hand side of (2.28) may take on any value. For this reason, we develop the following version of Stokes' theorem (for balls and spheres) which will hold simultaneously for almost all r as long as the function g is weakly differentiable. It will be extensively used in extending Stein's lemma to general spherically symmetric distributions in Chaps. 5 and 6. Interestingly, the proof is based on Stein's lemma and completeness of a certain exponential family. We provide an extension to general smooth open sets in Sect. A.5 of the Appendix.

Theorem 2.7 (Fourdrinier and Strawderman 2016) *Let g be a weakly differentiable function from $\mathbb{R}^p$ into $\mathbb{R}^p$. Then* (2.28) *holds for almost every r.*

Proof Since g is weakly differentiable, the functions $(X-\theta)^{\mathrm{T}}g$ and $\operatorname{div} g$ are locally integrable. The same is true for the functions $g_n = g\,h_n$ where, for $n \in \mathbb{N}$, h_n is a smooth cutoff function such that $h_n(x) = 1$ if $\|x\| < n$, $h_n(x) = 0$ if $\|x\| > n+1$, $h_n \in \mathscr{C}^\infty$, and $h_n(x) \leq 1$ for all x. Thus g_n is weakly differentiable and we have $E_\theta[|(X-\theta)^{\mathrm{T}}g_n(X)|] < \infty$ or $E_\theta[|\operatorname{div} g_n(X)|] < \infty$. Hence, Stein's lemma applies to g_n, so that (2.39) holds for g_n, that is,

$$E_\theta[(X-\theta)^{\mathrm{T}}g_n(X)] = \sigma^2\, E_\theta[\operatorname{div} g_n(X)]\,. \tag{2.35}$$

Then, as in (2.31), with ψ_{σ^2} given in (2.32),

$$E_{\theta,\sigma^2}[(X-\theta)^{\mathrm{T}}g_n(X)] = \int_0^\infty \int_{S_{r,\theta}} \left(\frac{x-\theta}{\|x-\theta\|}\right)^{\mathrm{T}} g_n(x)\, d\sigma_{r,\theta}(x)\, \psi_{\sigma^2}(r)\, dr \tag{2.36}$$

and, as in (2.33), we also have

$$\sigma^2\, E_{\theta,\sigma^2}[\operatorname{div} g_n(X)] = \int_0^\infty \int_{B_{r,\theta}} \operatorname{div} g_n(x)\, dx\, \psi_{\sigma^2}(r)\, dr\,. \tag{2.37}$$

Hence, it follows from (2.35), (2.36), and (2.37) that, for all σ^2,

$$\int_0^\infty \int_{S_{r,\theta}} \left(\frac{x-\theta}{\|x-\theta\|}\right)^{\mathrm{T}} g_n(x)\, d\sigma_{r,\theta}(x)\, \psi_{\sigma^2}(r)\, dr = \int_0^\infty \int_{B_{r,\theta}} \operatorname{div} g_n(x)\, dx\, \psi_{\sigma^2}(r)\, dr\,.$$

Therefore, since the family $\{\psi_{\sigma^2}(r)\}_{\sigma^2>0}$ defined in (2.32) is proportional to a family of densities that is complete as an exponential family, we have

$$\int_{S_{r,\theta}} \left(\frac{x-\theta}{\|x-\theta\|}\right)^{\mathrm{T}} g_n(x)\, d\sigma_{r,\theta}(x) = \int_{B_{r,\theta}} \operatorname{div} g_n(x)\, dx\,, \tag{2.38}$$

for almost every $0 < r < n$. Now, since $g_n(x) = g(x)$ for $\|x\| < n$, it follows that (2.38) holds for g for almost every $r > 0$. □

As a first corollary, it follows that the classical (ball-sphere) version of Stokes' theorem holds for every r when g is continuously differentiable.

Corollary 2.7 *Let g be a continuously differentiable function from $\mathbb{R}^p$ into $\mathbb{R}^p$. Then* (2.28) *holds for every $r > 0$.*

Proof Because g is continuously differentiable, both sides of (2.38) are continuous. Then, since the equality holds almost everywhere, it must hold for all $r > 0$. □

Note that the proof of Proposition 2.2 remains valid when (2.28) holds for almost every $r > 0$. Hence the following corollary follows from Theorem 2.7 and Proposition 2.2.

Corollary 2.8 (Stein's lemma) *Let $X \sim \mathcal{N}_p(\theta, \sigma^2 I_p)$ and let g be a weakly differentiable function from $\mathbb{R}^p$ into $\mathbb{R}^p$ such that either $E_\theta[|(X-\theta)^{\mathrm{T}} g(X)|] < \infty$ or $E_\theta[|\mathrm{div} g(X)|] < \infty$. Then Stein's identity in* (2.7) *holds, that is,*

$$E_\theta[(X-\theta)^{\mathrm{T}} g(X)] = \sigma^2\, E_\theta[\mathrm{div} g(X)]\,. \qquad (2.39)$$

Note that, as in Proposition 2.2, Corollary 2.8 uses the weaker condition (2.29) instead of Condition (2.8) which was used in Theorem 2.1.

We have seen for balls and spheres that Stokes' theorem can be directly derived from Stein's identity, for weakly differentiable functions. This result will be particularly important for proving Stein type identities for spherically symmetric distributions in Chaps. 5 and 6. Note that we have in fact obtained a stronger result. It is actually shown that, any time Stein's identity is valid, then the version of Stokes' theorem given in Theorem 2.7 holds as well. This result is particularly interesting when the weak differentiability assumption is not met. For example, Fourdrinier et al. (2006) noticed that this may be the case when dealing with a location parameter restricted to a cone; Stein's identity (2.7) holds but the weak differentiability of the functions at hand is not guaranteed (see also Sect. 7.3).

2.6 Differential Operators and Dimension Cut-Off When Estimating a Mean

In the previous sections, when estimating the mean θ in the normal case, the MLE X is admissible when $p \leq 2$, but inadmissible when $p \geq 3$. Although specific to the normal case, this result can be extended to other distributional settings (such as exponential families) so that this dimension cut-off should reflect a more fundamental mathematical phenomenon. Below, we give an insight into such phenomena in terms of nonlinear partial differential operators.

Indeed, when estimating θ under quadratic loss, improvements on X through unbiased estimation techniques often involve a nonlinear partial differential operator of the form

$$\mathscr{R}g(x) = k \operatorname{div} g(x) + \|g(x)\|^2 \tag{2.40}$$

for a certain constant k. A sufficient condition for improvement is typically

$$\mathscr{R}g(x) \le 0 \tag{2.41}$$

for all $x \in \mathbb{R}^p$ (with strict inequality on a set of positive Lebesgue measure). We will see that (2.41) does not have a nontrivial solution g (i.e. g is not equal to 0 almost everywhere) when the dimension $p \le 2$, even if we look for solutions with smoothness conditions as weak as possible. Consequently, a necessary dimension condition for (2.41) to have solutions $g \not\equiv 0$ is $p \ge 3$.

Here is a precise statement of this fact.

Theorem 2.8 *Let $k \in \mathbb{R}$ be fixed. When $p \le 2$, the only weakly differentiable solution g with $\|g\|^2 \in L^1_{loc}(\mathbb{R}^p)$ of*

$$\mathscr{R}g(x) = k \operatorname{div} g(x) + \|g(x)\|^2 \le 0\,, \tag{2.42}$$

for any $x \in \mathbb{R}^p$, is $g = 0$ (a.e.).

Note that, in Theorem 2.8, the search for solutions of (2.42) is addressed in the general setting of weakly differentiable functions. The proof will follow the development in Blanchard and Fourdrinier (1999). However, in that paper, the g's are sought in the much larger space of distributions $\mathscr{D}'(\mathbb{R}^p)$ introduced by Schwartz (see Schwartz 1973 for a full account). Note also that the condition $\|g\|^2 \in L^1_{loc}(\mathbb{R}^p)$ is not restrictive. Any estimator of the form $X + g(X)$ with finite risk must satisfy $E_\theta[\|g(X)\|^2] < \infty$ and hence $\|g\|^2$ must be in $L^1_{loc}(\mathbb{R}^p)$.

The proof of Theorem 2.8 is based on the use of the following sequence of so-called test functions. Let φ be a nonnegative infinitely differentiable function on $\mathbb{R}_+$ bounded by 1, identically equal to 1 on [0, 1], and with support on the interval [0, 2] ($\mathrm{supp}(\varphi) = [0, 2]$), which implies that its derivative is bounded. Associate to φ the sequence $\{\varphi_n\}_{n\ge1}$ of infinitely differentiable functions from $\mathbb{R}^p$ into [0, 1] defined through

$$\forall n \ge 1 \quad \forall x \in \mathbb{R}^p \quad \varphi_n(x) = \varphi\!\left(\frac{||x||}{n}\right). \tag{2.43}$$

Clearly, for any $n \ge 1$, the function φ_n has compact support B_{2n}, the closed ball of radius $2n$ and centered at zero in $\mathbb{R}^p$. Also, an interesting property that follows from the uniform boundedness of φ', is that, for any $\beta \ge 1$ and for any $j = 1, \ldots, p$, there exists a constant $K > 0$ such that

$$\left|\frac{\partial \varphi_n^\beta}{\partial x_j}(x)\right| \le \frac{K}{n}\, \varphi_n^{\beta-1}(x)\,. \tag{2.44}$$

Note that, as all the derivatives of φ vanish outside of the compact interval $[1, 2]$ and φ is bounded by 1, (2.44) implies

$$\left| \frac{\partial \varphi_n^\beta}{\partial x_j}(x) \right| \leq \frac{K}{n} \, \mathbb{1}_{C_n}(x) \,. \tag{2.45}$$

where $\mathbb{1}_{C_n}$ is the indicator function of the annulus $C_n = \{x \in \mathbb{R}^p \mid n \leq ||x|| \leq 2n\}$.

Proof of Theorem 2.8 Let g be a weakly differentiable function g, with $\|g\|^2 \in L^1_{loc}(\mathbb{R}^p)$, satisfying (2.42). Then, using the defining property (2.6) of weak differentiability (see also Sect. A.1), we have, for any $n \in \mathbb{N}^*$ and any $\beta > 1$,

$$\begin{aligned}
\int_{\mathbb{R}^p} \|g(x)\|^2 \, \varphi_n^\beta(x) \, dx &\leq -k \int_{\mathbb{R}^p} \operatorname{div} g(x) \, \varphi_n^\beta(x) \, dx \\
&= -k \sum_{i=1}^p \int_{\mathbb{R}^p} \frac{\partial}{\partial x_i} g_i(x) \, \varphi_n^\beta(x) \, dx \\
&= k \sum_{i=1}^p \int_{\mathbb{R}^p} g_i(x) \, \frac{\partial}{\partial x_i} \varphi_n^\beta(x) \, dx \\
&= k \int_{\mathbb{R}^p} g^{\mathrm{T}}(x) \, \partial \varphi_n^\beta(x) \, dx \\
&\leq k \int_{\mathbb{R}^p} \|g(x)\| \, \|\partial \varphi_n^\beta(x)\| \, dx \,.
\end{aligned} \tag{2.46}$$

Then, using (2.44), it follows from (2.46) that there exists a constant $C > 0$ such that

$$\begin{aligned}
\int_{\mathbb{R}^p} \|g(x)\|^2 \, \varphi_n^\beta(x) \, dx &\leq \frac{C}{n} \int_{\mathbb{R}^p} \|g(x)\| \, \varphi_n^{\beta-1}(x) \, dx \\
&\leq \frac{C}{n} \left(\int_{\mathbb{R}^p} \varphi_n^{\beta-2}(x) \, dx \right)^{1/2} \left(\int_{\mathbb{R}^p} \|g(x)\|^2 \, \varphi_n^\beta(x) \, dx \right)^{1/2} ,
\end{aligned} \tag{2.47}$$

when applying Schwarz's inequality with $\beta > 2$ and using

$$\|g(x)\| \, \varphi_n^{\beta-1}(x) = \varphi_n^{\beta/2-1}(x) \; \|g(x)\| \, \varphi_n^{\beta/2}(x) \,.$$

Clearly (2.47) is equivalent to

$$\int_{\mathbb{R}^p} \|g(x)\|^2 \, \varphi_n^\beta(x) \, dx \leq \frac{C^2}{n^2} \int_{\mathbb{R}^p} \varphi_n^{\beta-2}(x) \, dx \,. \tag{2.48}$$

Thus, since $\varphi_n = 1$ on B_n and $\varphi_n \geq 0$,

$$\int_{B_n} \|g(x)\|^2 \, dx = \int_{B_n} \|g(x)\|^2 \, \varphi_n^\beta(x) \, dx \leq \int_{\mathbb{R}^p} \|g(x)\|^2 \, \varphi_n^\beta(x) \, dx \,. \quad (2.49)$$

Then, since supp $(\varphi_n) = B_{2n}$ and $0 \leq \varphi_n \leq 1$, using (2.48) gives

$$\int_{B_n} \|g(x)\|^2 \, dx \leq \frac{C^2}{n^2} \int_{\mathbb{R}^p} \varphi_n^{\beta-2}(x) \, dx \leq \frac{C^2}{n^2} \int_{B_{2n}} dx = A\, n^{p-2} \quad (2.50)$$

for some constant $A > 0$. Letting n go to infinity in (2.50) shows that, when $p < 2$, $g = 0$ almost everywhere, which proves the theorem in that case. It also implies that $\|g\|^2 \in L^1(\mathbb{R}^p)$ when $p = 2$.

In the case $p = 2$, the result will follow by applying (2.45). Indeed, it follows from (2.45), (2.49) and the first inequality in (2.47) that, for some constant $C > 0$,

$$\begin{aligned} \int_{B_n} \|g(x)\|^2 \, dx &\leq \frac{C}{n} \int_{C_n} \|g(x)\| \, dx \\ &\leq \frac{C}{n} \left(\int_{C_n} dx \right)^{1/2} \left(\int_{C_n} \|g(x)\|^2 \, dx \right)^{1/2} \end{aligned} \quad (2.51)$$

by Schwarz's inequality. Now, since $p = 2$,

$$\int_{C_n} dx \leq \int_{B_{2n}} dx \propto n^2 \,. \quad (2.52)$$

Hence (2.51) and (2.52) imply that

$$\int_{B_n} \|g(x)\|^2 \, dx \leq A \left(\int_{C_n} \|g(x)\|^2 \, dx \right)^{1/2} , \quad (2.53)$$

for some constant $A > 0$. Since as noted above, $\|g\|^2 \in L^1(\mathbb{R}^p)$, we have

$$\lim_{n\to\infty} \int_{C_n} \|g(x)\|^2 \, dx = 0$$

and consequently (2.53) gives rise to

$$0 = \lim_{n\to\infty} \int_{C_n} \|g(x)\|^2 \, dx = \int_{\mathbb{R}^p} \|g(x)\|^2 \, dx \,.$$

Thus $g = 0$ almost everywhere and gives the desired result for $p = 2$ is obtained.

□

Such a dimension cut-off result implies that the usual Stein inequality $2\,\mathrm{div} g(x) + ||g(x)||^2 \leq 0$, for any $x \in \mathbb{R}^p$, has no nontrivial solution g, with $\|g\|^2 \in L^1_{loc}(\mathbb{R}^p)$ when $p \leq 2$. This reinforces the fact that the MLE X is admissible in dimension $p \leq 2$ when estimating a normal mean. Blanchard and Fourdrinier (1999) (to which we refer for a full account of the dimension cut-off phenomenon) also considered more general nonlinear partial differential inequalities. We will again use their technique in Chap. 8 (for loss estimation) to prove that, for an inequality of the form $k\,\Delta\gamma(x) + \gamma^2(x) \leq 0$, the same dimension cut-off phenomenon occurs for $p \leq 4$ (there is no nontrivial solution γ, with $\gamma^2 \in L^1_{loc}(\mathbb{R}^p)$, when $p \leq 4$).

Chapter 3
Estimation of a Normal Mean Vector II

As we saw in Chap. 2, the frequentist paradigm is well suited for risk evaluations, but is less useful for estimator construction. It turns out that the Bayesian approach is complementary, as it is well suited for the construction of possibly optimal estimators. In this chapter we take a Bayesian view of minimax shrinkage estimation. In Sect. 3.1 we derive a general sufficient condition for minimaxity of Bayes and generalized Bayes estimators in the known variance case, we also illustrate the theory with numerous examples. In Sect. 3.2 we extend these results to the case when the variance is unknown. Section 3.3 considers the case of a known covariance matrix under a general quadratic loss. The admissibility of Bayes estimators in discussed in Sect. 3.4. Interesting connections to MAP estimation, penalized likelihood methods, and shrinkage estimation are developed in Sect. 3.5. The fascinating connections between Stein estimation and estimation of a predictive density under Kullback-Leibler divergence are outlined in Sect. 3.6.

3.1 Bayes Minimax Estimators

In this section, we derive a general sufficient condition for minimaxity of Bayes and generalized Bayes estimators when $X \sim \mathcal{N}_p(\theta, \sigma^2 I_p)$, with known σ^2, and the loss function is $\|\delta - \theta\|^2$, due to Stein (1973, 1981). The condition depends only on the marginal distribution and states that a generalized Bayes estimator is minimax if the square root of the marginal distribution is superharmonic. Alternative (stronger) sufficient conditions are that the prior distribution or the marginal distribution is superharmonic. We establish these results in Sect. 3.1.1 and apply them in Sect. 3.1.2 to obtain classes of prior distributions which lead to minimax (generalized and proper) Bayes estimators. Section 3.1.3 will be devoted to minimax multiple shrinkage estimators.

D. Fourdrinier et al., *Shrinkage Estimation*, Springer Series in Statistics,
https://doi.org/10.1007/978-3-030-02185-6_3

Throughout this section, let $X \sim \mathcal{N}_p(\theta, \sigma^2 I_p)$ (with σ^2 known) and the loss be $L(\theta, \delta) = \|\delta - \theta\|^2$. Let θ have the (generalized) prior distribution π and let the marginal density, $m(x)$, of X be

$$m(x) = K \int_{\mathbb{R}^p} e^{-\frac{\|x-\theta\|^2}{2\sigma^2}} \, d\pi(\theta). \tag{3.1}$$

Recall from Sect. 1.4 that the Bayes estimator corresponding to $\pi(\theta)$ is given by

$$\delta_\pi(X) = X + \sigma^2 \frac{\nabla m(X)}{m(X)}. \tag{3.2}$$

Since the constant K in (3.1) plays no role in (3.2) we will typically take it to be equal to 1 for simplicity. It may happen that an estimator will have the form (3.2) where $m(X)$ does not correspond to a true marginal distribution. In this case we will refer to such an estimator as a pseudo-Bayes estimator, provided $x \mapsto \nabla m(x)/m(x)$ is weakly differentiable. Recall that, if $\delta_\pi(X)$ is generalized Bayes, $x \mapsto m(x)$ is a positive analytic function and so $x \mapsto \nabla m(x)/m(x)$ is automatically weakly differentiable.

3.1.1 A Sufficient Condition for Minimaxity of (Proper, Generalized, and Pseudo) Bayes Estimators

Stein (1973, 1981) gave the following sufficient condition for a generalized Bayes estimator to be minimax. This condition relies on the superharmonicity of the square root of the marginal. Recall from Corollary A.2 in Appendix A.8.3 that a function f from $\mathbb{R}^p$ into $\mathbb{R}$ which is twice weakly differentiable and lower semicontinuous is superharmonic if and only if, for almost every $x \in \mathbb{R}^p$, we have $\Delta f(x) \leq 0$, where Δf is the weak Laplacian of f. Note that, if the function f is analytic, the last inequality holds for any $x \in \mathbb{R}^p$.

Theorem 3.1 *Under the model of this section, an estimator of the form* (3.2) *has finite risk if* $E_\theta\big[\|\nabla m(X)/m(X)\|^2\big] < \infty$ *and is minimax provided* $x \mapsto \sqrt{m(x)}$ *is superharmonic (i.e.,* $\Delta\sqrt{m(x)} \leq 0$*, for any* $x \in \mathbb{R}^p$*).*

Proof First, note that, as noticed in Example 1.1, the marginal m is a positive analytic function, and so is $\sqrt{m}$.

Using Corollary 2.1 and the fact that $\delta_\pi(X) = X + \sigma^2 g(X)$ with $g(X) = \nabla m(X)/\, m(X)$, the estimator $\delta_\pi(X)$ has finite risk if $E_\theta\big[\|\nabla m(X)/m(X)\|^2\big] < \infty$. Also, it is minimax provided, for almost any $x \in \mathbb{R}^p$,

$$\mathcal{D}(x) = \frac{\|\nabla m(x)\|^2}{m^2(x)} + 2\,\mathrm{div}\frac{\nabla m(x)}{m(x)} \leq 0\,.$$

Now, for any $x \in \mathbb{R}^p$,

$$\mathscr{D}(x) = \frac{\|\nabla m(x)\|^2}{m^2(x)} + 2\,\frac{m(x)\,\Delta m(x) - \|\nabla m(x)\|^2}{m^2(x)}$$

where

$$\Delta m(x) = \sum_{i=1}^{p} \frac{\partial^2}{\partial x_i^2} m(x)$$

is the Laplacian of $m(x)$. Hence, by straightforward calculation,

$$\begin{aligned}\mathscr{D}(x) &= \frac{2\,m(x)\,\Delta m(x) - \|\nabla m(x)\|^2}{m^2(x)} \qquad (3.3)\\ &= 4\,\frac{\Delta\sqrt{m(x)}}{\sqrt{m(x)}}\,.\end{aligned}$$

Therefore $\mathscr{D}(x) \leq 0$ since $x \mapsto \sqrt{m(x)}$ is superharmonic. □

It is convenient to assemble the following results for the case of spherically symmetric marginals. The proof is straightforward and left to the reader.

Corollary 3.1 *Assume the prior density $\pi(\theta)$ is spherically symmetric around 0 (i.e., $\pi(\theta) = \pi(\|\theta\|^2)$). Then*

(1) *the marginal density m of X is spherically symmetric around 0 (i.e., $m(x) = m(\|x\|^2)$, for any $x \in \mathbb{R}^p$);*
(2) *the Bayes estimator equals*

$$\delta_\pi(X) = X + 2\,\sigma^2\,\frac{m'(\|X\|^2)}{m(\|X\|^2)}\,X$$

and has the form of a Baranchik estimator (2.19) *with*

$$a\,r(t) = -2\,\frac{m'(t)}{m(t)}\,t \qquad \forall t \geq 0;$$

(3) *the unbiased estimator of the risk difference between $\delta_\pi(X)$ and X is given by*

$$\mathscr{D}(X) = 4\,\sigma^4\left\{p\,\frac{m'(\|X\|^2)}{m(\|X\|^2)} + 2\,\|X\|^2\,\frac{m''(\|X\|^2)}{m(\|X\|^2)} - \|X\|^2\left(\frac{m'(\|X\|^2)}{m(\|X\|^2)}\right)^2\right\}.$$

While, in Theorem 3.1 minimaxity of $\delta_\pi(X)$ follows from the superharmonicity of $\sqrt{m(X)}$, it is worth noting that, in the setting of Corollary 3.1, it can be obtained from the concavity of $t \mapsto m^{1/2}(t^{2/(2-p)})$.

The following corollary is often useful. It shows that $\sqrt{m(X)}$ is superharmonic if $m(X)$ is superharmonic, which in turn follows if the prior density $\pi(\theta)$ is superharmonic.

Corollary 3.2

(1) *A finite risk (generalized, proper, or pseudo) Bayes estimator of the form* (3.2) *is minimax provided the marginal m is superharmonic (i.e. $\Delta m(x) \le 0$, for any $x \in \mathbb{R}^p$).*
(2) *If the prior distribution has a density, π, which is superharmonic, then a finite risk generalized or proper Bayes estimator of the form* (3.2) *is minimax.*

Proof Part (1) follows from the first equality in (3.3), which shows that superharmonicity of m implies superharmonicity of $\sqrt{m}$. Indeed, the superharmonicity of m implies the superharmonicity of any nondecreasing concave function of m.

Part (2) follows since, for any $x \in \mathbb{R}^p$,

$$\begin{aligned}
\Delta_x m(x) &= \Delta_x \int_{\mathbb{R}^p} \exp\left(-\frac{1}{2\sigma^2}\|x-\theta\|^2\right) \pi(\theta)\, d\theta \\
&= \int_{\mathbb{R}^p} \Delta_x \exp\left(-\frac{1}{2\sigma^2}\|x-\theta\|^2\right) \pi(\theta)\, d\theta \\
&= \int_{\mathbb{R}^p} \Delta_\theta \exp\left(-\frac{1}{2\sigma^2}\|x-\theta\|^2\right) \pi(\theta)\, d\theta \\
&= \int_{\mathbb{R}^p} \exp\left(-\frac{1}{2\sigma^2}\|x-\theta\|^2\right) \Delta_\theta \pi(\theta)\, d\theta
\end{aligned}$$

where the second equality follows from exponential family properties and the last equality is Green's formula (see also Sect. A.9). More generally, any mixture of superharmonic functions is superharmonic (Sect. A.8). □

Note that the condition of finiteness of risk is superfluous for proper Bayes estimators since the Bayes risk is bounded above by $p\,\sigma^2$, and Fubini's theorem assures that the risk function is finite a.e. (π). Continuity of the risk function implies finiteness for all θ in the convex hull of the support of π (see Berger (1985a) and Lehmann and Casella (1998) for more discussion on finiteness and continuity of risk).

As an example of a pseudo-Bayes estimator, consider $m(X)$ of the form

$$m(X) = \frac{1}{(\|X\|^2)^b}\,.$$

The case $b = 0$ corresponds to $m(X) = 1$ which is the marginal corresponding to the "uniform" generalized prior distribution $\pi(\theta) \equiv 1$, which in turn corresponds to the generalized Bayes estimator $\delta_0(X) = X$. If $b > 0$, $m(X)$ is unbounded in a neighborhood of 0 and consequently is not analytic. Thus, $m(X)$ cannot be a true marginal (for any generalized prior). However,

$$\nabla m(X) = \frac{-2\,b}{(\|X\|^2)^{b+1}}\,X$$

and

$$\frac{\nabla m(X)}{m(X)} = \frac{-2\,b}{\|X\|^2}\,X,$$

which is weakly differentiable if $p \geq 3$ (see Sect. 2.3). Hence, for $p \geq 3$, the James-Stein estimator

$$\delta_{2b}^{JS}(X) = \left(1 - \frac{2\,b\,\sigma^2}{\|X\|^2}\right) X$$

is a pseudo-Bayes estimator. Also, a simple calculation gives

$$\Delta m(X) = \frac{(-2\,b)[p - 2\,(b+1)]}{(\|X\|^2)^{b+1}}.$$

It follows that $m(X)$ is superharmonic for $0 \leq b \leq (p-2)/2$ and similarly that $\sqrt{m(X)}$ is superharmonic for $0 \leq b \leq p-2$. An application of Theorem 3.1 gives minimaxity for $0 \leq b \leq p-2$ which agrees with Theorem 2.2 (with $a = 2b$), while an application of Corollary 3.2 establishes minimaxity for only half of the interval, i.e. $0 \leq b \leq (p-2)/2$. Thus, while useful, the corollary is considerably weaker than the theorem.

Another interesting aspect of this example relates to the existence of proper Bayes minimax estimators for $p \geq 5$. Considering the behavior of $m(x)$ for $\|x\| \geq R$ for some positive R, note that

$$\int_{\|x\|\geq R} m(x)\,dx = \int_{\|x\|\geq R} \frac{1}{(\|X\|^2)^b}\,dX \propto \int_R^\infty \frac{r^{p-1}}{r^{2b}}\,dr = \int_R^\infty r^{p-2b-1}\,dr$$

and that this integral is finite if and only if $p - 2b < 0$. Thus, integrability of $m(x)$ for $\|x\| \geq R$ and minimaxity of the (James-Stein) pseudo-Bayes estimator corresponding to $m(X)$ are possible if and only if $p/2 < b \leq p-2$, which implies $p \geq 5$.

It is also interesting to note that superharmonicity of $m(X)$ (i.e. $0 \leq b \leq (p-2)/2$) is incompatible with integrability of $m(x)$ on $\|x\| \geq R$ (i.e. $b > p/2$). This is illustrative of a general fact that a generalized Bayes minimax estimator corresponding to a superharmonic marginal cannot be proper Bayes (see Theorem 3.2).

3.1.2 Construction of (Proper and Generalized) Minimax Bayes Estimators

Corollary 3.1 provides a method of constructing pseudo-Bayes minimax estimators. In this section, we concentrate on the construction of proper and generalized Bayes minimax estimators. The results in this section are primarily from Fourdrinier et al. (1998). Although Corollary 3.1 is helpful in constructing minimax estimators it cannot be used to develop proper Bayes minimax estimators as indicated in the example at the end of the previous section. The following result establishes that a superharmonic marginal (and consequently a superharmonic prior density) cannot lead to a proper Bayes estimator.

Theorem 3.2 *Let m be a superharmonic marginal density corresponding to a prior π. Then π is not a probability measure.*

Proof Assume π is a probability measure. Then it follows that m is an integrable, strictly positive, and bounded function in C^∞ (the space of functions which have derivatives of all orders). Recall from Example 1.1 of Sect. 1.4 that the posterior risk is given, for any $x \in \mathbb{R}^p$, by

$$p\sigma^2 + \sigma^4 \frac{m(x)\,\Delta m(x) - \|\nabla m(x)\|^2}{m^2(x)}.$$

Hence, the Bayes risk is

$$r(\pi) = E^m\left[p\sigma^2 + \sigma^4 \frac{m(X)\Delta m(X) - \|\nabla m(X)\|^2}{m^2(X)}\right],$$

where E^m is the expectation with respect to the marginal density m. Also, denoting by E^π the expectation with respect to the prior π, we may use the unbiased estimate of risk to express $r(\pi)$ as

$$\begin{aligned} r(\pi) &= E^\pi\left[E_\theta\left[p\sigma^2 + \sigma^4 \tfrac{2m(X)\Delta m(X) - \|\nabla m(X)\|^2}{m^2(X)}\right]\right] \\ &= E^m\left[p\sigma^2 + \sigma^4 \tfrac{2m(X)\Delta m(X) - \|\nabla m(X)\|^2}{m^2(X)}\right], \end{aligned}$$

since the unbiased estimate of risk does not depend on θ, by definition. Hence, by taking the difference,

$$E^m\left[\frac{\Delta m(X)}{m(X)}\right] = 0.$$

Now, since the marginal m is superharmonic ($\Delta m(x) \leq 0$ for any $x \in \mathbb{R}^p$), strictly positive and in C^∞, it follows that $\Delta m \equiv 0$. Finally, the strict positivity and harmonicity of m implies that $m \equiv C$ where C is a positive constant (see Doob 1984), and hence, that $\int_{\mathbb{R}^p} m(X)\,dx = \infty$, which contradicts the integrability of m. □

We now turn to the construction of Bayes minimax estimators. Consider prior densities of the form

$$\pi(\theta) = k \int_0^\infty \exp\left(-\frac{\|\theta\|^2}{2\sigma^2 v}\right) v^{-p/2}\, h(v)\, dv \tag{3.4}$$

for some constant k and some nonnegative function h on $\mathbb{R}^+$ such that the integral exists, i.e. $\pi(\theta)$ is a variance mixture of normal distributions. It follows from Fubini's theorem that, for any $x \in \mathbb{R}^p$,

$$m(x) = \int_0^\infty m_v(x)\, h(v)\, dv$$

where

$$m_v(x) = k \exp\left(-\frac{\|x\|^2}{2\sigma^2(1+v)}\right)(1+v)^{-p/2}.$$

Lebesgue's dominated convergence theorem ensures that we may differentiate under the integral sign and so

$$\nabla m(x) = \int_0^\infty \nabla m_v(x)\, h(v)\, dv \tag{3.5}$$

and

$$\Delta m(x) = \int_0^\infty \Delta m_v(x)\, h(v)\, dv \tag{3.6}$$

where

$$\nabla m_v(x) = -\frac{k}{\sigma^2} \exp\left(-\frac{\|x\|^2}{2\sigma^2(1+v)}\right)(1+v)^{-p/2-1}\, x$$

and

$$\Delta m_v(x) = -\frac{k}{\sigma^2}\left[p - \frac{\|x\|^2}{\sigma^2(1+v)}\right] \exp\left(-\frac{\|x\|^2}{2\sigma^2(1+v)}\right)(1+v)^{-p/2-1}.$$

Then the following integral

$$I_j(y) = \int_0^\infty \exp(-y/(1+v))\,(1+v)^{-j}\, h(v)\, dv$$

exists for $j \geq p/2$. Hence, with $y = \|x\|^2/2\sigma^2$, we have

$$
\begin{aligned}
m(x) &= k\, I_{p/2}(y) \qquad (3.7)\\
\nabla m(x) &= -\frac{k}{\sigma^2}\, I_{p/2+1}(y)\, x\\
\Delta m(x) &= -\frac{k}{\sigma^2}\left[p\, I_{p/2+1}(y) - 2\, y\, I_{p/2+2}(y)\right]\\
\|\nabla m(x)\|^2 &= 2\,\frac{k^2}{\sigma^2}\, y\, I^2_{\frac{p}{2}+1}(y).
\end{aligned}
$$

Note that

$$
\frac{\|\nabla m(x)\|^2}{m^2(x)} = \frac{2}{\sigma^2}\,\frac{I^2_{p/2+1}(y)}{I^2_{\frac{p}{2}}(y)}\, y \leq \frac{2\,y}{\sigma^2} = \frac{\|x\|^2}{\sigma^4}
$$

since $I_{j+p}(y) \leq I_j(y)$. Hence,

$$
E_0\!\left[\frac{\|\nabla m(x)\|^2}{m^2(x)}\right] \leq E_0\!\left[\frac{\|x\|^2}{\sigma^4}\right] < \infty\,,
$$

which, according to Theorem 3.1, guarantees the finiteness of the risk of the Bayes estimator $\delta_\pi(X)$ in (3.2). Furthermore, the unbiased estimator of risk difference (3.3) can be expressed as

$$
\begin{aligned}
\mathscr{D}(X) &= -\tfrac{2}{\sigma^2}\left[p\, I_{p/2+1}(y) - 2\, y\, I_{p/2+2}(y)\right]/I_{p/2}(y) \qquad (3.8)\\
&\quad -\tfrac{2}{\sigma^2}\left[y\, I^2_{p/2+1}(y)/I^2_{p/2}(y)\right]\\
&= \tfrac{2\, I_{p/2+1}(y)}{\sigma^2\, I_{p/2}(y)}\left[\tfrac{2\, y\, I_{p/2+2}(y)}{I_{p/2+1}(y)} - p - \tfrac{y\, I_{p/2+1}(y)}{I_{p/2}(y)}\right].
\end{aligned}
$$

Then the following intermediate result follows immediately from (3.2) and Theorem 3.1 since finiteness of risk has been guaranteed above.

Lemma 3.1 *The generalized Bayes estimator corresponding to the prior density* (3.4) *is minimax provided*

$$
\frac{2\, I_{p/2+2}(y)}{I_{p/2+1}(y)} - \frac{I_{p/2+1}(y)}{I_{p/2}(y)} \leq \frac{p}{y}\,. \qquad (3.9)
$$

The next theorem gives sufficient conditions on the mixing density $h(\cdot)$ so that the resulting generalized Bayes estimator is minimax.

Theorem 3.3 *Let h be a positive differentiable function such that the function $-(v+1)h'(v)/h(v) = l_1(v) + l_2(v)$ where $l_1(v) \leq A$ and is nondecreasing while*

$0 \le l_2 \le B$ with $A + 2B \le (p-2)/2$. Assume also that $\lim_{v\to\infty} h(v)/(v+1)^{p/2-1} = 0$ and that $\int_0^\infty \exp(-y/(1+v))\,(1+v)^{-p/2}\,h(v)\,dv < \infty$. Then the generalized Bayes estimator (3.2) for the prior density (3.4) corresponding to the mixing density h is minimax. Furthermore, if h is integrable, the resulting estimator is also proper Bayes.

Proof Via integration by parts, we first find an alternative expression for

$$I_k(y) = \int_0^\infty \exp(-y/(1+v))\,(1+v)^{-k}\,h(v)\,dv.$$

Letting $u = (1+v)^{-k+2}h(v)$ and $dw = (1+v)^{-2}\exp(-y/(1+v))\,dv$, so that $du = (-k+2)(1+v)^{-k+1}\,h(v) + (1+v)^{-k+2}h'(v)$ and $w = \exp(-y/(1+v))/y$, we have, for $k \ge p/2+1$,

$$\begin{aligned} I_k(y) &= \frac{(1+v)^{-k+2}\exp(-y/(1+v))\,h(v)}{y}\Big|_0^\infty \\ &\quad + \frac{k-2}{y}\int_0^\infty \exp\left(-\frac{y}{1+v}\right)(1+v)^{-k+1}\,h(v)\,dv \\ &\quad - \frac{1}{y}\int_0^\infty \exp\left(-\frac{y}{1+v}\right)(1+v)^{-k+2}\,h'(v)\,dv \\ &= -\frac{e^{-y}\,h(0)}{y} + \frac{k-2}{y}\,I_{k-1}(y) \\ &\quad - \frac{1}{y}\int_0^\infty \exp\left(-\frac{y}{1+v}\right)(1+v)^{-k+2}\,h'(v)\,dv\,. \end{aligned} \tag{3.10}$$

Applying (3.10) to both numerators in the left-hand side of (3.9) we have

$$\begin{aligned} &\frac{2}{I_{p/2+1}(y)}\left[\frac{-e^{-y}\,h(0)}{y} + \frac{p}{2\,y}\,I_{p/2+1}(y) - \frac{1}{y}\int_0^\infty \exp\left(-\frac{y}{1+v}\right)(1+v)^{-p/2}\,h'(v)\,dv\right] \\ &- \frac{1}{I_{p/2}(y)}\left[\frac{-e^{-y}\,h(0)}{y} + \frac{p-2}{2\,y}\,I_{p/2}(y) - \frac{1}{y}\int_0^\infty \exp\left(-\frac{y}{1+v}\right)(1+v)^{-p/2+1}\,h'(v)\,dv\right] \\ &\le \frac{p+2}{2\,y} - \frac{2\int_0^\infty \exp\left(-\frac{y}{1+v}\right)(1+v)^{-p/2+2}\,h'(v)\,dv}{y\,I_{p/2+1}(y)} \\ &+ \frac{\int_0^\infty \exp\left(-\frac{y}{1+v}\right)(1+v)^{-p/2+1}\,h'(v)\,dv}{y\,I_{p/2}(y)} \end{aligned}$$

since $I_{p/2+1}(y) < I_{p/2}(y)$. Then it follows from Lemma 3.1 that $\delta_\pi(X)$ is minimax provided, for any $y \ge 0$,

$$J_p^y \le p - \frac{p+2}{2} = \frac{p-2}{2},$$

where

$$J_p^y = -2\, E_{p/2+1}^y\left[(V+1)\frac{h'(V)}{h(V)}\right] + E_{p/2}^y\left[(V+1)\frac{h'(V)}{h(V)}\right]$$

and where $E_k^y[f(V)]$ is the expectation of $f(V)$ with respect to the random variable V with density $g_k^y(v) = \exp(-y/(1+v))\,(1+v)^{-k}\,h(v)/I_k(y)$. Now upon setting $-(v+1)\,h'(v)/h(v) = l_1(v)+l_2(v)$ and noting that $g_k^y(v)$ has monotone decreasing likelihood ratio in k, for fixed y, we have

$$\begin{aligned} J_p^y &= 2\, E_{p/2+1}^y\,[l_1(V)+l_2(V)] - E_{p/2}^y\,[l_1(V)+l_2(V)] \\ &\le 2\, E_{p/2+1}^y\,[l_1(V)] - E_{p/2}^y\,[l_1(V)] + 2\, E_{p/2+1}^y\,[l_2(V)] \end{aligned}$$

since $l_2 \ge 0$. Also

$$E_{p/2+1}^y\,[l_1(V)] \le E_{p/2}^y\,[l_1(V)]$$

since l_1 is nondecreasing. Then

$$J_p^y \le E_{p/2}^y\,[l_1(V)] + 2\, E_{p/2+1}^y\,[l_2(V)] \le A + 2\,B \le \frac{p-2}{2}.$$

since $l_1 \le A$ and $l_2 \le B$ and by the assumptions on A and B. The result follows. □

The following corollary allows the construction of mixing distributions so that the conditions of the theorem are met and the resulting (generalized or proper) Bayes estimators are minimax.

Corollary 3.3 *Let $\psi = \psi_1 + \psi_2$ be a continuous function such that $\psi_1 \le C$ and is nondecreasing, while $0 \le \psi_2 \le D$, and where $C \le -2D$. Define, for $v > 0$, $h(v) = \exp\left[-\frac{1}{2}\int_{v_0}^{v} \frac{2\,\psi(u)+p-2}{u+1}\,du\right]$ where $v_0 \ge 0$. Assume also that $\lim_{v\to\infty} h(v)/(1+v)^{p/2-1} = 0$ and that $I_{p/2}(y) = \int_0^\infty \exp(-y/(1+v))\,(1+v)^{-p/2}\,h(v)\,dv < \infty$.*

Then the Bayes estimator corresponding to the mixing density h is minimax. Furthermore if h is integrable the estimator is proper Bayes.

Proof A simple calculation shows that

$$-(v+1)\,\frac{h'(v)}{h(v)} = \psi_1(v) + \psi_2(v) + \frac{p-2}{2}.$$

Setting $l_1(v) = \psi_1(v) + (p-2)/2$ and $l_2(v) = \psi_2(v)$, the result follows from Theorem 3.1 with $A = (p-2)/2 + C$ and $B = D$. □

Note that finiteness of $I_{p/2}(y)$ in Corollary 3.2 is assured if we strengthen the limit condition to $\lim_{v\to\infty} h(v)/(1+v)^{p/2-1-\epsilon} = 0$ for some $\epsilon > 0$, since this implies that, for $h(v)/(1+v)^{p/2} \le M/(1+v)^{1+\epsilon}$ for some $M > 0$ and any $v > 0$. Thus

$$
\begin{aligned}
I_{p/2}(y) = \int_0^\infty \exp(-y/(1+v))\,(1+v)^{-p/2}\,h(v)\,dv &\le \int_0^\infty (1+v)^{-p/2}\,h(v)\,dv \\
&\le \int_0^\infty \frac{M}{(1+v)^{1+\epsilon}}\,dv \\
&< \infty .
\end{aligned}
$$

3.1.3 Examples

An interesting and useful class of examples results from the choice

$$
\psi(v) = \alpha + \beta/v + \gamma/v^2 \tag{3.11}
$$

for some $(\alpha, \beta, \gamma) \in \mathbb{R}^3$. A simple calculation shows

$$
\begin{aligned}
h(v) &= \exp\left[-\int_{v_0}^{v} \frac{\alpha + \beta/u + \gamma/u^2 + (p-2)/2}{u+1}\,du\right] \\
&\propto (v+1)^{\beta-\alpha-\gamma-\frac{p-2}{2}}\, v^{\gamma-\beta} \exp\left(\frac{\gamma}{v}\right).
\end{aligned}
\tag{3.12}
$$

Example 3.1 (The Strawderman 1971 prior) Suppose $\alpha \le 0$ and $\beta = \gamma = 0$ so that $h(v) \propto (v+1)^{-\alpha-(p-2)/2}$. Let $\psi_1(v) = \psi(v) \equiv \alpha$ and $\psi_2(v) \equiv 0$ so that $C = D = 0$. Then the minimaxity conditions of Corollary 3.1 require $\lim_{v\to\infty} h(v)/(1+v)^{p/2-1} = \lim_{v\to\infty}(v+1)^{-\alpha-(p-2)} = 0$ and this is satisfied if $\alpha > 2-p$. Also

$$
\begin{aligned}
I_{p/2}(y) &= \int_0^\infty \exp(-y/(1+v))\,(1+v)^{-p/2}\,h(v)\,dv \\
&\propto \int_0^\infty \exp(-y/(1+v))\,(1+v)^{-\alpha-p+1}\,h(v)\,dv \\
&\le \int_0^\infty (1+v)^{-\alpha-p+1}\,h(v)\,dv \\
&< \infty
\end{aligned}
$$

if $\alpha > 2-p$ as above. Hence in this case the corresponding generalized Bayes estimator is minimax if $2-p < \alpha \le 0$ (which requires $p \ge 3$).

Furthermore it is proper Bayes minimax if $\int_0^\infty (1+v)^{-\alpha-(p-2)/2}\,dv < \infty$ which is equivalent to $2-p/2 < \alpha \le 0$. This latter condition requires $p \ge 5$ and demonstrates the existence of proper Bayes minimax estimators for $p \geq 5$. We will see below that this is the class of priors studied in Strawderman (1971) under the alternative parametrization $\lambda = 1/(1+v)$.

Example 3.2 Consider $\psi(v)$ given by (3.11) with $\alpha \leq 0$, $\beta \leq 0$ and $\gamma \leq 0$. Here we take $\psi_1(v) = \psi(v)$, $\psi_2(v) = 0$, and $C = D = 0$. The minimaxity conditions of Corollary 3.2 require

$$\lim_{v\to\infty} h(v)/(1+v)^{p/2-1} = \lim_{v\to\infty} (v+1)^{\beta-\alpha-\gamma-p+2} v^{\gamma-\beta} \exp(\gamma/v) = 0.$$

This implies $2 - p < \alpha \leq 0$. The finiteness condition on

$$\begin{aligned} I_{p/2}(y) &= \int_0^\infty \exp(-y/(1+v))\,(1+v)^{-p/2}\,h(v)\,dv \\ &\propto \int_0^\infty e^{-\frac{y}{1+v}}\,(v+1)^{\beta-\alpha-\gamma-p+1} v^{\gamma-\beta} \exp(\gamma/v)\,dv \end{aligned}$$

also requires $2 - p < \alpha \leq 0$. Therefore, minimaxity is ensured as soon as $2 - p < \alpha \leq 0$.

Furthermore, the minimax estimator will be proper Bayes if

$$\int_0^\infty h(v)\,dv \propto \int_0^\infty (1+v)^{\beta-\alpha-\gamma-(p-2)/2}\, v^{\gamma-\beta}\, \exp(\gamma/v)\,dv < \infty.$$

This holds if $2 - \frac{p}{2} < \alpha \leq 0$ as in Example 3.1.

Example 3.3 Suppose $\alpha \leq 0$, $\beta > 0$, and $\gamma < 0$ and take

$$\psi_1(v) = \alpha + (\gamma/v)(1/+\beta/\gamma) I_{[0,-2\gamma/\beta]}(v)\,,$$

$$\psi_2(v) = (\gamma/v)(1/v + \beta/\gamma)\, \mathbb{1}_{[-2\gamma/\beta,\infty]}(v)\,,$$

for $C = \alpha$ and $D = -\beta^2/4\gamma$.

Note first that $\psi_1(v)$ is monotone nondecreasing and bounded above by α; also, $0 \leq \psi_2(v) \leq -\beta^2/4\gamma$. Therefore, we require $C = \alpha < -2D = \beta^2/2\gamma$. The conditions $\lim_{v\to\infty} h(v)/(1+v)^{p/2-1} = 0$ and $\int_0^\infty \exp(-y/(1+v))\,(1+v)^{-p/2}\,h(v)\,dv < \infty$ are, as in Example 3.2, $2 - p < \alpha \leq 0$.

Thus, $\delta_\pi(X)$ is minimax for $2 - p < \alpha \leq \beta^2/2\gamma < 0$. The condition for integrability of h is also, as in Example 3.2, i.e. $2 - \frac{p}{2} < \alpha \leq \beta^2/2\gamma < 0$.

In this example, $\psi(v)$ is not monotone but is increasing on $[0, -2\gamma/\beta)$ and decreasing thereafter. This typically corresponds to a non-monotone $r(\|X\|^2)$ in the Baranchik-type representation of $\delta_\pi(X)$.

For simplicity, in the following examples, we assume $\sigma^2 = 1$.

Example 3.4 (Student-t priors) In this example we take $\psi(v)$ as in Examples 3.2 and 3.3 with the specific choices $\alpha = (m-p+4)/2 \le 0$, $\beta = (m\,(1-\varphi)+2)/2$, and $\gamma = -m\,\varphi/2 \le 0$, where $m \ge 1$. In this case $h(v) = C\,v^{-(m+2)/2}\exp(-m\,\varphi/2\,v)$, an inverse gamma density. Hence, as is well known, $\pi(\theta)$ is a multivariate-t distribution with m-degrees of freedom and scale parameter φ if m is an integer (see e.g. Muirhead 1982, p.33 or Robert 1994, p.174). If $\sigma^2 \ne 1$, the scale of the t-distribution is $\varphi\,\sigma$.

For various different values of m and φ, either the conditions of Example 3.2 or the conditions of Example 3.3 apply. Both examples require $\alpha = (m-p+4)/2 \le 0$, or equivalently $1 \le m \le p-4$ (so that $p \ge 5$), and $\gamma = -m\,\varphi/2 \le 0$.

Example 3.2 requires $\beta = \big(m\,(1-\varphi)+2\big)/2 < 0$, or equivalently, $\varphi \ge (m+2)/m$. The condition for minimaxity $2-p < \alpha \le 0$ is satisfied since it is equivalent to $m > -p$. Furthermore the condition for proper Bayes minimaxity, $2-\frac{p}{2} < \alpha \le 0$, is satisfied as well since it reduces to $m > 0$. Hence, if $\varphi \ge (m+2)/m$, the scaled p-variate t prior distribution leads to a proper Bayes minimax estimator for $p \ge 5$ and $m \le p-4$.

On the other hand, when $\varphi < (m+2)/m$, or equivalently, $\beta > 0$, the conditions of Example 3.3 are applicable. Considering the proper Bayes case only, the condition for minimaxity of the Bayes estimator is

$$2-\frac{p}{2} < \alpha = \frac{m-p+4}{2} \le \frac{\beta^2}{2\gamma} \le \frac{\beta^2}{2\gamma} = -\frac{1}{4}\,\frac{\big(m\,(1-\varphi)+2\big)^2}{m\,\varphi}.$$

The first inequality is satisfied by the fact that $m > 0$. The second inequality can be satisfied only for certain φ since, when φ goes to 0, the last expression tends to $-\infty$. A straightforward calculation shows that the second inequality can hold only if

$$\varphi \ge \frac{p-2}{m}\left[1-\sqrt{1-\left(\frac{m+2}{p-2}\right)^2}\,\right] > 0\,.$$

In particular, if $\varphi = 1$ (the standard multivariate t), the condition becomes $2-p/2 < \frac{m-p+4}{2} \le -\frac{1}{m}$. As $m \ge 1$ this is equivalent to $m + 2/m \le p-4$, which requires $p \ge 7$ for $m = 1$ or 2, and $p \ge m+5$ for $m \ge 3$.

An alternative approach to the results of this section can be made using the techniques of Sect. 2.4.2 applied to Baranchik-type estimators of the form $\big(1 - a\,r(\|X\|^2)/\|X\|^2\big)\,X$. Indeed any spherically symmetric prior distribution will lead to an estimator of the form $\phi(\|X\|^2)X$. More to the point, for prior distributions of the form studied in this section, the $r(\cdot)$ function is closely connected to the function $v \mapsto -(v+1)h'(v)/h(v)$. To see this, note that

$$
\begin{aligned}
\delta_\pi(X) &= X + \sigma^2 \frac{\nabla m(X)}{m(X)} \\
&= \left(1 - \frac{I_{p/2+1}(y)}{I_{p/2}(y)}\right) X \qquad \text{from (3.2) with } y = \|X\|^2/2\sigma^2
\end{aligned}
$$

$$
= \left(1 - \frac{1}{y}\left(\frac{p-2}{2} - \frac{\int_0^\infty e^{-\frac{y}{1+v}}(1+v)^{-p/2}[(v+1)h'(v)/h(v)]\,dv - e^{-y}h(0)}{I_{p/2}(y)}\right)\right) X
$$

$$
= \left(1 - \frac{2\sigma^2}{\|X\|^2}\left(\frac{p-2}{2} + E^y_{p/2}\left[-\frac{(V+1)h'(V)}{h(V)}\right] - \frac{e^{-\frac{\|X\|^2}{2\sigma^2}} h(0)}{I_{p/2}(\frac{\|X\|^2}{2\sigma^2})}\right)\right) X,
$$

where $E^y_k(f)$ is as in the proof of Theorem 3.1, the second to last equality following from (3.4).

Hence, the Bayes estimator is of Baranchik form with

$$
ar(\|X\|^2) = 2\left(\frac{p-2}{2} + E^{\frac{\|X\|^2}{2\sigma^2}}_{p/2}\left[-\frac{(V+1)h'(V)}{h(V)}\right] - \frac{e^{-\frac{\|X\|^2}{2\sigma^2}} h(0)}{I_{p/2}(\frac{\|X\|^2}{2\sigma^2})}\right).
$$

□

Recall, as in the proof of Theorem 3.1, that the density $g^y_k(V)$ has a monotone decreasing likelihood ratio in k, but notice also that it has a monotone increasing likelihood ratio (actually as an exponential family) in y.

Hence, if $-\frac{(v+1)h'(v)}{h(v)}$ is nondecreasing, it follows that r is nondecreasing since $e^{-y}/I_{p/2}(y)$ is also nondecreasing. Then the following corollary is immediate from Theorem 3.3.

Corollary 3.4 *Suppose the prior is of the form* (3.4) *where* $-(v+1)\,h'(v)/h(v)$ *is nondecreasing and bounded above by* $A > 0$. *Then, the generalized Bayes estimator is minimax provided* $A \le \frac{p-2}{2}$.

Proof As noted, $r(\cdot)$ is nondecreasing and is bounded above by $p - 2 + 2A \le 2(p-2)$. □

Corollary 3.3 yields an alternative proof for the minimaxity of the generalized Bayes estimator in Example 3.1.

Finally, as indicated earlier in this section, an alternative parametrization has often been used in minimaxity proofs for the mixture of normal priors, namely $\lambda = \frac{1}{1+v}$, or equivalently, $v = \frac{1-\lambda}{\lambda}$.

Perhaps the easiest way to proceed is to reconsider the prior distribution as a hierarchical prior as discussed in Sect. 1.7. Here the distribution of $\theta \mid v \sim \mathcal{N}_p(0, v\sigma^2 X)$ and the unconditional density of v is the mixing density $h(v)$. The conditional distribution of θ given X and v is $\mathcal{N}_p(\frac{v}{1+v}X, \frac{V}{1+v}\sigma^2 I_p)$. The Bayes estimator is

$$
\begin{aligned}
\delta_\pi(X) &= E(\theta \mid X) \\
&= E[E(\theta \mid X, V) \mid X] \\
&= E[\tfrac{v}{1+v} X \mid X] \\
&= (1 - E[\tfrac{1}{1+v} \mid X])X \\
&= (1 - E[\lambda \mid X])X.
\end{aligned}
$$

Note also that the Bayes estimator for the first stage prior

$$
\theta \mid \lambda \sim \mathcal{N}(0, \frac{1-\lambda}{\lambda}\sigma^2 I) \tag{3.13}
$$

is $(1-\lambda)X$. Therefore, in terms of the λ parametrization, one may think of $E[\lambda \mid X]$ as the posterior mean of the shrinkage factor and of the (mixing) distribution on λ as the distribution of the shrinkage factor.

In particular, for the prior distribution of Example 3.1 where the mixing density on v is $h(v) = C\,(1+v)^{-\alpha-(p-2)/2}$, the corresponding mixture density on λ is given by $g(\lambda) = C\lambda^{\alpha+\frac{p-2}{2}-2} = C\lambda^\beta$ and $(\beta = \alpha + p/2 - 3)$. The resulting prior is proper Bayes minimax if $2 - p/2 < \alpha \le 0$ or equivalently, $-1 < \beta \le /2 - 3$ (and $p \ge 5$). Note that, if $p \ge 6$, $\beta = 0$ satisfies the conditions and consequently the mixing prior $g(\lambda) \equiv 1$ on $0 \le \lambda \le 1$, i.e. the uniform prior on the shrinkage factor λ gives a proper Bayes minimax estimator. This class of priors is often referred to as the Strawderman priors.

To formalize the above discussion further we present a version of Theorem 3.3 in terms of the mixing distribution on λ. The proof follows from Theorem 3.3 and the change of variable $\lambda = 1/(1+v)$.

Corollary 3.5 *Let θ have the hierarchical prior $\theta \mid \lambda \sim \mathcal{N}_p(0, (\{1-\lambda\}/\lambda)\,\sigma^2 I_p)$ where $\lambda \sim g(\lambda)$ for $0 \le \lambda \le 1$. Assume that $\lim_{\lambda\to 0} g(\lambda)\lambda^{p/2+1} = 0$ and that $\int_0^1 e^{-\lambda}\lambda^{p/2} g(\lambda) d\lambda < \infty$. Suppose $\lambda g'(\lambda)/g(\lambda)$ can be decomposed as $l_1^*(\lambda) + l_2^*(\lambda)$ where $l_1^*(\lambda)$ is monotone nonincreasing and $l_1^*(\lambda) \le A^*$, $0 \le l_2^*(\lambda) \le B^*$ with $A^* + 2B^* \le p/2 - 3$.*

Then the generalized Bayes estimator is minimax. Furthermore, if $\int_0^1 g(\lambda) d\lambda < \infty$, the estimator is also proper Bayes.

Example 3.5 (Beta priors) Suppose the prior $g(\lambda)$ on λ is a Beta (a, b) distribution, i.e. $g(\lambda) = K\lambda^{a-1}(1-\lambda)^{b-1}$. Note that the Strawderman (1971) prior is of this form if $b = 1$. An easy calculation shows $\frac{\lambda g'(\lambda)}{g(\lambda)} = a - 1 - (b-1)\frac{\lambda}{1-\lambda}$. Letting $l_1^*(\lambda) = \frac{\lambda g'(\lambda)}{g(\lambda)}$ and $l_2^*(\lambda) \equiv 0$, we see that the resulting proper Bayes estimator is minimax for $0 < a \le p/2 - 2$ and $b \ge 1$.

It is clear that our proof fails for $0 < b < 1$ since in this case $\lambda g'(\lambda)/g(\lambda)$ is not bounded from above (and is also monotone increasing). Maruyama (1998) shows, using a different proof technique involving properties of confluent hypergeometric functions, that the generalized Bayes estimator is minimax (in our notation) for $-p/2 < a \leq p/2 - 2$ and $b \geq (p + 2a + 2)(3p/2 + a)^{-1}$. This bound in b is in (0, 1) for $a < p/2 - 2$. Hence, certain Beta distributions with $0 < b < 1$ also give proper Bayes minimax estimators. The generalized Bayes minimax estimators of Alam (1973) are also in Maruyama's class.

3.1.4 Multiple Shrinkage Estimators

In this subsection, we consider a class of estimators that adaptively choose a point (or subspace) toward which to shrink. George (1986a,b) originated work in this area and the results in this section are largely due to him. The basic fact upon which the results rely is that a mixture of superharmonic functions is superharmonic (see the discussion in the Appendix), that is, if $m_\alpha(x)$ is superharmonic for each α, then $\int m_\alpha(x)\, dG(\alpha)$ is superharmonic if $G(\cdot)$ is a positive measure such that $\int m_\alpha(x)\, dG(\alpha) < \infty$. Using this property, we have the following result from Corollary 3.1.

Theorem 3.4 *Let $m_\alpha(x)$ be a family of twice weakly differentiable nonnegative superharmonic functions and $G(x)$ a positive measure such that $m(x) = \int m_\alpha(x)\, dG(x) < \infty$, for all $x \in \mathbb{R}^p$.*

Then the (generalized, proper, or pseudo) Bayes estimator

$$\delta(X) = X + \sigma^2 \frac{\nabla m(X)}{m(X)}$$

is minimax provided $E[\|\nabla m\|^2/m^2] < \infty$.

The following corollary for finite mixtures is useful.

Corollary 3.6 *Suppose that $m_i(x)$ is superharmonic and $E[\|\nabla m_i(X)\|^2/m_i^2(X)] < \infty$ for $i = 1, \ldots, n$. Then, if $m(x) = \sum_{i=1}^n m_i(x)$, the (generalized, proper, or pseudo) Bayes estimator*

$$\begin{aligned}\delta(X) &= X + \sigma^2 \tfrac{\nabla m(X)}{m(X)} \\ &= \textstyle\sum_{i=1}^n (X + \sigma^2 \tfrac{\nabla m_i(X)}{m_i(X)}) W_i(X)\end{aligned}$$

where $W_i(X) = m_i(X)/\sum_{i=1}^n m_i(X)$ for $0 < W_i(X) < 1, \sum_{i=1}^n W_i(X) = 1$ is minimax. (Note that $E_\theta[\|\nabla m(X)\|^2/m^2(X)] < \sum_{i=1}^n E_\theta[\|\nabla m_i(X)\|^2/m^2(X_i)] < \infty$.)

Example 3.6

(1) Multiple shrinkage James-Stein estimator. Suppose we have several possible points $X_1, X_2, \ldots, X_n$ toward which to shrink. Recall that $m_i(x) = (1/\|x - X_i\|^2)^{(p-2)/2}$ is superharmonic if $p \geq 3$ and the corresponding pseudo-Bayes estimator is $\delta_i(X) = X_i + \left(1 - (p-2)\,\sigma^2/\|X - X_i\|^2\right)(X - X_i)$. Hence, if $m(x) = \sum_{i=1}^n m_i(x)$, the resulting minimax pseudo Bayes estimator is given by

$$\delta(X) = \sum_{i=1}^{n} \left[X_i + (1 - \frac{(p-2)\sigma^2}{\|X - X_i\|^2})(X - X_i) \right] W_i(X)$$

where $W_i(X) \propto \left(1/\|X - X_i\|^2\right)^{(p-2)/2}$ and $\sum_{i=1}^n W_i(X) = 1$. Note that $W_i(X)$ is large when X is close to X_i and the estimator is seen to adaptively shrink toward X_i.

(2) Multiple shrinkage positive-part James-Stein estimators. Another possible choice for the $m_i(x)$ (leading to a positive-part James Stein estimator) is

$$m_i(x) = \begin{cases} C \, \exp\left(\frac{\|x - X_i\|^2}{2\sigma^2}\right) & \text{if } \|x - X_i\|^2 < (p-2)\,\sigma^2 \\ \left(\frac{1}{\|x - X_i\|^2}\right) & \text{if } \|x - X_i\|^2 \geq (p-2)\,\sigma^2 \end{cases}$$

where $C = \left(1/(p-2)\,\sigma^2\right)^{(p-2)/2} e^{(p-2)/2}$ so that $m_i(x)$ is continuous. This gives

$$\delta_i(X) = X_i + \left(1 - \frac{(p-2)\sigma^2}{\|X - X_i\|^2}\right)_+ (X - X_i)$$

since

$$\frac{\nabla m_i(X)}{m_i(X)} = \begin{cases} -\frac{X - X_i}{\sigma^2} & \text{if } \|X - X_i\|^2 < (p-2)\sigma^2, \\ -\frac{(p-2)}{\|X - X_i\|^2} & \text{otherwise.} \end{cases}$$

The adaptive combination is again minimax by the corollary and inherits the usual advantages of the positive-part estimator over the James-Stein estimator.

Note that a smooth alternative to the above is $m_i(x) = \left(\frac{1}{b + \|x - X_i\|^2}\right)^{\frac{p-2}{2}}$ for some $b > 0$.

In each of the above examples we may replace $(p-2)/2$ in the exponent by $a/2$ where $0 \leq a \leq p-2$ (and where $0 \leq \|x - X_i\|^2 < (p-2)\,\sigma^2$ is replaced by $0 \leq \|x - X_i\|^2 < a\,\sigma^2$ for the positive-part estimator). The choice of $p-2$ as an upper bound for a ensures superharmonicity of $m_i(x)$. A choice of a in the range of $p - 2 < a \leq 2\,(p-2)$ seems also quite natural since $\sqrt{m_i(x)}$ is superharmonic (but $m_i(x)$ is not) for a in this range so that each $\delta_i(X)$ is minimax. Unfortunately minimaxity of $\delta(X) = \sum_{i=1}^{n} W_i(X)\delta_i(X)$ does not follow from Corollary 3.3 for $p-2 < a \le 2(p-2)$ since it need not be true that $\sqrt{\sum_{i=1}^{n} m_i(x)}$ is superharmonic even though $\sqrt{m_i(x)}$ is superharmonic for each i.

(3) A generalized Bayes multiple shrinkage estimator. If $\pi_i(\theta)$ is superharmonic then $\pi(\theta) = \sum_{i=1}^{n} \pi_i(\theta)$ is also superharmonic as is $m(x) = \sum_{i=1}^{n} m_i(x)$.

For example, $\pi_i(\theta) = \left(1/b + \|\theta - X_i\|^2\right)^{a/2}$, for $b \ge 0$ and $0 \le a \le p-2$, is a suitable prior. Interestingly, according to a heuristic of Brown (1971), $m(x)$ in this case should behave for large $\|x\|^2$ as $\sum_{i=1}^{n} 1/\left(b + \|x - X_i\|^2\right)^{a/2}$, the "smooth" version of the adaptive positive-part multiple shrinkage pseudo-marginal in part (2) of this example.

By obvious modifications of the above, multiple shrinkage estimators may be constructed that shrink adaptively toward subspaces. Further examples can be found in George (1986a,b), Ki and Tsui (1990) and Wither (1991).

3.2 Bayes Estimators in the Unknown Variance Case

3.2.1 A Class of Proper Bayes Minimax Admissible Estimators

In this subsection, we give a class of hierarchical Bayes minimax estimators for the model

$$X \sim \mathcal{N}_p(\theta, \sigma^2 I_p) \quad S \sim \sigma^2 \chi_k^2 , \tag{3.14}$$

where S is independent of X, under scale invariant squared error loss

$$L(\theta, \delta(X, S)) = \frac{\|\delta(X, S) - \theta\|^2}{\sigma^2} . \tag{3.15}$$

We reparameterize σ^2 as $1/\eta$ and consider the following hierarchically, on the unknown parameters, structured prior(θ, η), which is reminiscent of the hierarchical version of the Strawderman prior in (3.13),

$$\begin{aligned} \theta | \lambda, \eta &\sim \mathcal{N}_p\left(0, \frac{1}{\eta}\frac{1-\lambda}{\lambda} I_p\right) \\ \eta &\sim Gamma\left(\frac{b}{2}, \frac{c}{2}\right) \\ \lambda &\sim (1+a)\,\lambda^a, \quad 0 < \lambda < 1 . \end{aligned} \tag{3.16}$$

Lemma 3.2 *For the model* (3.14) *and loss* (3.15), *the (generalized or proper) Bayes estimator of* θ *is given by*

$$\delta(X, S) = \left(1 - \frac{S}{\|X\|^2} r(\|X\|^2, S)\right) X \tag{3.17}$$

where

$$r(\|X\|^2, S) = \frac{\|X\|^2}{\|X\|^2 + c} \frac{\int_0^{(\|X\|^2+c)/S} u^{A+1} \left(\frac{1}{u+1}\right)^{B+1} du}{\int_0^{(\|X\|^2+c)/S} u^{A} \left(\frac{1}{u+1}\right)^{B+1} du} \tag{3.18}$$

where

$$A = \frac{p+a+b}{2} \quad \text{and} \quad B = \frac{p+k+b-2}{2} \tag{3.19}$$

provided $A > -1$, $A - B < 0$, *and* $c > 0$.

Proof Under the loss in (3.15) the Bayes estimator for the model in (3.16) is given by

$$\delta(X, S) = \frac{E[\theta\,\eta|X, S]}{E[\eta|X, S]}. \tag{3.20}$$

Expressing the expectation in the numerator of (3.20) gives

$$\begin{aligned}
E[\theta\,\eta|X, S] = &\int_0^\infty \int_0^1 \int_{\mathbb{R}^p} \theta\,\eta^{p/2+1} \left(\frac{\lambda\,\eta}{1-\lambda}\right)^{p/2} \\
&\times \exp\left(-\frac{\eta}{2}\left[\|x-\theta\|^2 + \frac{\lambda}{1-\lambda}\|\theta\|^2\right]\right) \eta^{(k+b-2)/2} \\
&\times \lambda^{(b+a)/2} \exp\left(-\frac{\eta}{2}(S+\lambda\, c)\right) d\theta\, d\eta\, d\lambda \\
= &\int_0^\infty \int_0^1 (1-\lambda)\lambda^A \eta^B \exp\left(-\frac{\eta}{2}(S+\lambda(\|x\|^2+c))\right) d\eta\, d\lambda
\end{aligned} \tag{3.21}$$

upon integrating with respect to θ and evaluating with the constants in (3.19). Similarly, for the denominator in (3.20)

$$\begin{aligned}
E[\eta|X, S] = &\int_0^\infty \int_0^1 \int_{\mathbb{R}^p} \eta^{p/2+1} \left(\frac{\lambda\,\eta}{1-\lambda}\right)^{p/2} \\
&\times \exp\left(-\frac{\eta}{2}\left[\|x-\theta\|^2 + \frac{\lambda}{1-\lambda}\|\theta\|^2\right]\right) \eta^{(k+b-2)/2}
\end{aligned}$$

$$\times\lambda^{(b+a)/2} \exp\left(-\frac{\eta}{2}(S+\lambda c)\right) d\theta\, d\eta\, d\lambda$$

$$= \int_0^\infty \int_0^1 \eta^B \lambda^A \exp\left(-\frac{\eta}{2}(S+\lambda(\|x\|^2+c))\right) d\eta d\lambda. \tag{3.22}$$

Therefore from (3.21) and (3.22) the Bayes estimator in (3.20) has the form

$$\delta(X,S) = \left(1 - \frac{S}{\|X\|^2} r(\|X\|^2, S)\right) X$$

where

$$r(\|X\|^2, S) = \frac{\|X\|^2}{S} \frac{\int_0^\infty \int_0^1 \eta^B \lambda^{A+1} \exp\left(-\frac{\eta S}{2}\left(1+\lambda \frac{\|x\|^2+c}{S}\right)\right) d\eta\, d\lambda}{\int_0^\infty \int_0^1 \eta^B \lambda^{A} \exp\left(-\frac{\eta S}{2}\left(1+\lambda \frac{\|x\|^2+c}{S}\right)\right) d\eta\, d\lambda}$$

$$= \frac{\|X\|^2/S}{(\|X\|^2+c)\, S} \frac{\int_0^{(\|X\|^2+c)/S} \int_0^\infty \eta^B u^{A+1} \exp\left(-\frac{\eta S}{2}(1+u)\right) d\eta\, du}{\int_0^{(\|X\|^2+c)/S} \int_0^\infty \eta^B u^{A} \exp\left(-\frac{\eta S}{2}(1+u)\right) d\eta\, du}$$

$$= \frac{\|X\|^2}{\|X\|^2+c} \frac{\int_0^{(\|X\|^2+c)/S} u^{A+1} \left(\frac{1}{u+1}\right)^{B+1} du}{\int_0^{(\|X\|^2+c)/S} u^{A} \left(\frac{1}{u+1}\right)^{B+1} du},$$

with the change of variable $u = \lambda\,(\|X\|^2+c)/S$ is made in the next to last step. □

The properties of $r(\|X\|^2, S)$ in Lemma 3.2 are given in the following result.

Lemma 3.3 *The function* $r(\|X\|^2, S)$ *given in* (3.18) *satisfies the following properties:*

(i) $r(\|X\|^2, S)$ *is nondecreasing in* $\|X\|^2$ *for fixed* S*;*
(ii) $r(\|X\|^2, S)$ *is nonincreasing in* S *for fixed* $\|X\|^2$*; and*
(iii) $0 \le r(\|X\|^2, S) \le (A+1)/(B-A-1) = (p+a+b+2)/(k-a-4)$

provided the conditions of Lemma 3.2 *hold.*

Proof Note first that $\int_0^t u\, f(u)\, du / \int_0^t f(u)\, du$ is nondecreasing in t for any integrable nonnegative function $f(\cdot)$. Hence Part (i) follows since $r(\|X\|^2, S)$ is the product of two nonnegative nondecreasing functions $\|X\|^2/\|X\|^2 + c$ and $\int_0^{(\|X\|^2+c)/S} u\, f(u)\, du / \int_0^{(\|X\|^2+c)/S} f(u)\, du$ for $f(u) = u^A\,(1+u)^{-(B+1)}$.

Part (ii) follows from a similar reasoning since the first term is constant in S and $(\|X\|^2+c)/S$ is decreasing in S.

To show Part (iii) note that, by Parts (i) and (ii),

$$
\begin{aligned}
0 \le r(\|X\|^2, S) &\le \lim_{\substack{\|X\|^2 \to \infty \\ S \to 0}} r(\|X\|^2, S) \\
&\le \frac{\int_0^\infty u^{A+1} \left(\frac{1}{u+1}\right)^{B+1} du}{\int_0^\infty u^{A} \left(\frac{1}{u+1}\right)^{B+1} du} \\
&= \frac{\int_0^1 \lambda^{B-A-2} (1-\lambda)^{A+1}}{\int_0^1 \lambda^{B-A-1} (1-\lambda)^{A}} \\
&= \frac{A+1}{B-A-1} \\
&= \frac{p+a+b+2}{k-a-4},
\end{aligned}
$$

expressing the beta functions and according to the values of A and B. □

We also need the following straightforward generalization of Corollary 2.6. The proof is left to the reader.

Corollary 3.7 *Under model* (3.14) *and loss* (3.15) *an estimator of the form*

$$
\delta(X, S) = \left(1 - \frac{S}{\|X\|^2} r(\|X\|^2, S)\right) X
$$

is minimax provided

(i) $r(\|X\|^2, S)$ *is nondecreasing in* $\|X\|^2$ *for fixed* S;
(ii) $r(\|X\|^2, S)$ *is nonincreasing in* S *for fixed* $\|X\|^2$; *and*
(iii) $0 \le r(\|X\|^2, S) \le 2\,(p-2)/(k+2)$.

Combining Lemmas 3.2 and 3.3 and Corollary 3.7 gives the following result.

Theorem 3.5 *For the model* (3.14), *loss* (3.15) *and hierarchical prior* (3.16), *the generalized or proper Bayes estimator in Lemma* 3.2 *is minimax provided*

$$
\frac{p+a+b+2}{k-a-4} \le \frac{2\,(p-2)}{k+2}. \tag{3.23}
$$

Furthermore, if $p \ge 5$, *there exist values of* $a > -2$ *and* $b > 0$ *which satisfy* (3.23), *i.e. such that the estimator is proper Bayes, minimax and admissible.*

Proof The first part is immediate. To see the second part, note that it suffices, if $a = -2 + \epsilon\ b = \delta$, for $\epsilon, \delta > 0$, that

$$
\frac{p}{k-2} < \frac{p+\epsilon+\delta}{k-2-\epsilon} \le \frac{2\,(p-2)}{k+2}
$$

equivalently $p > 4\frac{k-2}{k-6}$. Hence, for $p \geq 5$ and k sufficiently large, $k > 2(3p - 4)/(p - 4)$, there are values of a and b such that the priors are proper. □

Note that there exist values of a and b satisfying (3.23) and the assumptions of Lemma 3.2 whenever $p \geq 3$.

Strawderman (1973) gave the first example of a generalized and proper Bayes minimax estimators in the unknown variance setting. Zinodiny et al. (2011) also give classes of generalized and proper Bayes minimax estimators along somewhat similar lines as the above. The major difference is that the prior distribution on η $(= 1/\sigma^2)$ in the above development is also hierarchical, as it also depends on λ.

3.2.2 The Construction of a Class of Generalized Bayes Minimax Estimators

In this subsection we extend the generalized Bayes results of Sect. 3.1.2, using the ideas in Maruyama and Strawderman (2005) and Wells and Zhou (2008), to consider point estimation of the mean of a multivariate normal when the variance is unknown. Specifically, we assume the following model in (3.14) and the scaled squared loss function in (3.15).

In order to derive the (formal) Bayes estimator we reparameterize the model in (3.14) by replacing σ by η^{-1}. The model then becomes

$$\begin{aligned} X &\sim \mathcal{N}_p(\theta, \eta^{-2} I_p), \quad S \sim s^{k/2-1}\, \eta^k \exp(s\,\eta^2/2), \\ \theta &\sim \mathcal{N}_p(0, \nu\, \eta^{-2} I_p), \quad \nu \sim h(\nu), \quad \eta \sim \eta^d\,, \eta > 0\,, \end{aligned} \tag{3.24}$$

for some constant d. Under this model, the prior for θ is a scale mixture of normal distributions. Note that the above class of priors cannot be proper due to the impropriety of the distribution of η. However, as a consequence of the form of this model, the resulting generalized Bayes estimator is of the Baranchik form (3.17), with $r(\|X\|^2, S) = r(F)$, where $F = ||X||^2/S$.

We develop sufficient conditions on k, p, and $h(\nu)$ such that the generalized Bayes estimators with respect to the class of priors in (3.24) are minimax under the invariant loss function in (3.15). Maruyama and Strawderman (2005) and Wells and Zhou (2008) were able to obtain such sufficient conditions by applying the bounds and monotonicity results of Baranchik (1970), Efron and Morris (1976), and Fourdrinier et al. (1998).

Before we derive the formula for the generalized Bayes estimator under the model (3.24), we impose three regularity conditions on the parameters of priors. These conditions are easily satisfied by many hierarchical priors. These three conditions are assumed throughout this section.

C1: $A > 1$ where $A = \frac{d+k+p+3}{2}$;

C2: $\int_0^1 \lambda^{\frac{p}{2}-2} h\left(\frac{1-\lambda}{\lambda}\right) d\lambda < \infty$; and

C3: $\lim_{\nu \to \infty} \frac{h(\nu)}{(1+\nu)^{p/2-1}} = 0$.

Now, as in Sect. 3.1, we will first find the form of the Bayes estimator and then show that it satisfies some sufficient conditions for minimaxity. We start with the following lemma that corresponds to (3.2) in the known variance case and (3.18) in the previous subsection.

Lemma 3.4 *Under the model in* (3.24), *the generalized Bayes estimator can be written as*

$$\delta(X, S) = X - R(F)\, X = X - \frac{r(F)}{F}\, X, \tag{3.25}$$

where $F = ||X||^2/S$,

$$R(F) = \frac{\int_0^1 \lambda^{p/2-1}\, (1 + \lambda\, F)^{-A}\, h\left(\frac{1-\lambda}{\lambda}\right)\, d\lambda}{\int_0^1 \lambda^{p/2-2}\, (1 + \lambda\, F)^{-A}\, h\left(\frac{1-\lambda}{\lambda}\right)\, d\lambda}, \tag{3.26}$$

and

$$r(F) = F\, R(F)\,. \tag{3.27}$$

Proof Under the loss function (3.15), the generalized Bayes estimator for the model (3.24) is

$$\begin{aligned}
\delta(X, S) &= \frac{E(\frac{\theta}{\sigma^2}|X, S)}{E(\frac{1}{\sigma^2}|X, S)} \\
&= \frac{\int_0^\infty h(\nu) \int_0^\infty [(\eta^2)^{A-\frac{1}{2}} e^{-\frac{1}{2}\eta^2 S} \int_{\mathbb{R}^p} (\frac{1}{2\pi \nu \eta^{-2}})^{\frac{p}{2}} \theta e^{-\frac{1}{2}\eta^2(\frac{||\theta||^2}{\nu} + ||X-\theta||^2)} d\theta] d\eta\, d\nu}{\int_0^\infty h(\nu) \int_0^\infty [(\eta^2)^{A-\frac{1}{2}} e^{-\frac{1}{2}\eta^2 S} \int_{\mathbb{R}^p} (\frac{1}{2\pi \nu \eta^{-2}})^{\frac{p}{2}} e^{-\frac{1}{2}\eta^2(\frac{||\theta||^2}{\nu} + ||X-\theta||^2)} d\theta] d\eta d\nu} \\
&= \left(1 - \frac{\int_0^\infty [(\frac{1}{1+\nu}) h(\nu) (\frac{1}{1+\nu})^{\frac{p}{2}} \int_0^\infty (\eta^2)^{A-\frac{1}{2}} e^{-\frac{1}{2}\eta^2(S + \frac{||X||^2}{1+\nu})}\, d\eta]\, d\nu}{\int_0^\infty [h(\nu) (\frac{1}{1+\nu})^{\frac{p}{2}} \int_0^\infty (\eta^2)^{A-\frac{1}{2}} e^{-\frac{1}{2}\eta^2(S + \frac{||X||^2}{1+\nu})}\, d\eta]\, d\nu}\right) X \\
&= \left(1 - \frac{\int_0^\infty (\frac{1}{1+\nu}) h(\nu) (\frac{1}{1+\nu})^{\frac{p}{2}} (1 + \frac{F}{1+\nu})^{-A}\, d\nu}{\int_0^\infty h(\nu) (\frac{1}{1+\nu})^{\frac{p}{2}} (1 + \frac{F}{1+\nu})^{-A}\, d\nu}\right) X.
\end{aligned} \tag{3.28}$$

Letting $\lambda = (1 + \nu)^{-1}$, $\delta(X, S) = (1 - R(F))X$, which gives the form of the generalized Bayes estimator. □

Recall from Stein (1981) that when σ^2 is known the Bayes estimator under squared error loss and corresponding to a prior $\pi(\theta)$ is given by (3.2), that is, $\delta^{\pi}(X) = X + \sigma^2 \frac{\nabla m(X)}{m(X)}$.

The form of the Bayes estimator given in (3.25) gives an analogous form with the unknown variance replaced by a multiple of the usual unbiased estimator. In particular, define the "quasi-marginal"

$$\mathbf{M}(x, s) = \int \int f_X(x)\, f_S(s)\, \pi(\theta, \sigma^2)\, d\theta\, d\sigma^2$$

where

$$f_X(x) = \left(\frac{1}{2\pi\sigma^2}\right)^{p/2} e^{-\frac{1}{2\sigma^2}||x-\theta||^2}$$

and

$$f_S(s) = \frac{1}{2^{k/2}\Gamma(k/2)} s^{k/2-1} (\sigma^2)^{-k/2} e^{-\frac{s}{2\sigma^2}}.$$

A straightforward calculation shows $\mathbf{M}(x, s)$ is proportional to

$$\int_0^{\infty} h(v) \int_0^{\infty} [(\eta^2)^{A-\frac{3}{2}} e^{-\frac{1}{2}\eta^2 s} \int_{\mathbb{R}^p} (\frac{1}{2\pi v \eta^{-2}})^{\frac{p}{2}} e^{-\frac{1}{2}\eta^2(\frac{||\theta||^2}{v} + ||x-\theta||^2)} d\theta] d\eta dv.$$

It is interesting to note the unknown variance analog of (3.2) is

$$\delta(X, S) = X - \frac{1}{2} \frac{\nabla_X \mathbf{M}(X, S)}{\nabla_S \mathbf{M}(X, S)}.$$

Lastly, note that the exponential term in the penultimate expression in the representation of $\delta(X, S)$ in (3.28) (that comes from the normal sampling distribution assumption) cancels. Hence there is a sort of robustness with respect to the sampling distribution. We will develop this theme in greater detail in Chap. 6 in the setting of spherically symmetric distributions.

3.2.2.1 Preliminary Results

The minimax property of the generalized Bayes estimator is closely related to the behavior of the $r(F)$ and $R(F)$ functions, which is in turn closely related to the behavior of

$$g(v) = -(v + 1)\frac{h'(v)}{h(v)}. \tag{3.29}$$

Fourdrinier et al. (1998) gave a detailed analysis of the type of function in (3.29). However, their argument was deduced from the superharmonicity of the square root of a marginal condition. Baranchik (1970) and Efron and Morris (1976) gave certain regularity conditions on the shrinkage function $r(\cdot)$ such that an estimator

$$\widehat{\theta}(X, S) = X - \frac{r(F)}{F} X \tag{3.30}$$

is minimax under the loss function (3.15) for the model (3.14). Both results require an upper bound on $r(F)$ and a condition on how fast $R(F) = r(F)/F$ decreases with F. Both theorems follow from a general result for spherically symmetric distributions given in Chap. 6 (Proposition 6.1), or by applying Theorem 2.5 in a manner similar to that in Corollary 2.3. The proofs are left to the reader.

Theorem 3.6 (Baranchik 1970) *Assume that $r(F)$ is increasing in F and $0 \le r(F) \le 2\,(p-2)/(k+2)$. Then any point estimator of the form* (3.30) *is minimax.*

Theorem 3.7 (Efron and Morris 1976) *Define $c_k = \frac{p-2}{k+2}$. Assume that $0 \le r(F) \le 2\,c_k$, that for all F with $r(F) < 2c_k$,*

$$\frac{F^{p/2-1}\, r(F)}{(2 - r(F)/c_k)^{1+2c_k}} \text{ is increasing in } F, \tag{3.31}$$

and that, if an F_0 exists such that $r(F_0) = 2c_k$, then $r(F) = 2\,c_k$ for all $F \ge F_0$. With the above assumptions, the estimator $\widehat{\theta}(X, S) = X - r(F)/F\ X$ is minimax.

Consequently, to apply these results one has to establish an upper bound for $r(F)$ in (3.27) and the monotonicity property for some variant of $r(F)$. The candidate we use is $\widetilde{r}(F) = F^c r(F)$ with a constant c. Note that the upper bound $2\,c_k$ is exactly the same upper bound needed in Corollary 3.7(iii). We develop the needed results below.

First note that if $h(\nu)$ is a continuously differentiable function on $[0, \infty)$, and regularity conditions C1, C2 and C3 hold, then the integrations by parts used in Lemmas 3.5 and 3.6 are valid.

Lemma 3.5 *Assume the regularity conditions C1, C2 and C3, and that $g(\nu) \le M$, where M is a positive constant and $g(\nu)$ is defined as in* (3.29)*. Then, for the $r(F)$ function* (3.27)*, we have*

$$0 \le r(F) \le \frac{\frac{p}{2} - 1 + M}{A - \frac{p}{2} - M},$$

where A is defined in condition C1.

Proof By the definition in (3.26), $R(F) \ge 0$. Then $r(F) = F R(F) \ge 0$. Note that

$$r(F) = F \frac{\int_0^1 \lambda^{\frac{p}{2}-1}(1+\lambda F)^{-A} h(\frac{1-\lambda}{\lambda})\, d\lambda}{\int_0^1 \lambda^{\frac{p}{2}-2}(1+\lambda F)^{-A} h(\frac{1-\lambda}{\lambda})\, d\lambda} = F \frac{I_{\frac{p}{2}-1,A,h}(F)}{I_{\frac{p}{2}-2,A,h}(F)},$$

where we are using the notation

$$I_{\alpha,A,h}(F) = \int_0^1 \lambda^\alpha (1+\lambda F)^{-A} h(\frac{1-\lambda}{\lambda})\, d\lambda\,.$$

Using integration by parts, we obtain

$$\begin{aligned}
F I_{\frac{p}{2}-1,A,h}(F) &= \int_0^1 \lambda^{p/2-1} h\left(\frac{1-\lambda}{\lambda}\right) d\left[\frac{(1+\lambda F)^{1-A}}{1-A}\right] \\
&= \lambda^{\frac{p}{2}-1} h\left(\frac{1-\lambda}{\lambda}\right) \frac{(1+\lambda F)^{1-A}}{1-A}|_0^1 + \frac{1}{A-1}\int_0^1 (1+\lambda F)^{-A}(1+\lambda F) \\
&\left[\left(\frac{p}{2}-1\right)\lambda^{\frac{p}{2}-2} h\left(\frac{1-\lambda}{\lambda}\right) - \frac{1}{\lambda^2}\lambda^{\frac{p}{2}-1} h'\left(\frac{1-\lambda}{\lambda}\right)\right] d\lambda.
\end{aligned}$$

By C1 and C3, we know that the first term of the right hand side is nonpositive. The second term of the right hand side can be written as $N_1 + N_2 + N_3 + N_4$ where

$$N_1 = \frac{1}{A-1}\int_0^1 (1+\lambda F)^{-A}\left(\frac{p}{2}-1\right)\lambda^{\frac{p}{2}-2} h\left(\frac{1-\lambda}{\lambda}\right) d\lambda = \frac{\frac{p}{2}-1}{A-1} I_{\frac{p}{2}-2,A,h}(F),$$

$$\begin{aligned}
N_2 &= \frac{1}{A-1}\int_0^1 (1+\lambda F)^{-A}\lambda^{\frac{p}{2}-2} h'\left(\frac{1-\lambda}{\lambda}\right)\left(\frac{-\lambda}{\lambda^2}\right) d\lambda \\
&= \frac{I_{\frac{p}{2}-2,A,h}(F)}{A-1} \frac{\int_0^1 \lambda^{\frac{p}{2}-2}(1+\lambda F)^{-A} g(\frac{1-\lambda}{\lambda}) h(\frac{1-\lambda}{\lambda})\, d\lambda}{\int_0^1 \lambda^{\frac{p}{2}-2}(1+\lambda F)^{-A} h(\frac{1-\lambda}{\lambda})\, d\lambda} \\
&\le \frac{M}{A-1} I_{\frac{p}{2}-2,A,h}(F),
\end{aligned}$$

$$N_3 = \frac{\frac{p}{2}-1}{A-1} F I_{\frac{p}{2}-1,A,h}(F) = \frac{(\frac{p}{2}-1) r(F)}{A-1} I_{\frac{p}{2}-2,A,h}(F),$$

and

$$\begin{aligned}
N_4 &= \frac{I_{\frac{p}{2}-2,A,h}(F)}{A-1} \frac{F\int_0^1 \lambda^{\frac{p}{2}-1}(1+\lambda F)^{-A} h'(\frac{1-\lambda}{\lambda})(\frac{-1}{\lambda}) d\lambda}{I_{\frac{p}{2}-2,A,h}(F)} \\
&= \frac{I_{\frac{p}{2}-2,A,h}(F)}{A-1} \frac{F\int_0^1 (1+\lambda F)^{-A}\lambda^{\frac{p}{2}-1} g(\frac{1-\lambda}{\lambda}) h(\frac{1-\lambda}{\lambda}) d\lambda}{I_{\frac{p}{2}-2,A,h}(F)} \\
&\le \frac{M r(F)}{A-1} I_{\frac{p}{2}-2,A,h}(F).
\end{aligned}$$

Combining all the terms, we get the following inequality

$$(A-1)r(F) \le \left(\frac{p}{2}-1\right)+M+\left(\frac{p}{2}-1\right)r(F)+Mr(F) \Rightarrow r(F) \le \frac{\frac{p}{2}-1+M}{A-\frac{p}{2}-M}.$$

Therefore, we have the needed bound on the $r(F)$ function. □

We will now show that under certain regularity conditions on $g(\nu)$, we have the monotonicity property for $\tilde{r}(F) = F^c r(F)$ with a constant c. This monotonicity property enables us to establish the minimaxity of the generalized Bayes estimator. The following lemma is analogous to Theorem 3.3 in the known variance case.

Lemma 3.6 *If* $g(\nu) = -(\nu+1)\frac{h'(\nu)}{h(\nu)} = l_1(\nu) + l_2(\nu)$ *such that* $l_1(\nu)$ *is increasing in* ν *and* $0 \le l_2(\nu) \le c$*, then* $\tilde{r}(F) = F^c r(F)$ *is nondecreasing.*

Proof By taking the derivative, we only need to show (since $r(F) = FR(F)$)

$$0 \le FR'(F) + (1+c)R(F), \tag{3.32}$$

which is equivalent to

$$0 \le F\frac{I'_{\frac{p}{2}-1,A,h}(F)I_{\frac{p}{2}-2,A,h}(F) - I'_{\frac{p}{2}-2,A,h}(F)I_{\frac{p}{2}-1,A,h}(F)}{I^2_{\frac{p}{2}-2,A,h}(F)} + (1+c)\frac{I_{\frac{p}{2}-1,A,h}(F)}{I_{\frac{p}{2}-2,A,h}(F)}.$$

This in turn equivalent to

$$\begin{aligned} &-FI'_{\frac{p}{2}-1,A,h}(F)I_{\frac{p}{2}-2,A,h}(F) \\ &\le -FI'_{\frac{p}{2}-2,A,h}(F)I_{\frac{p}{2}-1,A,h}(F) + (1+c)I_{\frac{p}{2}-2,A,h}(F)I_{\frac{p}{2}-1,A,h}(F). \end{aligned} \tag{3.33}$$

Now note that

$$-FI'_{a,A,h}(F) = \int_0^1 \lambda^a(1+\lambda F)^{-A} h\left(\frac{1-\lambda}{\lambda}\right)\frac{A\lambda F}{1+\lambda F}d\lambda.$$

Define the intergral operator

$$J_a\left(f(u)\right) = \int_0^F u^a(1+u)^{-A} f(u)\, du.$$

Therefore,

$$J_a\left(h\left(\frac{F-u}{u}\right)\right) = \int_0^F u^a(1+u)^{-A} h\left(\frac{F-u}{u}\right) du$$

and

$$J_a\left(\frac{Au}{1+u}h\left(\frac{F-u}{u}\right)\right)=\int_0^F u^a(1+u)^{-A}\frac{Au}{1+u}h\left(\frac{F-u}{u}\right)du.$$

Also, note that

$$J_a\left(\frac{Au}{1+u}h\left(\frac{F-u}{u}\right)\right)=F^{a+1}\int_0^1 \lambda^a(1+\lambda F)^{-A}h\left(\frac{1-\lambda}{\lambda}\right)\frac{A\lambda F}{1+\lambda F}d\lambda,$$

and

$$J_a\left(h\left(\frac{F-u}{u}\right)\right)=F^{a+1}I_{a,A,h}(F).$$

Now, with this new notation, it follows that (3.33) is equivalent to

$$\frac{J_{\frac{p}{2}-1}(\frac{Au}{1+u}h(\frac{F-u}{u}))}{J_{\frac{p}{2}-1}(h(\frac{F-u}{u}))}\leq\frac{J_{\frac{p}{2}-2}(\frac{Au}{1+u}h(\frac{F-u}{u}))}{J_{\frac{p}{2}-2}(h(\frac{F-u}{u}))}+(1+c). \tag{3.34}$$

Using integration by parts, we have

$$J_a\left(\frac{Au}{1+u}h\left(\frac{F-u}{u}\right)\right)=\int_0^F u^a(1+u)^{-A}h\left(\frac{F-u}{u}\right)\frac{Au}{1+u}du$$

$$=-u^{a+1}h\left(\frac{F-u}{u}\right)(1+u)^{-A}|_0^F$$

$$+\int_0^F(1+u)^{-A}\left[(a+1)u^a h\left(\frac{F-u}{u}\right)+u^{a+1}h'\left(\frac{F-u}{u}\right)\left(\frac{-F}{u^2}\right)\right]du.$$

Hence, (3.34) is equivalent to

$$\frac{-F^{\frac{p}{2}}h(0)(1+F)^{-A}}{J_{\frac{p}{2}-1}(h(\frac{F-u}{u}))}+\left(\frac{p}{2}\right)$$

$$+\frac{\int_0^F u^{\frac{p}{2}-1}(1+u)^{-A}h(\frac{F-u}{u})\left[\frac{h'(\frac{F-u}{u})}{h(\frac{F-u}{u})}(\frac{-F}{u})\right]du}{\int_0^F u^{\frac{p}{2}-1}(1+u)^{-A}h(\frac{F-u}{u})\,du}$$

$$\leq\frac{-F^{\frac{p}{2}-1}h(0)(1+F)^{-A}}{J_{\frac{p}{2}-2}(h(\frac{F-u}{u}))}+\left(\frac{p}{2}-1\right)$$

$$+\frac{\int_0^F u^{\frac{p}{2}-2}(1+u)^{-A}h(\frac{F-u}{u})\left[\frac{h'(\frac{F-u}{u})}{h(\frac{F-u}{u})}(\frac{-F}{u})\right]du}{\int_0^F u^{\frac{p}{2}-2}(1+u)^{-A}h(\frac{F-u}{u})\,du}+(1+c). \tag{3.35}$$

Since $-(v+1)h'(v)/h(v) = l_1(v) + l_2(v)$ (3.35) is equivalent to

$$\begin{aligned}&\frac{-h(0)(1+F)^{-A}}{I_{\frac{p}{2}-1,A,h}(F)}+\frac{J_{\frac{p}{2}-1}(h(\frac{F-u}{u})l_1(\frac{F-u}{u}))}{J_{\frac{p}{2}-1}(h(\frac{F-u}{u}))}+\frac{J_{\frac{p}{2}-1}(h(\frac{F-u}{u})l_2(\frac{F-u}{u}))}{J_{\frac{p}{2}-1}(h(\frac{F-u}{u}))}\\&\le\frac{-h(0)(1+F)^{-A}}{I_{\frac{p}{2}-2,A,h}(F)}+\frac{J_{\frac{p}{2}-2}(h(\frac{F-u}{u})l_1(\frac{F-u}{u}))}{J_{\frac{p}{2}-2}(h(\frac{F-u}{u}))}+\frac{J_{\frac{p}{2}-2}(h(\frac{F-u}{u})l_2(\frac{F-u}{u}))}{J_{\frac{p}{2}-2}(h(\frac{F-u}{u}))}+c. \end{aligned}\tag{3.36}$$

It is clear that $I_{\frac{p}{2}-1,A,h}(F) \le I_{\frac{p}{2}-2,A,h}(F)$, so we then have

$$\frac{-h(0)(1+F)^{-A}}{I_{\frac{p}{2}-1,A,h}(F)} \le \frac{-h(0)(1+F)^{-A}}{I_{\frac{p}{2}-2,A,h}(F)}$$

which accounts for the first terms on the left and right hand sides of (3.36). As for the second term on each side of (3.36) note that the hypothesis $l_1(\nu)$ is increasing in ν implies that for all fixed F, $l_1(\frac{F-u}{u})$ is decreasing in u. When $t < u$, we have

$$\frac{(1+u)^{-A}u^{\frac{p}{2}-2}h(\frac{F-u}{u})\,\mathbb{1}\{u\le F\}}{(1+t)^{-A}t^{\frac{p}{2}-2}h(\frac{F-t}{t})\,\mathbb{1}\{t\le F\}} \le \frac{(1+u)^{-A}u^{\frac{p}{2}-1}h(\frac{F-u}{u})\,\mathbb{1}\{u\le F\}}{(1+t)^{-A}t^{\frac{p}{2}-1}h(\frac{F-t}{t})\,\mathbb{1}\{t\le F\}}.$$

By a monotone likelihood ratio argument, we have

$$\begin{aligned}\frac{J_{\frac{p}{2}-1}(h(\frac{F-u}{u})l_1(\frac{F-u}{u}))}{J_{\frac{p}{2}-1}(h(\frac{F-u}{u}))}&=\frac{\int_0^F u^{\frac{p}{2}-1}(1+u)^{-A}h(\frac{F-u}{u})l_1(\frac{F-u}{u})}{\int_0^F u^{\frac{p}{2}-1}(1+u)^{-A}h(\frac{F-u}{u})\,du}\\&\le\frac{\int_0^F u^{\frac{p}{2}-2}(1+u)^{-A}h(\frac{F-u}{u})l_1(\frac{F-u}{u})\,du}{\int_0^F u^{\frac{p}{2}-2}(1+u)^{-A}h(\frac{F-u}{u})\,du}=\frac{J_{\frac{p}{2}-2}(h(\frac{F-u}{u})l_1(\frac{F-u}{u}))}{J_{\frac{p}{2}-2}(h(\frac{F-u}{u}))}.\end{aligned}$$

Finally, note that since $0 \le l_2(v) \le c$ for the third term on each side of (3.36) we have

$$0 \le \frac{J_{\frac{p}{2}-i}(l_2(\frac{F-u}{u})h(\frac{F-u}{u}))}{J_{\frac{p}{2}-i}(h(\frac{F-u}{u}))} \le c \quad \text{for } i = 1, 2.$$

Therefore we established the inequality (3.36) and the proof is complete. □

3.2.2.2 Minimaxity of the Generalized Bayes Estimators

In this subsection we apply Lemmas 3.4, 3.5, 3.6 and Theorems 3.6 and 3.7 to show minimaxity of the generalized Bayes estimator (3.25).

Theorem 3.8 *Assume that $g(v) = -(v+1)\,h'(v)/h(v)$ is increasing in v, $g(v) \le M$, where M is a positive constant, and*

$$\frac{p-2+2M}{k+3+d-2M} \le 2\frac{p-2}{k+2}.$$

Then $\delta(X, S)$ in (3.25) is minimax.

Proof Let $l_2(v) = 0$ and $l_1(v) = g(v)$. By applying Lemma 3.6 to the case $c = 0$, we have $r(F)$ increasing in F. Applying the bound in Lemma 3.5, we can get $0 \le r(F) \le 2\frac{p-2}{m+2}$. Therefore, by Lemma 3.4, $\delta(X, S)$ is minimax. □

It is interesting to make connections to the result in Faith (1978). Faith (1978) considered generalized Bayes estimator for $\mathcal{N}_p(\theta, I_p)$ and showed that when $g(v)$ is increasing in v and $M \le \frac{p-2}{2}$, the generalized Bayes estimator would be minimax. By taking $k \to \infty$, we deduce the same conditions as Faith (1978). The next lemma is a variant of Alam (1973) for the known variance case.

Theorem 3.9 *Define $c_k = \frac{p-2}{k+2}$. If there exists $b \in (0, 1]$ and $c = \frac{b(p-2)}{4+4(2-b)c_k}$, such that $0 \le r(F) \le (2-b)c_k$, and $F^c r(F)$ is increasing in F, then the generalized Bayes estimator $\delta(X, S)$ in (3.25) is minimax.*

Proof By taking the derivative of the Efron and Morris' condition, (3.31) can be satisfied by requiring

$$0 \le 2\left(\frac{p}{2} - 1\right) R(F)\left(2 - \frac{r(F)}{c_m}\right) + 4r'(F)(1 + r(F)). \tag{3.37}$$

Since $r(F) \le (2-b)c_k$, then (3.37) is satisfied at the point where $r'(F) \ge 0$. Since $r(F) \le (2-b)c_k$ with $\beta = (2-b)c_k$

$$4r'(F)(1+\beta) \le 4r'(F)(1 + r(F)), \tag{3.38}$$

at the point where $r'(F) < 0$. We now have

$$\begin{aligned}
0 &\le (4+4\beta)(cR(F) + R(F) + FR'(F)) \\
&= 2b\left(\frac{p}{2} - 1\right) R(F) + 4r'(F)(1+\beta) \\
&\le 2\left(\frac{p}{2} - 1\right) R(F)\left(2 - \frac{r(F)}{c_k}\right) + 4r'(F)(1 + r(F))
\end{aligned}$$

since $F^c r(F)$ is increasing in F. Thus, for all values of F, we have proven (3.37), and combining with the bound on the $r(F)$ function, we have proven the minimaxity of the generalized Bayes estimator. □

It is interesting to observe that by requiring a tighter upper bound on $r(F)$, we can relax the monotonicity requirement on $r(F)$. The tighter the upper bound, the more flexible $r(F)$ can be. This result enriches the class of priors whose generalized Bayes estimators are minimax. Direct application of Lemmas 3.4, 3.5, 3.6, and 3.9 gives the following theorem.

Theorem 3.10 *If there exists $b \in (0, 1]$ such that $g(v) = l_1(v) + l_2(v) \leq M$, and $l_1(v)$ is increasing in v, $0 \leq l_2(v) \leq c = \frac{b(p-2)}{4+4(2-b)\frac{p-2}{k+2}}$, and $\frac{p-2+2M}{k+3+d-2M} \leq \frac{(2-b)(p-2)}{k+2}$, then the generalized Bayes estimator $\delta(X, S)$ in (3.25) is minimax.*

3.2.2.3 Examples of the Priors in (3.24)

In this subsection, we will give several examples to which our results can be applied and make some connection to the existing literature found in Maruyama and Strawderman (2005) and Fourdrinier et al. (1998).

Example 3.7 Maruyama and Strawderman (2005) considered the priors with $h(v) \propto v^b(1+v)^{-a-b-2}$ for $b > 0$ and show that $r(F) \leq \frac{\frac{p}{2}+a+1}{\frac{k}{2}+\frac{d}{2}-a-\frac{1}{2}}$ (in terms of the Maruyama and Strawderman (2005) notation $d = 2e + 1$). Condition C1 is equivalent to the condition that $d + k + p > -1$. C2 and C3 are equivalent here, and both are equivalent to the condition that $a + \frac{p}{2} + 1 > 0$. Then, using Theorem 3.8, we have $g(v) = a + 2 - bv^{-1}$. The condition that $g(v)$ is increasing in v is equivalent to the condition that $b \geq 0$. Clearly, we can let $M = a + 2$. Then the condition of Theorem 3.8 is that

$$\frac{k}{2} + \frac{d}{2} - \frac{1}{2} > a \quad \text{and} \quad \frac{\frac{p}{2}+a+1}{\frac{k}{2}+\frac{d}{2}-a-\frac{1}{2}} \leq 2c_k.$$

A close examination of the Maruyama and Strawderman (2005) proof shows that their upper bound on $r(F)$ is sharp. This implies that our bound in Lemma 3.5 cannot be relaxed.

Example 3.8 Generalized Student-t priors correspond to a mixing distribution of the form

$$h(v) = c(v+1)^{\beta-\alpha-\gamma-\frac{p-2}{2}} v^{\gamma-\beta} e^{\frac{\gamma}{v}}.$$

Consider the following two cases. The first case where $\alpha \leq 0$, $\beta \leq 0$ and $\gamma < 0$ involves the construction of a monotonic $r(\cdot)$ function. The second case where $\alpha \leq 0$, $\beta > 0$ and $\gamma < 0$ does not require the $r(\cdot)$ function to be monotonic. In both cases,

$$\ln h(\nu) = (\beta - \alpha - \gamma - \frac{p-2}{2}) \ln(1+\nu) + (\gamma - \beta) \ln \nu + \frac{\gamma}{\nu}$$

and

$$g(\nu) = \left(\frac{p-2}{2} + \alpha + \gamma - \beta\right) + \frac{(1+\nu)(\beta-\gamma)}{\nu} + \frac{\gamma(1+\nu)}{\nu^2} = \frac{p-2}{2} + \alpha + \frac{\beta}{\nu} + \frac{\gamma}{\nu^2}.$$

Clearly, $g(\nu)$ is monotonic in the first case, and minimaxity of the generalized Bayes estimator follows when

$$0 \le \frac{p-2+\alpha}{\frac{k}{2}+\frac{1}{2}+\frac{d}{2}-\frac{p}{2}-\alpha} \le \frac{p-2}{\frac{k}{2}+1}$$

in addition to the conditions C1, C2, and C3. In the limiting case where $m \to \infty$, C1 holds trivially. Both C2 and C3 can be satisfied by $\alpha > 2-p$. The upper bound on $R(F)$ can be satisfied by any $\alpha \le 0$. Consequently, the conditions reduce to those in Example 3.4 for the case of known variance.

Next we consider spherical multivariate Student-t priors with f degrees of freedom and a scale parameter τ and with $\alpha = \frac{f-p+4}{2}$, $\beta = \frac{f(1-\tau)+2}{2}$, and $\gamma = -\frac{f\tau}{2}$. The case of $\tau = 1$ is of particular interest but does not necessarily give a monotonic $r(\cdot)$ function. However, we can use the result in Theorem 3.10 to show that the generalized Bayes estimator is minimax under the following conditions for $f \le p-4$, suppose there exists a constant $b \in (0, 1]$ such that

$$\frac{p+f+\frac{1}{f}}{k+1+d-f-\frac{1}{f}} \le (2-b)\frac{p-2}{k+2},$$

$$\frac{1}{2f} \le c = \frac{b(p-2)}{4+4(2-b)\frac{p-2}{k+2}}. \tag{3.39}$$

Condition (3.39) can be established by observing that for this case,

$$g(\nu) = \frac{p-2}{2} + \alpha + \frac{\beta}{\nu} + \frac{\gamma}{\nu^2} = \frac{f}{2} + 1 + \frac{1}{\nu} - \frac{f}{2\nu^2}$$

is clearly nonmonotonic. We then let $M = \frac{f}{2} + 1 + \frac{1}{2f}$ and apply Lemma 3.5 to get the upper bound on $r(\cdot)$. We define $l_1(\nu) = g(\nu) - \frac{1}{2f}$ when $\nu \le f$ and $l_1(\nu) = \frac{f}{2}+1$ otherwise. We also define $l_2(\nu) = \frac{1}{2f}$ when $\nu \le f$ and $l_2(\nu) = \frac{1}{\nu} - \frac{f}{2\nu^2}$ otherwise. By applying Lemma 3.6, we get condition (3.39).

The spherical multivariate Cauchy prior corresponds to the case $f = 1$. If $k = O(p)$ and $d = 3$, then condition (3.39) reduces to $p \ge 5$, $\frac{p+2}{k+2} \le (2-b)\frac{p-2}{k+2}$, and $\frac{1}{2} \le \frac{b(p-2)}{4+8-4b}$.

3.3 Results for Known Σ and General Quadratic Loss

3.3.1 Results for the Diagonal Case

Much of this section is based on the review in Strawderman (2003). We begin with a discussion of the multivariate normal case where $\Sigma = \text{diag}(\sigma_1^2, \ldots, \sigma_p^2)$ is diagonal, which we assume throughout this subsection. Let

$$X \sim \mathcal{N}_p(\theta, \Sigma) \tag{3.40}$$

and the loss be equal to a weighted sum of squared errors loss

$$L(\theta, \delta) = (\delta - \theta)^{\mathrm{T}} D(\delta - \theta) = \sum_{i=1}^{p} (\delta_i - \theta_i)^2 d_i \,. \tag{3.41}$$

The results in Sects. 2.3, 2.4 and 3.1 extend by the use of Stein's lemma in a straightforward way to give the following basic theorem.

Theorem 3.11 *Let X have the distribution* (3.40) *and let the loss be given by* (3.41).

(1) *If $\delta(X) = X + \Sigma g(X)$, where $g(X)$ is weakly differentiable and $E||g||^2 < \infty$, then the risk of δ is*

$$\begin{aligned} R(\delta, \boldsymbol{\theta}) &= E_{\boldsymbol{\theta}}((\delta - \boldsymbol{\theta})^{\mathrm{T}} D(\delta - \boldsymbol{\theta})) \\ &= tr(\Sigma D) + E_\theta \left[\sum_{i=1}^{p} \sigma_i^4 d_i \left(g_i^2(X) + 2 \frac{\partial g_i(X)}{\partial X_i} \right) \right]. \end{aligned}$$

(2) *If $\boldsymbol{\theta} \sim \pi(\theta)$, then the Bayes estimator of θ is $\delta_\Pi(X) = X + \Sigma \frac{\nabla m(X)}{m(X)}$, where $m(X)$ is the marginal distribution of X.*

(3) *If $\boldsymbol{\theta} \sim \pi(\theta)$, then the risk of a proper (generalized, pseudo-) Bayes estimator of the form $\delta_m(X) = X + \Sigma \frac{\nabla m(X)}{m(X)}$ is given by*

$$\begin{aligned} R(\delta_m, \theta) &= \text{tr}(\Sigma D) \\ &\quad + E_\theta \left[\frac{2m(X) \sum_{i=1}^{p} \sigma_i^4 d_i \partial m^2(X)/\partial^2 X_i}{m^2(X)} - \frac{\sum_{i=1}^{p} \sigma_i^4 d_i \left(\partial m(X)/\partial X_i\right)^2}{m^2(X)} \right] \\ &= \text{tr}(\Sigma D) + 4\, E_\theta \left[\frac{\sum_{i=1}^{p} \sigma_i^4 d_i \partial^2 \sqrt{m(X)}/\partial^2 X_i}{\sqrt{m(X)}} \right]. \end{aligned}$$

(4) *If* $\frac{\sum_{i=1}^{p} \sigma_i^4 d_i \partial^2 \sqrt{m(X)}/\partial^2 X_i}{\sqrt{m(X)}}$ *is nonpositive, the proper (generalized, pseudo) Bayes* $\delta_m(X)$ *is minimax.*

The proof follows closely to that of corresponding results in Sects. 2.3, 2.4 and 3.1. The result is essentially from Stein (1981).

A key observation that allows us to construct Bayes minimax procedures for this situation, based on the procedures for the case $\Sigma = D = I$, is the following straightforward result from Strawderman (2003).

Lemma 3.7 *Suppose* $\eta(X)$ *is such that* $\Delta\eta(X) = \sum_{i=1}^{p} \partial^2\eta(X)/\partial^2 X_i^2 \leq 0$ *(i.e.* $\eta(X)$ *is superharmonic). Then* $\eta^*(X) = \eta(\Sigma^{-1}D^{-1/2}X)$ *is such that* $\sum_{i=1}^{p} \sigma_i^4 d_i \partial^2\eta^*(X)/\partial^2 X_i \leq 0$.

Note, that for any scalar a, if $\eta(X)$ is superharmonic, then so is $\eta(aX)$. This leads to the following result.

Theorem 3.12 *Suppose* X *has the distribution* (3.40) *and the loss is given by* (3.41).

(1) *Suppose* $\sqrt{m(X)}$ *is superharmonic* ($m(X)$ *is a proper, generalized, or pseudo-marginal for the case* $\Sigma = D = I$). *Then*

$$\delta_m(X) = X + \Sigma\left(\frac{\nabla m(\Sigma^{-1}D^{-1/2}X)}{m(\Sigma^{-1}D^{-1/2}X)}\right)$$

is a minimax estimator.

(2) *If* $\sqrt{m(\|X\|^2)}$ *is spherically symmetric and superharmonic, then*

$$\delta_m(X) = X + \frac{2m'(X^{\mathrm{T}}\Sigma^{-1}D^{-1}\Sigma^{-1}X)D^{-1}\Sigma^{-1}X}{m(X^{\mathrm{T}}\Sigma^{-1}D^{-1}\Sigma^{-1}X)}$$

is minimax.

(3) *Suppose the prior distribution* $\pi(\theta)$ *has the hierarchical structure* $\theta|\lambda \sim \mathcal{N}_p(0, A_\lambda)$ *for* $\lambda \sim h(\lambda)$, $0 < \lambda < 1$, *where* $A_\lambda = (c/\lambda)\Sigma D\Sigma - \Sigma$, *c is such that* A_1 *is positive definite, and* $h(\lambda)$ *satisfies the conditions of Theorem* 3.12. *Then*

$$\delta_\pi(X) = X + \Sigma\frac{\nabla m(X)}{m(X)}$$

is minimax.

(4) *Suppose* $m_i(X), i = 1, 2 \ldots k$ *are superharmonic. Then the multiple shrinkage estimator*

$$\delta_m(X) = X + \Sigma \left[\frac{\sum_{i=1}^{k} \nabla m_i(\Sigma^{-1} D^{-1/2} X)}{\sum_{i=1}^{k} m_i(\Sigma^{-1} D^{-1/2} X)} \right]$$

is a minimax multiple shrinkage estimator.

Proof Part (1) follows directly from Parts (3) and (4) of Theorem 3.11 and Lemma 3.7. Part (2) follows from Part (1) and Part (2) of Theorem 3.11 with a straightforward calculation.

For Part (3), first note that $\theta|\lambda \sim \mathcal{N}_p(0, A_\lambda)$ and $X - \theta|\lambda \sim \mathcal{N}_p(0, \Sigma)$. Thus, $X - \theta$ and θ are conditionally independent given λ. Hence we have $X|\lambda \sim \mathcal{N}_p(0, A_\lambda + \Sigma)$. It follows that

$$m(X) \propto \int_0^1 \lambda^{p/2} \exp\left[-\frac{\lambda}{c} \left(X^{\mathrm{T}} \Sigma^{-1} D^{-1} \Sigma^{-1} X \right) \right] h(\lambda)\, d\lambda$$

but $m(X) = \eta\left(X^{\mathrm{T}} \Sigma^{-1} D^{-1} \Sigma^{-1} X / c \right)$, where $\sqrt{\eta\,(X^{\mathrm{T}} X)}$ is superharmonic by Theorem 3.11. Hence, by Part (2), $\delta_\pi(X)$ is minimax (and proper or generalized Bayes depending on whether $h(\lambda)$ is integrable or not).

Since superharmonicity of $\eta(X)$ implies the superharmonicity of $\sqrt{\eta\,(X)}$, Part (4) follows from Part (1) and the superharmonicity of mixtures of superharmonic functions. □

Example 3.9 (Pseudo-Bayes minimax estimators) When $\Sigma = D = \sigma^2 I$, we saw in Sect. 3.3 that by choosing $m(X) = \frac{1}{\|X\|^{2b}}$, the pseudo-Bayes estimator was the James-Stein estimator $\delta_m(X) = (1 - \frac{2b\sigma^2}{\|X\|^2})X$. It now follows from this and part (2) of Theorem 3.12 that $m(X^{\mathrm{T}} \Sigma^{-1} D^{-1} \Sigma^{-1} X) = (1/X^{\mathrm{T}} \Sigma^{-1} D^{-1} \Sigma^{-1} X)^b$ has associated with it the pseudo-Bayes estimator $\delta_m(X) = (1 - \frac{2bD^{-1}\Sigma^{-1}}{(X^{\mathrm{T}} \Sigma^{-1} D^{-1} \Sigma^{-1} X)})X$. This estimator is minimax for $0 < b \le 2(p-2)$.

Example 3.10 (Hierarchical proper Bayes minimax estimator) As suggested by Berger (1976) suppose the prior distribution has the hierarchical structure $\theta|\lambda \sim \mathcal{N}_p(0, A_\lambda)$ where $A_\lambda = c\Sigma D \Sigma - \Sigma$, $c > 1/\min(\sigma_i^2 d_i)$ and $h(\lambda) = (1+b)\lambda^b$ for $0 < \lambda < 1$ and $-1 < b \le \frac{(p-6)}{2}$. The resulting proper Bayes estimator will be minimax for $p \ge 5$ by part (3) of Theorem 3.12 and Example 3.9. For $p \ge 3$, the estimator $\delta_\pi(X)$ given in part (3) of Theorem 3.12 is a generalized Bayes minimax estimator provided $-\frac{(p+2)}{2} < b \le \frac{(p-6)}{2}$.

It can be shown to be admissible if the lower bound is replaced by -2, by the results of Brown (1971). Also see the development in Berger and Strawderman (1996) and Kubokawa and Strawderman (2007).

Example 3.11 (Multiple shrinkage minimax estimators) It follows from Example 3.9 and Theorem 3.12 that $m(X) = \sum_{i=1}^{k} \left[\frac{1}{(X-\nu_i)^{\mathrm{T}} \Sigma^{-1} D^{-1} \Sigma^{-1} (X-\nu_i)} \right]^b$ satisfies the conditions of Theorem 3.12 (4) for $0 < b \le (p-2)/2$. and hence

$$\delta_m(X) = X - \frac{2b \sum_{i=1}^{k} \left[D^{-1} \Sigma^{-1} (X - \nu_i) \right] \Big/ \left[(X - \nu_i)^{\mathrm{T}} \Sigma^{-1} D^{-1} \Sigma^{-1} (X - \nu_i) \right]^{b+1}}{\sum_{i=1}^{k} 1 \Big/ \left[(X - \nu_i)^{\mathrm{T}} \Sigma^{-1} D^{-1} \Sigma^{-1} (X - \nu_i) \right]^{b}} \tag{3.42}$$

is a minimax multiple shrinkage (pseudo-Bayes) estimator.

If, as in Example 3.11 we used the generalized prior

$$\pi(\theta) = \sum_{i=1}^{k} \left[\frac{1}{(\theta - \nu_i)^{\mathrm{T}} \Sigma^{-1} D^{-1} \Sigma^{-1} (\theta - \nu_i)} \right]^b,$$

the resulting generalized Bayes (as opposed to pseudo-Bayes) estimators is minimax for $0 < b \le (p-2)/2$.

3.3.2 General Σ and General Quadratic Loss

In this section, we generalize the above results to the case of

$$X \sim \mathcal{N}_p(\theta, \Sigma), \tag{3.43}$$

where Σ is a general positive definite covariance matrix and the loss is given by

$$L(\theta, \delta) = (\delta - \theta)^{\mathrm{T}} Q (\delta - \theta), \tag{3.44}$$

where Q is a general positive definite matrix. We will see that this case can be reduced to the canonical form $\Sigma = I$ and $Q = \mathrm{diag}(d_1, d_2, \ldots, d_p) = D$. We continue to follow the development in Strawderman (2003).

The following well known fact will be used repeatedly to obtain the desired generalization.

Lemma 3.8 *For any pair of positive definite matrices, Σ and Q, there exits a non-singular matrix A such that $A \Sigma A^{\mathrm{T}} = I$ and $(A^{\mathrm{T}})^{-1} Q A^{-1} = D$ where D is diagonal.*

Using this fact we can now present the canonical form of the estimation problem.

Theorem 3.13 *Let $X \sim \mathcal{N}_p(\theta, \Sigma)$ and suppose that the loss is $L_1(\delta, \theta) = (\delta - \theta)^{\mathrm{T}} Q (\delta - \theta)$. Let A and D be as in Lemma* 3.8 *and let $Y = AX \sim \mathcal{N}_p(\upsilon, I_p)$, where $\upsilon = A\theta$ and $L_2(\delta, \upsilon) = (\delta - \upsilon)^{\mathrm{T}} D (\delta - \upsilon)$.*

(1) *If $\delta_1(X)$ is an estimator with risk function $R_1(\delta_1, \boldsymbol{\theta}) = E_\theta L_1(\delta_1(X), \theta)$, then the estimator $\delta_2(Y) = A\delta_1(A^{-1}Y)$ has risk function $R_2(\delta_2, \upsilon) = R_1(\delta_1, \theta) = E_\theta L_2(\delta_2(Y), \upsilon)$.*
(2) *$\delta_1(X)$ is proper or generalized Bayes with respect to the proper prior distribution $\pi_1(\theta)$ (or pseudo-Bayes with respect to the pseudo-marginal $m_1(X)$) under loss L_1 if and only if $\delta_2(Y) = A\delta_1(A^{-1}Y)$ is proper or generalized Bayes with respect to $\pi_2(\upsilon) = \pi_1(A^{-1}\upsilon)$ (or pseudo-Bayes with respect to the pseudo-marginal $m_2(Y) = m_1(A^{-1}Y)$).*
(3) *$\delta_1(X)$ is admissible (or minimax or dominates $\delta_1^*(X)$) under L_1 if and only if $\delta_2(Y) = A\delta_1(A^{-1}Y)$ is admissible (or minimax or dominates $\delta_2^*(Y) = A\delta_1^*(A^{-1}Y)$ under L_2).*

Proof To establish Part (1) note that the risk function

$$\begin{aligned}
R_2(\delta_2, \upsilon) &= E_\theta L_2[\delta_2(Y), \upsilon] \\
&= E_\theta[(\delta_2(Y) - \upsilon)^{\mathrm{T}} D (\delta_2(Y) - \upsilon)] \\
&= E_\theta[(A\delta_1(A^{-1}(AX)) - A\theta)^{\mathrm{T}} D (A\delta_1(A^{-1}(AX)) - A\theta)] \\
&= E_\theta[(\delta_1((X) - \theta)^{\mathrm{T}} A^{\mathrm{T}} D A (\delta_1(X) - \theta)] \\
&= E_\theta[(\delta_1((X) - \theta)^{\mathrm{T}} Q (\delta_1(X) - \theta)] \\
&= R_1(\delta_1, \theta).
\end{aligned}$$

Since the Bayes estimator for any quadratic loss is the posterior mean and $\theta \sim \pi_1(\theta)$ and $\upsilon = A\theta \sim \pi_2(\upsilon) = \pi_1(A^{-1}\upsilon)$ (ignoring constants), then Part (2) follows by noting that

$$\delta_2(Y) = E[\upsilon|Y] = E[A\theta|Y] = E[A\theta|AX] = A\ E[\theta|X] = A\ \delta_1(X) = A\delta_1(A^{-1}Y).$$

Lastly, Part (3) follows directly from Part (1). □

Note that if $\Sigma^{1/2}$ is the positive definite square root of Σ and $A = P\Sigma^{-1/2}$ where P is orthogonal and diagonalizes $\Sigma^{1/2} Q \Sigma^{1/2}$, then this A and $D = P\Sigma^{1/2} Q \Sigma^{1/2} P^{\mathrm{T}}$ satisfy the requirements of the theorem.

Example 3.12 Proceeding as we did in Example 3.9 and applying Theorem 3.13, $m(X^{\mathrm{T}} \Sigma^{-1} Q^{-1} \Sigma^{-1} X) = (X^{\mathrm{T}} \Sigma^{-1} Q^{-1} \Sigma^{-1} X)^{-b}$ has associated with it, the pseudo-Bayes minimax James-Stein estimators is

$$\delta_m(X) = \left(1 - \frac{2\,b\,Q^{-1}\Sigma^{-1}}{(X^{\mathrm{T}} \Sigma^{-1} Q^{-1} \Sigma^{-1} X)}\right) X,$$

for $0 < b \leq 2\,(p-2)$.

Generalizations of Example 3.10 to hierarchical Bayes minimax estimators and generalizations of Example 3.11 to multiple shrinkage estimators are straightforward. We omit the details.

3.4 Admissibility of Bayes Estimators

Recall from Sect. 2.4 that an admissible estimator is one that cannot be dominated in risk, i.e. $\delta(X)$ is admissible if there does not exist an estimator $\delta'(X)$ such that $R(\theta, \delta') \leq R(\theta, \delta)$ for all θ, with strict inequality for some θ. We have already derived classes of minimax estimators in the previous sections.

In this section, we study their possible admissibility or inadmissibility. One reason that admissibility of these minimax estimators is interesting is that, as we have already seen, the usual estimator $\delta_0(X) = X$ is minimax but inadmissible if $p \geq 3$. Actually, we have seen that it is possible to dominate X with a minimax estimator (e.g., $\delta^{JS}_{(p-2)}(X)$) that has a substantially smaller risk at $\theta = 0$. Hence, it is of interest to know if a particular (dominating) estimator is admissible.

Note that a unique proper Bayes estimator is automatically admissible (see Lemma 2.6), so we already have examples of admissible minimax estimators for $p \geq 5$.

We also note that the class of generalized Bayes estimators contains all admissible estimators if loss is quadratic (i.e., it is a complete class; see e.g., Sacks 1963; Brown 1971; Berger and Srinivasan 1978). It follows that if an estimator is not generalized Bayes, it is not admissible. Further, in order to be generalized Bayes, an estimator must be everywhere differentiable by properties of the Laplace transform . In particular, the James-Stein estimators and the positive-part James-Stein estimators (for $a \neq 0$) are not generalized Bayes and therefore not admissible.

In this section, we will study the admissibility of estimators corresponding to priors which are variance mixtures of normal distributions for the case of $X \sim \mathcal{N}_p(\theta, I)$ and quadratic loss $\|\delta - \theta\|^2$ as in Sect. 3.1.2. In particular, we consider prior densities of the form (3.4) and establish a connection between admissibility and the behavior of the mixing (generalized) density $h(v)$ at infinity. The analysis will be based on Brown (1971), Theorem 1.2. An Abelian Theorem (see, e.g., Widder (1946), Corollary 1.a, p. 182) along with Brown's theorem are our main tools. We use the notation $f(x) \sim g(x)$ as $x \to a$ to mean $\lim_{x\to a} f(x)/g(x) = 1$. Here is an adaptation of the Abelian theorem in Widder that meets our needs.

Theorem 3.14 *Assume* $g : \mathbb{R}^+ \to \mathbb{R}$ *has a Laplace transform* $f(s) = \int_0^\infty g(t)e^{-st}\,dt$ *that is finite for* $s \geq 0$. *If* $g(t) \sim t^\gamma$ *as* $t \to 0_+$ *for some* $\gamma > -1$, *then* $f(s) \sim s^{-(\gamma+1)}\Gamma(\gamma + 1)$ *as* $s \to \infty$.

The proof is essentially as in Widder (1946) but the assumption of finiteness of the Laplace transform at $s = 0$ allows the extension from $\gamma \geq 0$ to $\gamma > -1$.

We first give a lemma which relates the tail behavior of the mixing density $h(v)$ to the tail behavior of $\pi(\|\theta\|^2)$ and $m(\|x\|^2)$ and also shows that $\|\delta(x) - x\|$ is bounded whenever $h(v)$ has polynomial tail behavior.

Lemma 3.9 *Suppose* $X \sim \mathcal{N}_p(\theta, I_p)$, $L(\theta, \delta) = \|\delta - \theta\|^2$ *and* $\pi(\theta)$ *is given by* (3.4) *where* $h(v) \sim K\, v^a$ *as* $v \to \infty$ *with* $a < (p-2)/2$ *and where* $v^{-p/2} h(v)$ *is integrable in a neighborhood of* 0. *Then*

(1) $\pi(\theta) \sim K\, (\|\theta\|^2)^{a-(p-2)/2}\, \Gamma((p-2)/2 - a)$ *as* $\|\theta\|^2 \to \infty$,
$m(x) \sim K(\|x\|^2)^{a-(p-2)/2}\, \Gamma((p-2)/2 - a)$ *as* $\|x\|^2 \to \infty$,
and therefore $\pi(\|x\|^2) \sim m(\|x\|^2)$ *as* $\|x\|^2 \to \infty$,

(2) $\|\delta(x) - x\|$ *is uniformly bounded, where* δ *is the generalized Bayes estimator corresponding to* π.

Proof First note that (with $t = 1/v$)

$$\pi(\theta) = \pi^*(\|\theta\|^2) = \int_0^\infty \exp\left\{-\frac{\|\theta\|^2}{2} t\right\} t^{\frac{p}{2}-2} h(1/t)\, dt$$

and $g(t) = t^{\frac{p}{2}-2} h(1/t) \sim K t^{\frac{p-4}{2}-a}$ as $t \to 0_+$. Therefore, by Theorem 3.14, $\pi(\theta) \sim K(\|\theta\|^2)^{a-\frac{p-2}{2}}\, \Gamma\left(\frac{p-2}{2} - a\right)$ as $\|\theta\|^2 \to \infty$. Similarly

$$\begin{aligned} m(x) &= \int_0^\infty e^{-\frac{\|\theta\|^2}{2(1+v)}} (1+v)^{-\frac{p}{2}}\, h(v)\, dv \left(\text{for } t = \frac{1}{1+v}\right) \\ &= \int_1^\infty e^{-\frac{\|\theta\|^2}{2} t}\, t^{\frac{p}{2}-2} h\left(\frac{1-t}{t}\right) dt. \end{aligned}$$

We note that as $t \to 0_+$, $t^{\frac{p}{2}-2} h\left(\frac{1-t}{t}\right) \sim t^{\frac{p-4}{2}} \left(\frac{1-t}{t}\right)^a \sim t^{\frac{p-4}{2}-a}$. Thus, again by Theorem 3.14,

$$m(x) \sim K(\|x\|^2)^{a-\frac{p-2}{2}}\, \Gamma\left(\frac{p-2}{2} - a\right) \text{ as } \|x\|^2 \to \infty,$$

and Part (1) follows.

To prove Part (2) note that

$$\begin{aligned} \delta(x) - x &= \frac{\nabla m(x)}{m(x)} \\ &= -\frac{-\int_0^\infty \exp\left\{-\frac{\|x\|^2}{2(1+v)}\right\}(1+v)^{-(\frac{p}{2}+1)}\, h(v)\, dv}{\int_0^\infty \exp\left\{-\frac{\|x\|^2}{2(1+v)}\right\}(1+v)^{\frac{p}{2}}\, h(v)\, dv} x. \end{aligned}$$

The above argument applied to the numerator and denominator shows

$$\|\delta(x) - x\|^2 \sim \left[\frac{(\|x\|^2)^{a-\frac{p}{2}}\, \Gamma(\frac{p}{2}-a)}{\|x\|^2)^{a-\frac{p-2}{2}}\, \Gamma(\frac{p-2}{2}-a)} \right]^2 \|x\|^2$$

$$\sim \left(\tfrac{p-2}{2} - a\right)^2 \frac{1}{\|x\|^2} \text{ as } \|x\|^2 \to \infty.$$

Since $\delta(x) - x$ is in $\mathscr{C}^\infty$ and tends to zero as $\|x\|^2 \to \infty$, the function is uniformly bounded. □

The following result characterizes admissibility and inadmissibility for generalized Bayes estimators when the mixing density $h(v) \sim v^a$ as $v \to \infty$.

Theorem 3.15 *For priors $\pi(\theta)$ of the form* (3.4) *with mixing density $h(v) \sim v^a$ as $v \to \infty$, the corresponding generalized Bayes estimator δ is admissible if and only if $a \leq 0$.*

Proof (Admissibility if $a \leq 0$) By Lemma 3.9, we have $\bar{m}(r) = m^*(r^2) \sim K^* (r^2)^{a-(p-2)/2}$, with $m(x) = m^*(\|x\|^2)$. Thus, for any $\epsilon > 0$, there is an $r_0 > 0$ such that, for $r > r_0$, $\bar{m}(r) \leq (1+\epsilon)K^* r^{2a-(p-2)}$. Since $\|\delta(x) - x\|$ is uniformly bounded,

$$\int_{r_0}^{\infty} (r^{p-1}\bar{m}(r))^{-1}\, dr \geq (K^*(1+\epsilon))^{-1} \int_{r_0}^{\infty} r^{-(2a+1)}\, dr = \infty$$

if $a \geq 0$. Hence, $\delta(x)$ is admissible if $a \leq 0$, by Theorem 1.2.
(Inadmissibility if $a > 0$) Similarly, we have, for $r \geq r_0$,

$$\underline{m}(r) = \frac{1}{m^*(r^2)} \sim \frac{1}{K^*}(r^2)^{\frac{p-2}{2}-a},$$

$$\underline{m}(r) \leq \frac{1}{(1-\epsilon)K^*} r^{p-2-2a},$$

and

$$\int_0^{\infty} r^{1-p}\underline{m}\,(r)\, dr \leq \frac{1}{K^*} \int_{r_0}^{\infty} r^{-(1+2a)}\, dr < \infty$$

if $a > 0$. Thus $\delta(x)$ is inadmissible if $a > 0$. □

Example 3.13 (Continued) Recall for the Strawderman prior that $h(v) = C(1+v)^{-\alpha-(\frac{p-2}{2})} \sim v^a$ as $v \to \infty$ for $a = -(\alpha + \frac{p-2}{2})$.

The above theorem implies that the generalized Bayes estimator is admissible if and only if $\alpha + \frac{p-2}{2} \geq 0$ or $1 - \frac{p}{2} \leq \alpha$. We previously established minimaxity when $2 - p < \alpha \leq 0$ for $p \geq 3$ and propriety of the prior when $2 - \frac{p}{2} < \alpha \leq 0$ for $p \geq 5$.

Note in general that for a mixing distribution of the form $h(v) \sim K v^a$ as $v \to \infty$, the prior distribution $\pi(\theta)$ will be proper if and only if $a < -1$ by the same argument as in the proof of Theorem 3.15. Hence the bound for admissibility, $a \leq 0$, differs from the bound for propriety, $a < -1$, by 1.

3.5 Connections to Maximum a Posteriori Estimation

3.5.1 *Hierarchical Priors*

As we have seen in previous sections of this chapter, the classical Stein estimate and its positive-part modification can be motivated in a number of ways, perhaps most commonly as empirical Bayes estimates (i.e., posterior means) under a normal hierarchical model in which $\theta \sim \mathcal{N}_p(0, \psi I_p)$ where ψ, viewed as a hyperparameter, is estimated. In this section we look at shrinkage estimation through the lens of maximum a posteriori (MAP) estimation. The development of this section follows Strawderman and Wells (2012).

The class of proper Bayes minimax estimators constructed in Sect. 3.1 relies on the use of a hierarchically specified class of proper prior distributions $\pi_S(\theta, \kappa)$. In particular, for the prior in Strawderman (1971), $\pi_S(\theta, \kappa)$ is specified according to

$$\theta|\kappa \sim \mathcal{N}_p(0, g(\kappa)I_p), \quad \pi_S(\kappa) = \kappa^{-a}(1-a)^{-1}\, \mathbb{1}_{[0<\kappa<1]}, \tag{3.45}$$

where $g(\kappa) = (1-\kappa)/\kappa$ and the constant a satisfies $0 \leq a < 1$, i.e., $\pi_S(\kappa)$ is a Beta$(1-a, 1)$ probability distribution. Suppose $a = 1/2$; then, utilizing the transformation $\psi = g(\kappa) > 0$ in (3.45), we obtain the equivalent specification

$$\theta|\psi \sim \mathcal{N}_p(0, \psi I_p), \quad \pi_S(\psi) \propto \left(\frac{1}{1+\psi}\right)^{\frac{3}{2}} \mathbb{1}_{[\psi>0]}. \tag{3.46}$$

Two interesting alternative formulations of (3.46) are given below for the case $p = 1$ and generalized later for arbitrary p. In what follows, we let Gamma(τ, ξ) denote a random variable with probability density function

$$g(x|\tau, \xi) = \frac{\xi^\tau}{\Gamma(\tau)} x^{\tau-1} e^{-x\xi}\, \mathbb{1}_{[x>0]} \quad \text{for } \tau > 0 \quad \text{and} \quad \xi > 0$$

and Exp(ξ) corresponds to the choice $\tau = 1$ (i.e., an exponential random variable in its rate parametrization).

For $p = 1$, the marginal prior distribution on θ induced by (3.46) is equivalent to that obtained under the specification

$$\theta|\psi, \lambda \sim \mathcal{N}(0, \psi), \quad \psi|\lambda \sim \text{Exp}\left(\frac{\lambda^2}{2}\right), \quad \lambda|\alpha \sim \text{HN}(\alpha^{-1}), \tag{3.47}$$

where $\alpha = 1$ and HN(ζ) denotes the half-normal density

$$f(x|\zeta) = \sqrt{\frac{2}{\pi\,\zeta}} \exp\left\{-\frac{x^2}{2\zeta}\right\} 1\!\!1_{[x>0]} \quad \text{for} \quad \zeta > 0.$$

The marginal prior distribution on θ induced by (3.46) is also equivalent to that obtained under the alternative specification

$$\theta|\lambda \sim \text{Laplace}(\lambda), \quad \lambda|\alpha \sim \text{HN}(\alpha^{-1}), \tag{3.48}$$

where $\alpha = 1$ and Laplace(λ) denotes a random variable with the Laplace (double exponential) probability density function

$$f(y|\lambda) = \frac{\lambda}{2} e^{-\lambda|y|}\, 1\!\!1_{[y\in\mathbb{R}]}.$$

This result follows from Griffin and Brown (2010). Define

$$\theta|\psi, \omega \sim \mathcal{N}(0, \psi), \quad \psi|\omega \sim \text{Exp}(\omega), \quad \omega|\delta, \alpha \sim \text{Gamma}(1/2, \alpha) \tag{3.49}$$

as a hierarchically specified prior distribution for θ, ψ and ω. The resulting marginal prior distribution for θ, obtained by integrating out ψ and ω, is exactly the quasi-Cauchy distribution of Johnstone and Silverman (2004); see Griffin and Brown (2010) for details. Carvalho et al. (2010) showed that this distribution also coincides with the marginal prior distribution for θ induced by taking $a = 1/2$ in (3.45). The transformation $\lambda = \sqrt{2\omega}$ in (3.49) leads directly to (3.47) upon setting $\alpha = 1$; (3.48) is then obtained by integrating out ψ in (3.47).

3.5.2 The Positive-Part Estimator and Extensions as MAP Estimators

Takada (1979) showed that a positive-part type minimax estimator

$$\delta^c_{JS+}(X) = \left(1 - \frac{c}{\|X\|_2^2}\right)_+ X, \tag{3.50}$$

where $(t)_+ = \max(t, 0)$, is also the MAP estimator under a certain class of hierarchically specified generalized prior distributions, say $\pi_T(\theta, \kappa) = \pi(\theta|\kappa)\pi_T(\kappa)$. For the specific choice $c = p - 2$ in (3.50), Takada's prior reduces to

$$\theta|\kappa \sim \mathcal{N}_p(0, g(\kappa)I_p), \quad \pi_T(\kappa) \propto (1-\kappa)^{p/2}\kappa^{-1}\, \mathbb{1}_{[0<\kappa<1]}. \tag{3.51}$$

The improper prior (3.51) evidently behaves similarly to Strawderman's proper prior (3.45) (i.e., for $a = 1/2$). Notably, the numerator $(1-\kappa)^{p/2}$ in $\pi_T(\kappa)$ explicitly offsets the contribution of $(1-\kappa)^{-p/2}$ arising from the determinant of the variance matrix $g(\kappa)\, I_p$ in the conditional prior specification $\theta|\kappa$. Under the monotone decreasing variable transformation $\psi = g(\kappa) > 0$, (3.51) implies an alternative representation that is analogous to (3.46):

$$\theta|\psi \sim \mathcal{N}_p(0, \psi I_p), \quad \pi_T(\psi) \propto \psi^{p/2}\left(\frac{1}{1+\psi}\right)^{p/2+1} \mathbb{1}_{[\psi>0]}. \tag{3.52}$$

We observe that the proper prior (3.46) and improper prior (3.52) (almost) coincide when $p = 1$; in particular, multiplying the former by $\psi^{1/2}$ yields the latter. In view of the fact that (3.46) and (3.47) lead to the same marginal prior on $\boldsymbol{\theta}$ when $p = 1$, one is led to question whether a deeper connection between these two prior specifications might exist. Supposing $p \geq 1$, consider the following straightforward generalization of (3.47):

$$\theta|\psi, \lambda \sim \mathcal{N}_p(0, \psi I_p), \quad \psi|\lambda \sim \text{Gamma}\left(\frac{p+1}{2}, \frac{\lambda^2}{2}\right), \quad \lambda|\alpha \sim \text{HN}(\alpha^{-1}). \tag{3.53}$$

Integrating λ out of the higher level prior specification the resulting marginal (proper) prior for ψ reduces to

$$\pi(\psi|\alpha) \propto \psi^{-1/2}\psi^{p/2}\left(\frac{1}{1+\frac{\psi}{\alpha}}\right)^{\frac{p}{2}+1} \mathbb{1}_{[\psi>0]}. \tag{3.54}$$

For $\alpha = 1$ and any $p \geq 1$, we now observe that the proper prior (3.54) is simply the improper prior $\pi_T(\psi)$ in (3.52) multiplied by $\psi^{-1/2}$ and it reduces to Strawderman's prior (3.46) for $p = 1$.

3.5.3 *Penalized Likelihood and Hierarchical Priors*

Expressed in modern terms of penalization, Takada (1979) proved that the positive-part estimator (3.50) is the solution to a certain penalized likelihood estimation problem in which the penalty (or regularization) term is determined by the prior (3.51). Penalized likelihood estimation, and more generally problems of regularized estimation, have become a very important conceptual paradigm in both statistics and machine learning. Such methods suggest principled estimation and model selection procedures for a variety of high-dimensional problems. The statistical literature on penalized likelihood estimators has exploded, in part due to success in constructing procedures for regression problems in which one can simultaneously select variables and estimate their effects. The class of penalty functions leading to procedures with good asymptotic frequentist properties have singularities at the origin; important examples of separable penalties include the least absolute shrinkage and selection operator (LASSO) , Tibshirani (1996), smoothly clipped absolute deviation (SCAD), Fan and Li (2001), and minimax concave penalties (MCP) Zhang (2010). In fact, most such penalties utilized in the literature behave similarly to the LASSO penalty near the origin, differing more in their respective behaviors away from the origin, where control of estimation bias for those parameters not estimated to be zero becomes the driving concern. Generalizations of the LASSO penalty have been proposed to deal with correlated groupings of parameters, such as those that might arise in problems with features that can be sensibly ordered, as in the fused LASSO in Tibshirani et al. (2005), or separated into distinct subgroups as in the group LASSO in Yuan and Lin (2006). In such problems, the use of these penalties serves a related purpose.

The LASSO was initially formulated as a least squares estimation problem subject to a ℓ_1 constraint on the parameter vector. The more well-known penalized likelihood version arises from a Lagrange multiplier formulation of a convex relaxation of a ℓ_0 non-convex optimization problem. Since the underlying objective function is separable in the parameters, the underlying estimation problem is evidently directly related to the now-classical problem of estimating a bounded normal mean. From a decision theoretic point of view, if $X \sim \mathcal{N}(\theta, 1)$ for $|\theta| \leq \lambda$, then the projection of the usual estimator dominates the unrestricted MLE, but cannot be minimax for quadratic loss because it is not a Bayes estimator. Casella and Strawderman (1981) showed that the unique minimax estimator of θ is the Bayes estimator corresponding to a two-point prior on $\{-\lambda, \lambda\}$ for λ sufficiently small. Casella and Strawderman (1981) further showed that the uniform boundary Bayes estimator, $\lambda \tanh(\lambda x)$, is the unique minimax estimator if $\lambda < \lambda_0 \approx 1.0567$. They also considered three-point priors supported on $\{-\lambda, 0, \lambda\}$ and obtained sufficient conditions for such a prior to be least favorable. Marchand and Perron (2001) considered the multivariate extension, $X \sim \mathcal{N}_p(\theta, I_p)$ with $\|\theta\|_2 \leq \lambda$ and showed that the Bayes estimator with respect to a boundary uniform prior dominates the MLE whenever $\lambda \leq \sqrt{p}$ under squared error loss.

It has long been recognized that the class of penalized likelihood estimators also has a Bayesian interpretation. For example, in the canonical version of the LASSO problem, minimizing

$$\frac{1}{2}\|X - \theta\|_2^2 + \lambda\|\theta\|_1, \quad ||\theta||_1 = \sum_{i=1}^{p} |\theta_i| \tag{3.55}$$

with respect to θ is easily seen to be equivalent to computing the MAP estimator of θ under a model specification in which $X \sim \mathcal{N}_p(\theta, I_p)$ and θ has a prior distribution satisfying $\theta_i \overset{iid}{\sim}$ Laplace(λ). It is easily shown that the solution to (3.55) is $\widehat{\theta}_i(X) = \text{sign}(X_i)(|X_i| - \lambda)_+$, $i = 1, \ldots, p$. The critical hyperparameter λ, though regarded as fixed for the purposes of estimating θ, is typically estimated in some ad hoc manner (e.g., cross validation), resulting in an estimator with an empirical Bayes flavor.

The Laplace prior inherent in the LASSO minimization problem (3.55) has broad connections to estimation under hierarchical prior specifications that lead to scale mixtures of normal distributions. As pointed out above, the conditional prior distribution of $\theta|\lambda$ obtained by integrating out ψ in (3.47) is exactly Laplace(λ). More generally, the conditional distribution for $\theta|\lambda$ under the hierarchical prior specification (3.53) is a special case of the class of multivariate exponential power distributions in Gomez-Sanchez-Manzano et al. (2008); in particular, we obtain

$$\pi(\theta|\lambda) \propto \lambda^p \exp\{-\lambda\|\theta\|_2\}, \tag{3.56}$$

a direct generalization of the Laplace distribution that arises when $p = 1$. Treating λ as fixed hyperparameter, computation of the resulting MAP estimator under the previous model specification $X \sim \mathcal{N}_p(\theta, I_p)$ reduces to determining the value of θ that minimizes

$$\frac{1}{2}\|X - \theta\|_2^2 + \lambda\|\theta\|_2. \tag{3.57}$$

The resulting estimator is easily shown to be

$$\delta_{GL}(X) = \left(1 - \frac{\lambda}{\|X\|_2}\right)_+ X, \tag{3.58}$$

an estimator that coincides with the solution to the canonical version of the grouped LASSO problem involving a single group of parameters (see Yuan and Lin 2006) and equals $\widehat{\theta}(X) = \text{sign}(X)(|X| - \lambda)_+$ for the case where $p = 1$.

Consider the problem of estimating θ in the canonical setting $X \sim \mathcal{N}_p(\theta, I_p)$. In view of the fact that (3.53) leads to (3.56) upon integrating out ψ, our starting point is the (possibly improper) generalized class of joint prior distributions $\pi(\theta, \lambda|\alpha, \beta)$, which we define in the following hierarchical fashion

$$\begin{aligned} \pi(\theta|\lambda, \alpha, \beta) &\propto \lambda^p \exp\{-\lambda\|\theta\|_2\}, \\ \pi(\lambda|\alpha, \beta) &\propto \lambda^{-p} \exp\{-\alpha(\lambda - \beta)^2\}, \end{aligned} \tag{3.59}$$

where $\alpha, \beta > 0$ are hyperparameters. Equivalently,

$$\pi(\theta, \lambda|\alpha, \beta) \propto \exp\{-\lambda\|\theta\|_2\} \exp\{-\alpha(\lambda - \beta)^2\}. \tag{3.60}$$

The prior on λ is an improper modification of that given in (3.53), in which a location parameter β is introduced and the factor λ^{-p} is introduced to offset the contribution λ^p in (3.56). This construction mimics the idea underlying the prior used by Takada (1979) to motivate (3.50) as a MAP estimator.

Considering (3.60) as motivation for defining a new class of hierarchical penalty functions, Strawderman and Wells (2012) propose deriving the MAP estimator for (θ, λ) through minimizing the objective function

$$G(\theta, \lambda) = \frac{1}{2}\|X - \theta\|_2^2 + \lambda\|\theta\|_2 + \alpha(\lambda - \beta)^2 \tag{3.61}$$

jointly in $\theta \in \mathbb{R}^p$ and $\lambda > 0$, where $\alpha > 1/2$ and $\beta > 0$ are fixed. The resulting estimator for θ takes the closed form

$$\delta^{(\alpha,\beta)}(X) = w_{\alpha,\beta}(\|X\|_2)X, \tag{3.62}$$

where

$$w_{\alpha,\beta}(s) = \begin{cases} 0 & s \le \beta \\ \nu_\alpha\left(1 - \frac{\beta}{s}\right) & \beta < s \le 2\alpha\beta \\ 1 & s > 2\alpha\beta \end{cases}$$

for $\nu_\alpha = 2\alpha/(2\alpha - 1)$. Equivalently, we may write

$$w_{\alpha,\beta}(s) = \begin{cases} \nu_\alpha\left(1 - \frac{\beta}{s}\right)_+ & s \le 2\alpha\beta \\ 1 & s > 2\alpha\beta \end{cases}$$

demonstrating that (3.62) has the flavor of a range-modified positive-part estimator. A detailed derivation of this estimator is in Strawderman and Wells (2012).

Some interesting special cases of the estimator (3.62) arise when considering specific values of α, β and p. For example, letting $\alpha \to \infty$, we obtain (for $\beta > 0$)

$$\delta^{(\beta)}(X) = \left(1 - \frac{\beta}{\|X\|_2}\right)_+ X; \tag{3.63}$$

upon setting $\beta = \lambda$, we evidently recover (3.58); subsequently, setting $\lambda = \sqrt{p-2}$, one then obtains an obvious modification of (3.50) for the case where $c = p - 2$:

$$\delta^*_{PP}(X) = \left(1 - \frac{\sqrt{p-2}}{\|X\|_2}\right)_+ X \tag{3.64}$$

In the special case $p = 1$, the estimator (3.62) reduces to

$$\delta^M(X) = \begin{cases} 0 & \text{if } |X| \le \beta \\ \frac{2\alpha}{2\alpha-1}(X - \text{sign}(X)\beta) & \text{if } \beta < |X| \le 2\alpha\beta \\ X & \text{if } |X| > 2\alpha\beta \end{cases}. \tag{3.65}$$

As shown in Strawderman et al. (2013), (3.65) is also the solution to the penalized minimization problem

$$\frac{1}{2}(X-\theta)^2 + \rho(\theta;\alpha,\beta),$$

where $\beta > 0, \alpha > 1/2$ and

$$\rho(t;\alpha,\beta) = \beta \int_0^{|t|} (1 - \frac{z}{2\alpha\beta})_+ \, dz, \quad t \in \mathbb{R}.$$

This optimization problem is the univariate equivalent of the penalized likelihood estimation problem considered in Zhang (2010), who referred to $\rho(t;\alpha,\beta)$ as MCP. It follows that (3.65) is equivalent to the univariate MCP thresholding operator; consequently, (3.62) may be regarded as a generalization of this operator for thresholding a vector of parameters. Zhang (2010) showed that the LASSO, SCAD, and MCP belong to a family of quadratic spline penalties with certain sparsity and continuity properties. MCP turns out to be the simplest penalty that results in an estimator that is nearly unbiased, sparse and continuous. As demonstrated above, MCP also has an interesting Bayesian motivation under a hierarchical modeling strategy. Strawderman et al. (2013) undertook a more detailed study of the connections between MCP, the hierarchically penalized estimator, and proximal operators for the case of $p = 1$. They also compared this estimator to several others through consideration of frequentist and Bayes risks.

3.6 Estimation of a Predictive Density

Consider a parametric model $\{\mathscr{Y}, (\mathscr{P}'_\mu)_{\mu\in\Omega}\}$ where $\mathscr{Y}$ is the sample space, Ω is the parameter space and $\mathscr{P}' = \{p(y|\mu) : \mu \in \Omega\}$ is a class of densities of $\mathscr{P}'_\mu$ with respect to a σ -finite measure. In addition, suppose an observed value x of the random variable X follows a model $\{\mathscr{X}, (\mathscr{P}_\mu)_{\mu\in\Omega}\}$ indexed by the same parameter. In this section, we examine the problem of estimating the true density $p'(.|\mu) \in \mathscr{P}'$ of a random variable Y. In this context $p'(\cdot|\mu)$ is referred to as the predictive density of Y.

Let the density $\hat{q}(y|x)$ (belonging to some class of models $\mathscr{C} \supset \mathscr{P}'$) be an estimate, based on the observed data x, of the true density $p(y|\mu)$. Aitchison (1975) proposed using the Kullback and Leibler (1951) divergence, defined in (3.66) below, as a loss function for estimating $p(y|\mu)$.

The class of estimates $\mathscr{C}$ can be identical to the class $\mathscr{P}'$, that is, for any $y \in \mathscr{Y}$

$$\hat{q}(y|x) = p(y|\mu = \hat{\mu}(x))$$

where $\hat{\mu}$ is some estimate of μ. This type of density estimator is called the "plug-in density estimate" associated with the estimate $\hat{\mu}$. Alternatively, one may choose

$$\hat{q}(y|x) = \int_{\Omega} p(y|\mu)\, d\pi(\mu|x)$$

where $d\pi(\mu|x)$ may be a weight function (measure) or an *a posteriori* density associated with a priori measure $\pi(\mu)$. In this case, the class $\mathscr{C}$ will be broader than the class of the models $\mathscr{P}'$. Aitchison (1975) showed that this latter method is preferable to the plug-in approach for several families of probability distributions by comparing their risks induced by the Kullback-Leibler divergence.

3.6.1 The Kullback-Leibler Divergence

First, recall the definition of the Kullback-Leibler divergence and some of its properties.

Lemma 3.10 *The Kullback-Leibler divergence (relative entropy) $D_{KL}(p, q)$ between two densities p and q is defined by*

$$D_{KL}(p, q) = E_p\left[\log \frac{p}{q}\right] = \int \log\left[\frac{p(x)}{q(x)}\right] p(x)\, dx \geq 0 \tag{3.66}$$

and equality is achieved if and only if $p = q$, $p-$almost surely.

Note that the divergence can be finite only if the support of the density p is contained in the support of the density q. By convention, we define $0 \log \frac{0}{0} = 0$.

Proof By definition of the Kullback-Leibler divergence we can write

$$\begin{aligned}
-D_{\mathrm{KL}}(p, q) &= \int \log\left[\frac{q(x)}{p(x)}\right] p(x)\, dx \\
&\leq \log\left[\int \frac{q(x)}{p(x)}\, p(x)\, dx\right] \text{ (by Jensen's inequality)} \\
&= \log\left[\int q(x)\, dx\right] \\
&= 0.
\end{aligned}$$

We have equality, using Jensen's inequality, if and only if $p = q$, p -almost surely. Note that the lemma is true if q is assumed only to be a subdensity (mass less than or equal to 1). □

The Kullback-Leibler divergence is not a true distance since it is not symmetric and it does not satisfy the triangle inequality. But it appears as the natural discrepancy measure in information theory. An important property, given in the following lemma, is that it is strictly convex.

Lemma 3.11 *The Kullback-Leibler divergence is strictly convex, that is to say, if* (p_1, p_2) *and* (q_1, q_2) *are two pairs of densities then, for any* $0 \leq \lambda \leq 1$,

$$D_{KL}(\lambda\, p_1 + (1-\lambda)\, p_2, \lambda\, q_1 + (1-\lambda)\, q_2) \leq \lambda D_{KL}(p_1, q_1) + (1-\lambda) D_{KL}(p_2, q_2)\,, \tag{3.67}$$

with strict inequality unless $(p_1, p_2) = (q_1, q_2)$ *a.e. with respect to* $p_1 + p_2$.

Proof Note that $f(t) = t\, \log(t)$ is strictly convex on $(0, \infty)$. Let

$$\alpha_1 = \frac{\lambda q_1}{\lambda q_1 + (1-\lambda) q_2}, \; \alpha_2 = \frac{(1-\lambda) q_2}{\lambda q_1 + (1-\lambda) q_2}, \; t_1 = \frac{p_1}{q_1} \text{ and } t_2 = \frac{p_2}{q_2}.$$

From the convexity of the function f it follows that

$$f(\alpha_1 t_1 + \alpha_2 t_2) \leq \alpha_1 f(t_1) + \alpha_2 f(t_2)$$

and consequently

$$(\alpha_1 t_1 + \alpha_2 t_2) \log(\alpha_1 t_1 + \alpha_2 t_2) \leq t_1 \alpha_1 \log(t_1) + t_2 \alpha_2 \log(t_2)\,.$$

Substituting the above values of α_1, α_2, t_1 and t_2 gives

$$(\lambda p_1 + (1-\lambda) p_2) \log \frac{\lambda p_1 + (1-\lambda) p_2}{\lambda q_1 + (1-\lambda) q_2} \leq \lambda p_1 \log \frac{p_1}{q_1} + (1-\lambda) p_2 \log \frac{p_2}{q_2}.$$

Finally, by integrating the latter term, (3.67) and the strict convexity follow from the strict convexity of the function f. □

3.6.2 *The Bayesian Predictive Density*

Assume in the rest of this subsection that $p(x|\mu)$ and $p'(y|\mu)$ are densities with respect to the Lesbegue measure. For any estimator $\hat{p}(\cdot|x)$ of the density $p'(y|\mu)$, define the Kullback-Leibler loss by

$$\text{KL}(\mu, \hat{p}(\cdot|x)) = \int p'(y|\mu) \log \left[\frac{p'(y|\mu)}{\hat{p}(y|x)} \right] dy \tag{3.68}$$

and its corresponding risk as

$$\mathscr{R}_{\mathrm{KL}}(\mu,\hat{p}) = \int p(x|\mu)\left[\int p'(y|\mu)\log\left[\frac{p'(y|\mu)}{\hat{p}(y|x)}\right]dy\right]dx. \tag{3.69}$$

We say that the density estimate $\hat{p}_2$ dominates the density estimate $\hat{p}_1$ if, for any $\mu \in \Omega$, $\mathscr{R}_{\mathrm{KL}}(\mu,\hat{p}_1) - \mathscr{R}_{\mathrm{KL}}(\mu,\hat{p}_2) \le 0$, with strict inequality for at least some value of μ.

In the Bayesian framework we will compare estimates using Bayes risk. We will consider the class, more general than Aitchison (1975), of all subdensities,

$$\mathscr{D} = \left\{q(\cdot|x)| \int q(y|x)\,dy \le 1 \quad \text{for all } x\right\}.$$

Lemma 3.12 (Aitchison 1975) *The Bayes risk*

$$r_\pi(\hat{p}) = \int \mathscr{R}_{\mathrm{KL}}(\mu,\hat{p})\,\pi(\mu)\,d\mu$$

is minimized by

$$\hat{p}_\pi(y|x) = \int p'(y|\mu)\,p(\mu|x)\,d\mu = \frac{\int p'(y|\mu)\,p(x|\mu)\pi(\mu)\,d\mu}{\int p(x|\mu)\,\pi(\mu)\,d\mu}. \tag{3.70}$$

We call $\hat{p}_\pi$ the Bayesian predictive density.

Proof The difference between the Bayes risks of $\hat{p}_\pi$ and another competing subdensity estimator $\hat{q}$ is

$$\begin{aligned}
r_\pi(\hat{q}) - r_\pi(\hat{p}_\pi) &= \int_\Omega\left[\int_{\mathscr{X}}\left\{\int_{\mathscr{Y}} p'(y|\mu)\log\frac{\hat{p}_\pi(y|x)}{\hat{q}(y|x)}\,dy\right\}p(x|\mu)\,dx\right]\pi(\mu)\,d\mu\\
&= \int_\Omega\left[\int_{\mathscr{X}}\left\{\int_{\mathscr{Y}} p'(y|\mu)\log\frac{\hat{p}_\pi(y|x)}{\hat{q}(y|x)}\,dy\right\}p(x|\mu)\,\pi(\mu)\,dx\right]d\mu\\
&= \int_\Omega\left[\int_{\mathscr{X}}\left\{\int_{\mathscr{Y}} p'(y|\mu)\log\frac{\hat{p}_\pi(y|x)}{\hat{q}(y|x)}\,dy\right\}p(\mu|x)\,m(x)\,dx\right]d\mu\,.
\end{aligned}$$

Rearranging the order of integration thanks to Fubini'sTheorem gives

$$\begin{aligned}
r_\pi(\hat{q}) - r_\pi(\hat{r}) &= \int_{\mathscr{X}}\left[\int_{\mathscr{Y}}\left\{\int_\Omega p(\mu|x)\,p'(y|\mu)\,d\mu\right\}\log\frac{\hat{p}_\pi(y|x)}{\hat{q}(y|x)}\,dy\right]m(x)\,dx\\
&= \int_{\mathscr{X}}\left[\int_{\mathscr{Y}} \hat{p}_\pi(y|x)\log\frac{\hat{p}_\pi(y|x)}{\hat{q}(y|x)}\,dy\right]m(x)\,dx\\
&= \int_{\mathscr{X}} D_{\mathrm{KL}}(\hat{p}_\pi(.|x),\hat{q}(.|x))\,m(x)\,dx \ge 0.
\end{aligned}$$

□

3.6.3 Sufficiency Reduction in the Normal Case

Let $X_{(n)} = (X_1, \ldots, X_n)$ and $Y_{(m)} = (Y_1, \ldots, Y_m)$ be independent *iid* samples from p-dimensional normal distributions $\mathcal{N}_p(\mu, \Sigma_1)$ and $\mathcal{N}_p(\mu, \Sigma_2)$ with unknown common mean μ and known positive definite covariance matrices Σ_1 and Σ_2. On the basis of an observation $x_{(n)} = (x_1, \ldots, x_n)$ from $X_{(n)}$, consider the problem of estimating the true predictive density $p'(y_{(m)}|\mu)$ of $y_{(m)} = (y_1, \ldots, y_m)$, under the Kullback-Leibler loss. For a prior density $\pi(\mu)$, the Bayesian predictive density is given by

$$\hat{p}_\pi(y_{(m)}|x_{(n)}) = \frac{\displaystyle\int_\Omega p'(y_{(m)}|\mu)\, p(x_{(n)}|\mu)\, \pi(\mu)\, d\mu}{\displaystyle\int_\Omega p(x_{(n)}|\mu)\, \pi(\mu)\, d\mu}. \tag{3.71}$$

For simplicity, we consider the case where $\Sigma_1 = \Sigma_2 = I_p$. According to Komaki (2001) the Bayesian predictive densities satisfy

$$\int_{\mathbb{R}^{pm}} p'(y_{(m)}|\mu) \log \frac{p'(y_{(m)}|\mu)}{\hat{p}_\pi(y_{(m)}|x_{(n)})}\, dy_{(m)} = \int_{\mathbb{R}^p} p'(\bar{y}_m|\mu) \log \frac{p'(\bar{y}_m|\mu)}{\hat{p}_\pi(\bar{y}_m|\bar{x}_n)}\, d\bar{y}_m \tag{3.72}$$

where, denoting by $\phi_p(\cdot, |\mu, \Sigma)$ the density of $\mathcal{N}_p(\mu, \Sigma)$, in the left-hand side of (3.72),

$$p'(y_{(m)}|\mu) = \prod_{i=1}^m \phi_p(y_i, |\mu, I_p)$$

while, in the right-hand side of (3.72),

$$p'(\bar{y}_m|\mu) = \phi_p(\bar{y}_m|\mu, I_p/m)$$

with $\bar{y}_m = \sum_{j=1}^m y_j/m$. Similarly, $\hat{p}_\pi(y_{(m)}|x_{(n)})$ corresponds to the conditional density of the $p \times m$ matrix $y_{(m)}$ given the $p \times m$ matrix $x_{(n)}$ while $\hat{p}_\pi(\bar{y}_m|\bar{x}_m)$ corresponds to the conditional density of the $p \times 1$ vector $\bar{y}_m$ given the $p \times 1$ vector $\bar{x}_n = \sum_{i=1}^n x_i/n$.

To see this sufficiency reduction, use the fact that

$$\sum_{i=1}^m \|y_i - \mu\|^2 = \sum_{i=1}^m \|y_i - \bar{y}_m\|^2 + m\,(\|\bar{y}_m - \mu\|)^2.$$

Then we can express $p'(y_{(m)}|\mu)$ as

$$\begin{aligned} p'(y_{(m)}|\mu) &= \frac{1}{(2\pi)^{mp/2}} \exp\left(-\frac{1}{2}\sum_{i=1}^{m} \|y_i - \bar{y}_m\|^2\right) \exp\left(-\frac{m}{2}(\|\bar{y}_m - \mu\|)^2\right) \\ &= \frac{m^{p/2}}{(2\pi)^{(m-1)p/2}} \exp\left(-\frac{1}{2}\sum_{i=1}^{m} \|y_i - \bar{y}_m\|^2\right) p(\bar{y}_m|\mu). \end{aligned} \tag{3.73}$$

Similarly, it follows that

$$p(x_{(n)}|\mu) = \frac{n^{p/2}}{(2\pi)^{(n-1)p/2}} \exp\left(-\frac{1}{2}\sum_{i=1}^{n} \|x_i - \bar{x}_m\|^2\right) p(\bar{x}_m|\mu).$$

By replacing these expressions in the form of the predictive density in (3.71), we get

$$\begin{aligned} &\hat{p}_\pi(y_{(m)}|x_{(n)}) \\ &= \left\{\frac{m^{p/2}}{(2\pi)^{(m-1)p/2}} \exp\left(-\frac{1}{2}\sum_{i=1}^{m} \|y_i - \bar{y}_m\|^2\right)\right\} \frac{\int p'(\bar{y}_m|\mu)\, p(\bar{x}_m|\mu)\, \pi(\mu)\, d\mu}{\int p(\bar{x}_m|\mu)\, \pi(\mu)\, d\mu} \\ &= \left\{\frac{m^{p/2}}{(2\pi)^{(m-1)p/2}} \exp\left(-\frac{1}{2}\sum_{i=1}^{m} \|y_i - \bar{y}_m\|^2\right)\right\} \hat{p}_\pi(\bar{y}_m|\bar{x}_m). \end{aligned} \tag{3.74}$$

Finally, for (3.73) and (3.74), it follows that

$$\begin{aligned} \int p'(y_{(m)}|\mu) \log \frac{p'(y_{(m)}|\mu)}{\hat{p}(y_{(m)}|x_{(n)})} dy_{(m)} &= \int p'(y_{(m)}|\mu) \log \frac{p'(\bar{y}_m|\mu)}{\hat{p}(\bar{y}_m|\bar{x}_m)} dy_{(m)} \\ &= \int p'(\bar{y}_m|\mu) \log \frac{p'(\bar{y}_m|\mu)}{\hat{p}(\bar{y}_m|\bar{x}_m)} d\bar{y}_m. \end{aligned}$$

Therefore, for any prior π, the risk of the Bayesian predictive density estimator is equal to the risk of the Bayesian predictive density associated to π in the reduced model $X \sim \mathcal{N}_p(\mu, \frac{1}{n}I_p)$ and $Y \sim \mathcal{N}_p(\mu, \frac{1}{m}I_p)$. Thus, for the Bayesian predictive densities, it is sufficient to consider the reduced model.

Now we will compare two plug-in density estimators, $\hat{p}_1$ and $\hat{p}_2$ associated with the two different estimators of μ, δ_1 and δ_2. That is, for $i = 1, 2$, define

$$\hat{p}_i(y_{(m)}|x_{(n)}) = p'(y_{(m)}|\mu = \delta_i(x_{(n)})). \tag{3.75}$$

The difference in risk between $\hat{p}_2$ and $\hat{p}_1$ is given by

$$\begin{aligned}\Delta\mathscr{R}_{\mathrm{KL}}(\hat{p}_2,\hat{p}_1) &= \mathscr{R}_{\mathrm{KL}}(\mu,\hat{p}_2)-\mathscr{R}_{\mathrm{KL}}(\mu,\hat{p}_1)\\ &= \int p(x_{(n)}|\mu)\int p(y_{(m)}|\mu)\log\frac{\hat{p}_1(y_{(m)}|x_{(n)})}{\hat{p}_2(y_{(m)}|x_{(n)})}\,dy_{(m)}\,dx_{(n)}\\ &= \int p(x_{(n)}|\mu)\int p(y_{(m)}|\mu)\Big(\frac{1}{2}\sum_{i=1}^{m}\|\delta_2(x_{(n)})-y_i\|^2\\ &\quad -\frac{1}{2}\sum_{i=1}^{m}\|\delta_1(x_{(n)})-y_i\|^2\Big)\,dy_{(m)}\,dx_{(n)}\,.\end{aligned}$$

By the independence of $X_{(n)}$ and $Y_{(m)}$ this can be reexpressed in terms of expectations as

$$\begin{aligned}&\Delta\mathscr{R}_{\mathrm{KL}}(\hat{p}_2,\hat{p}_1)\\ &=\frac{1}{2}\sum_{i=1}^{m}E_{X_{(n)},Y_{(m)}}\Big(\|\delta_2(X_{(n)})-\mu+\mu-Y_i\|^2-\|\delta_1(X_{(n)})-\mu+\mu-Y_i\|^2\Big)\\ &=\frac{m}{2}E_{X_{(n)},Y_{(m)}}\Big[\|\delta_2(X_{(n)})-\mu\|^2-\|\delta_1(X_{(n)})-\mu\|^2\Big]\\ &\quad+\sum_{i=1}^{m}E_{X_{(n)},Y_{(m)}}\Big(\big[(\delta_2(X_{(n)})-\mu)(\mu-Y_i)\big]-\big[(\delta_1(X_{(n)})-\mu)(\mu-Y_i)\big]\Big)\\ &=\frac{m}{2}\Big(E_{X_{(n)}}\Big[\|\delta_2(X_{(n)})-\mu\|^2\Big]-E_{X_{(n)}}\Big[\|\delta_1(X_{(n)})-\mu\|^2\Big]\Big)\\ &=\frac{m}{2}\Big[\mathscr{R}_Q(\delta_2,\mu)-\mathscr{R}_Q(\delta_1,\mu)\Big],\end{aligned}$$

which shows that the risk difference between $\hat{p}_2$ and $\hat{p}_1$ is proportional to the risk difference between δ_2 and δ_1.

Note that, by completeness of the statistics $\bar{X}_n$, it suffices to consider only estimates of μ that depend only on $\bar{X}_n$.

3.6.4 Properties of the Best Invariant Density

In this subsection, we restrict our attention to location models. We assume $X \sim p(x|\mu) = p(x-\mu)$ and $Y \sim p'(y|\mu) = p'(y-\mu)$, where p and p' are two known possibly different densities. A density $\hat{q}$ is called invariant (equivariant) with respect to a location parameter if, for any $a \in \mathbb{R}^p$, $x \in \mathbb{R}^p$, and $y \in \mathbb{R}^p$ $q(y|x+a) = q(y-a|x)$. This is equivalent to $q(y+a|x+a) = q(y|x)$. The following result shows that the risk of an invariant predictive density is constant.

Lemma 3.13 *The invariant predictive densities with respect to the location group of translations have constant risk.*

Proof By the property of invariance, the risk of an invariant density $\hat{q}$ is equal to

$$
\begin{aligned}
\mathscr{R}(\mu, \hat{q}) &= \int \log \frac{p'(y-\mu)}{\hat{q}(y|x)}\, p(x-\mu)\, p'(y-\mu)\, dy\, dx \\
&= \int \log \frac{p'(y-\mu)}{\hat{q}(y-\mu|x-\mu)}\, p(x-\mu)\, p'(y-\mu)\, dy\, dx \\
&= \int \log \frac{p(z')}{\hat{q}(z'|z)}\, p(z)\, p'(z')\, dz'\, dz, \qquad (3.76)
\end{aligned}
$$

by the change of variables $z = x - \mu$ and $z' = z - \mu$. Therefore, the risk $\mathscr{R}(\mu, \hat{q})$ does not depend on μ and it is constant. □

Any invariant predictive density which minimizes this risk is known as the best invariant predictive density.

Lemma 3.14 *The best invariant predictive density is the Bayesian predictive density $\hat{p}_U$ associated with the Lebesgue measure on $\mathbb{R}^p$, $\pi(\mu) = 1$, is given by*

$$
\hat{p}_U(y|x) = \frac{\int_{\mathbb{R}^p} p'(y|\mu)\, p(x|\mu)\, d\mu}{\int_{\mathbb{R}^p} p(x|\mu)\, d\mu}. \qquad (3.77)
$$

Proof Let $Z = X - \mu$, $Z' = Y - \mu$, and $T = Y - X = Z' - Z$. We will show that $\hat{p}(t)$, the density of T, which is independent of μ, is the best invariant density. As noted in the previous section, if $\hat{q}$ is an invariant predictive density, $\hat{q}(y|x) = \hat{q}(y-x|0) = \hat{q}(y-x)$, by an abuse of notation. Hence,

$$
\begin{aligned}
\mathscr{R}(\mu, \hat{q}) - \mathscr{R}(\mu, \hat{p}) &= \int_{\mathbb{R}^p} \int_{\mathbb{R}^p} \left[\log \frac{\hat{p}(y-x)}{\hat{q}(y-x)}\right] p(x-\mu) p'(y-\mu)\, dx\, dy \\
&= \int_{\mathbb{R}^p} \int_{\mathbb{R}^p} \left[\log \frac{\hat{p}(z'-z)}{\hat{q}(z'-z)}\right] p(z) p'(z')\, dz\, dz' \\
&= \int_{\mathbb{R}^p} \left[\log \frac{\hat{p}(t)}{\hat{q}(t)}\right] \hat{p}(t)\, dt, \qquad (3.78)
\end{aligned}
$$

which is always positive by the inequality in (3.66). The result of the equality in (3.78), and hence the lemma, follows from the fact that $\hat{p}(t) = \hat{p}(y-x) = \hat{p}_U(y|x)$, that is,

$$\begin{aligned}\hat{p}(t) &= \int_{\mathbb{R}^p} p(z)\, p'(z+t)\, dz \\ &= \int_{\mathbb{R}^p} p(z)\, p'(z+y-x)\, dz \\ &= \int_{\mathbb{R}^p} p(x-\mu)\, p'(y-\mu)\, d\mu \\ &= \frac{\int_{\mathbb{R}^p} p'(y|\mu)\, p(x|\mu)\, d\mu}{\int_{\mathbb{R}^p} p(x|\mu)\; d\mu} \end{aligned} \tag{3.79}$$

which is the expression of $\hat{p}_U$ given in (3.70) with $\pi(\mu) = 1$. □

Murray (1977) showed that $\hat{p}_U$ is the best invariant density under the action of translations and of linear transformations for a Gaussian model. Ng (1980) has generalized this result. Liang and Barron (2004), without the hypothesis of independence between X and Y, for the estimation of $p'(y|x,\mu)$ showed that $\hat{p}_U = \dfrac{\int_{\mathbb{R}^p} p'(y|x,\mu)\, p(x|\mu)\, d\mu}{\int_{\mathbb{R}^p} p(x|\mu)\; d\mu}$ is the best invariant density.

We will now show that $\hat{p}_U$ is minimax in location problems.

Lemma 3.15 *Let $X \sim p(x|\mu) = p(x-\mu)$ and $Y \sim p(y|\mu) = p'(y-\mu)$, with unknown location parameter $\mu \in \mathbb{R}^p$. Assuming that $E_0\left[\|X\|^2\right] < \infty$, then the best predictive invariant density $\hat{p}_U$ is minimax.*

Proof We show minimaxity using Lemma 1.8. Consider a sequence $\{\pi_k\}$ of normal $\mathcal{N}_p(0, k\, I_p)$ priors. The difference of Bayes risk between $\hat{p}_U$ and $\hat{p}_{\pi_k}$, is given by

$$\begin{aligned} r(\hat{p}_U, \pi_k) - r(\hat{p}_{\pi_k}, \pi_k) &= \int_{\mathbb{R}^p} \left[\mathscr{R}(\mu, \hat{p}_U) - \mathscr{R}(\mu, \hat{p}_{\pi_k})\right] \pi_k(\mu)\; d\mu \\ &= \int_{\mathbb{R}^p}\int_{\mathbb{R}^p}\int_{\mathbb{R}^p} \log \frac{\hat{p}_{\pi_k}(y|x)}{\hat{p}_U(y|x)}\, p(y|\mu)\, p(x|\mu) \pi_k(\mu)\; dy\; dx\; d\mu \\ &= \int_{\mathbb{R}^p}\int_{\mathbb{R}^p} \log \frac{\hat{p}_{\pi_k}(y|x)}{\hat{p}_U(y|x)} \left[\int_{\mathbb{R}^p} p(y|\mu)\, p(x|\mu)\, \pi_k(\mu)\; d\mu\right] dy\; dx \\ &= E^{X,Y}_{\pi_k} \log \frac{\hat{p}_{\pi_k}(Y|X)}{\hat{p}_U(Y|X)} \end{aligned} \tag{3.80}$$

where $E^{x,y}_{\pi_k}$ denotes the expectation with respect to the joint marginal of (X, Y),

$$m_{\pi_k}(x, y) \;=\; \int_{\mathbb{R}^p} p(y|\mu)\, p(x|\mu)\, \pi_k(\mu)\; d\mu.$$

Since $r(\hat{p}_U, \pi_k) = \mathscr{R}(\mu, \hat{p}_U)$ ($\hat{p}_U$ has constant risk) it suffices to show (3.80) tends to 0 as k tends to infinity. By simplifying one gets

$$r(\hat{p}_U, \pi_k) - r(\hat{p}_{\pi_k}, \pi_k)$$
$$= E_{\pi_k}^{X,Y}\left[\log\left(\frac{\int p(x, y|\mu)\,\pi_k(\mu)\,d\mu}{\int p(x|\mu)\,\pi_k(\mu)\,d\mu}\frac{1}{\int p(x, y|\mu)\,d\mu}\right)\right]$$
$$= E_{\pi_k}^{X,Y}\left[-\log\frac{\int p(x, y|\mu)\,\pi_k(\mu)\frac{1}{\pi_k(\mu)}\,d\mu}{\int p(x, y|\mu)\,\pi_k(\mu)\,d\mu} - \log\left(\int p(x|\mu)\,\pi_k(\mu)\,d\mu\right)\right]$$
$$= E_{\pi_k}^{X,Y}\left[-\log E_{\mu|X,Y}\frac{1}{\pi_k(\mu)} - \log\left(\int p(x|\mu)\,\pi_k(\mu)\,d\mu\right)\right]$$

where $E_{\mu|X,Y}$ denotes the expectation with respect to the posterior of μ given (X, Y). An application of Jensen's inequality gives

$$r(\hat{p}_U, \pi_k) - r(\hat{p}_{\pi_k}, \pi_k)$$
$$\leq E_{\pi_k}^{X,Y} E_{\mu|X,Y}\log\pi_k(\mu) - E_{\pi_k}^{X,Y}\left[\int p(X|\mu)\,\log\pi_k(\mu)\,d\mu\right]. \tag{3.81}$$

By developing the expectations, it follows that

$$E_{\pi_k}^{X,Y} E_{\mu|X,Y}\log\pi_k(\mu) = \iint m_{\pi_k}(x, y)\frac{\int p(x, y|\mu)\pi_k(\mu)\log(\pi_k(\mu))d\mu}{m_{\pi_k}(x, y)}dxdy$$
$$= \iiint \pi_k(\mu)\log(\pi_k(\mu))\,d\mu\,dxdy$$
$$= \int \pi_k(\mu)\log(\pi_k(\mu))d\mu. \tag{3.82}$$

Similarly, by integrating with respect to y and by interchanging between μ and μ' we have

$$E_{\pi_k}^{X,Y}\left[\int p(X|\mu)\,\log\pi_k(\mu)\,d\mu\right]$$
$$= \iiiint p(x|\mu')p(y|\mu')\pi_k(\mu')p(x|\mu)\log\pi_k(\mu)\,d\mu' d\mu dxdy$$
$$= \iiint \pi_k(\mu')p(x|\mu)p(x|\mu')\log\pi_k(\mu)d\mu'\,dx\,d\mu$$
$$= \iiint \pi_k(\mu)p(x|\mu)p(x|\mu')\log\pi_k(\mu')d\mu\,dxd\mu'. \tag{3.83}$$

By grouping the expressions (3.81), (3.83) and (3.84) and making the changes of variables $z = x - \mu$ and $z' = x - \mu'$ it follows that

$$
\begin{aligned}
r(\hat{p}_U, \pi_k) - r(\hat{p}_{\pi_k}, \pi_k) &\\
&\le \iiint p(x|\mu)p(x|\mu')\pi_k(\mu)\left[\log(\pi_k(\mu)) - \log(\pi_k(\mu'))\right] d\mu d\mu' \, dx \\
&= \iiint \pi_k(\mu)p(x-\mu)p(x-\mu')\log\left(\frac{\pi_k(\mu)}{\pi_k(\mu')}\right) d\mu \, dz \, dz' \\
&= \iiint \pi_k(\mu)p(z)p(z')\log\left(\frac{\pi_k(\mu)}{\pi_k(\mu+z-z')}\right) d\mu \, dz \, dz'. \qquad (3.84)
\end{aligned}
$$

In view of the form $\pi_k(\mu)$, the term on the right in (3.84) can be written as

$$
\begin{aligned}
& E_{\pi_k} E_{Z,Z'} \log\left(\frac{\pi_k(\mu)}{\pi_k(\mu+Z-Z')}\right) \\
&= E_{\pi_k} E_{Z,Z'} \frac{1}{2k}\left(\|\mu+Z-Z'\| - \|\mu^2\|\right) \\
&= E_{\pi_k} E_{Z,Z'} \left[\frac{1}{2k}\left(\|Z\|^2 + \|Z'\|^2 + 2\langle \mu, Z-Z'\langle\right)\right] \\
&= E_{Z,Z'} \left[\frac{1}{2k}\left(\|Z\|^2 + \|Z'\|^2\right)\right],
\end{aligned}
$$

since $E(Z) = E(Z') = E_0(X)$ (here, $E_{Z,Z'}$ denotes the expectation with respect to $p(z, z') = p(z)p(z')$). We then see that the limit of the difference of Bayes risks tends toward zero when $k \to \infty$. Therefore, $\hat{p}_U$ is minimax by Lemma 1.8. □

This result is in Liang and Barron (2004), a more direct proof for the Gaussian case can be found in George et al. (2006) and is given in the next section.

3.6.5 An Explicit Expression for $\hat{p}_U$ and Its Risk in the Normal Case

We now give an explicit expression of $\hat{p}_U$, described the previous subsections, in the Gaussian setting. Let $X \sim \mathcal{N}_p(\mu, v_x I_p)$ and $Y \sim \mathcal{N}_p(\mu, v_y I_p)$.

Lemma 3.16 *The Bayesian predictive density associated with the uniform prior on* $\mathbb{R}^p$, $\pi(\mu) \equiv 1$, *is given by the following expression*

$$
\hat{p}_U(y|x) = \frac{1}{((2\pi)(v_y + v_x))^{p/2}} \exp\left(-\frac{\|y-x\|^2}{2(v_x+v_y)}\right). \qquad (3.85)
$$

Proof For $W = (v_y X + v_x Y)/(v_x + v_y)$ and $v_w = (v_x v_y)/(v_x + v_y)$ it is clear that $W \sim \mathcal{N}_p(\mu, v_w I_p)$, by the independence of X and Y. Further, note that

$$\frac{\|x-\mu\|^2}{2v_x}+\frac{\|y-\mu\|^2}{2v_y}=\frac{\|\mu-w\|^2}{2v_w}+\frac{\|y-x\|^2}{2(v_x+v_y)}. \tag{3.86}$$

By definition, and through the previous representation, it follows that

$$\begin{aligned}
\hat{p}_U(y|x) &= \frac{\int_{\mathbb{R}^p} p(y|\mu, v_y)\, p(x|\mu, v_x)\, d\mu}{\int_{\mathbb{R}^p} p(x|\mu, v_x)\, d\mu} \\
&= \int_{\mathbb{R}^p} \frac{1}{(2\pi)^p (v_y\, v_x)^{p/2}} \exp\left(-\frac{\|x-\mu\|^2}{2\, v_x} - \frac{\|y-\mu\|^2}{2\, v_y}\right) d\mu \\
&= \int_{\mathbb{R}^p} \frac{1}{(2\pi)^p (v_y\, v_x)^{p/2}} \exp\left(-\frac{\|\mu-w\|^2}{2\, v_w}\right) \exp\left(-\frac{\|y-x\|^2}{2\,(v_x+v_y)}\right) d\mu \\
&= \frac{(2\pi\, v_w)^{p/2}}{(2\pi)^p (v_y\, v_x)^{p/2}} \exp\left(-\frac{\|y-x\|^2}{2(v_x+v_y)}\right) \\
&= \frac{1}{\big((2\pi)\,(v_y+v_x)\big)^{p/2}} \exp\left(-\frac{\|y-x\|^2}{2\,(v_x+v_y)}\right).
\end{aligned}$$

□

Note that the risk of $\hat{p}_U$ is constant, as we have previously seen for invariant densities. Given the form of $\hat{p}_U(.|x)$ it follows that the Kullback-Liebler divergence is

$$\begin{aligned}
&\mathrm{KL}(\hat{p}_U(.|x), \mu) \\
&= \int p(y|\mu, v_y) \log \frac{p(y|\mu, v)}{\hat{p}_U(y|x)}\, dy \\
&= E^Y\left[\log \frac{p(Y|\mu, v)}{\hat{p}_U(Y|x)}\right] \\
&= E^Y\left[-\frac{p}{2}\log\frac{v_y}{v_x+v_y} - \frac{1}{2v_y}\|Y-\mu\|^2 + \frac{1}{2(v_x+v_y)}\|Y-x\|^2\right] \\
&= -\frac{p}{2}\log\frac{v_y}{v_x+v_y} - \frac{p}{2} + E^Y\left[\frac{1}{2(v_x+v_y)}\left(\|Y-\mu\|^2+\|\mu-x\|^2\right)\right] \\
&= \left[-\frac{p}{2}\log\frac{v_y}{v_x+v_y} - \frac{p}{2} + \frac{p v_y}{2(v_x+v_y)}\right] + \frac{1}{2(v_x+v_y)}\|\mu-x\|^2. \qquad (3.87)
\end{aligned}$$

Hence, we can conclude that the risk of $\hat{p}_U$ is

$$\begin{aligned}
\mathscr{R}_{\mathrm{KL}}(\hat{p}_U, \mu) &= E^X\left[\mathrm{KL}(\hat{p}_U, \mu, X)\right] \\
&= \left[-\frac{p}{2}\log\frac{v_y}{v_x+v_y} - \frac{p}{2} + \frac{p v_y}{2(v_x+v_y)}\right] + \frac{p v_x}{2(v_x+v_y)} \\
&= -\frac{p}{2}\log\left(\frac{v_y}{v_x+v_y}\right) = \frac{p}{2}\log\left(1+\frac{v_x}{v_y}\right). \qquad (3.88)
\end{aligned}$$

In the framework of the *iid* sampling model presented in Sect. 3.6.3 with $\Sigma_1 = \Sigma_2 = I_p$, we can express the risk as

$$\mathscr{R}_{\mathrm{KL}}(\hat{p}_U, \mu) = \frac{p}{2}\log\left(1+\frac{m}{n}\right).$$

A predictive density is called the plug-in relative to an estimator δ if it has the form

$$\hat{p}_\delta(y|x) = \frac{1}{(2\pi v_y)^{p/2}}\exp\left(-\frac{1}{2}\frac{\|y-\delta(x)\|^2}{v_y}\right).$$

The predictive plug-in density, which corresponds to the standard estimator of the mean, μ, $\delta_0(X) = X$, is

$$\hat{p}_\delta(y|x) = \frac{1}{\left(2\pi v_y\right)^{p/2}}\exp\left[-\frac{1}{2}\frac{\|y-x\|^2}{v_y}\right].$$

We can directly verify that the predictive density $\hat{p}_U$ dominates the plug-in density $\hat{p}_{\delta_0}$ for any $\mu \in \mathbb{R}^p$. In fact, their difference in risk is

$$\begin{aligned}
\Delta\mathscr{R}_{\mathrm{KL}}(\hat{p}_U, \hat{p}_{\delta_0}) &= E^{X,Y}\left(\log\frac{\hat{p}_U(Y|X)}{\hat{p}_{\delta_0}(Y|X)}\right) \\
&= -\frac{p}{2}\log\left(\frac{v_x+v_y}{v_y}\right) - \frac{1}{2}\left[\frac{1}{v_x+v_y} - \frac{1}{v_y}\right]E^{X,Y}\left(\|Y-X\|^2\right).
\end{aligned}$$

Since $E^{X,Y}\left(\|Y-X\|^2\right)$ equals

$$\begin{aligned}
E^{X,Y}\left(\|Y-\mu\|^2\right) + E^{X,Y}\left(\|X-\mu\|^2\right) - 2\left\langle E^{X,Y}(Y-\mu), E^{X,Y}(X-\mu)\right\rangle \\
= p(v_x+v_y),
\end{aligned}$$

we have

$$\Delta\mathscr{R}_{\mathrm{KL}}(\hat{p}_U, \hat{p}_{\delta_0}) = -\frac{p}{2}\left[\log\left(1+\frac{v_x}{v_y}\right) - \frac{v_x}{v_y}\right] > 0 .$$

Surprisingly, the predictive density $\hat{p}_U$ has similar properties to the standard estimator, $\delta_0(X) = X$, for the estimation of the mean under quadratic loss. Komaki (2001) showed that the density $\hat{p}_U$ is dominated by the Bayesian predictive density using the harmonic prior, $\pi(\mu) = \|\mu\|^{2-p}$. George et al. (2006) extended the analogy with point estimation. We give some of this development next.

Lemma 3.17 (George et al. 2006, Lemma 2) *For $W = (v_y X + v_x Y)/(v_x + v_y)$ and $v_w = (v_x v_y)/(v_x + v_y)$, let $m_\pi(W; v_w)$ and $m_\pi(X; v_x)$ be the marginals of W and X, respectively, relative to the a prior π. Then*

$$\hat{p}_\pi(y|X) = \frac{m_\pi(W; v_w)}{m_\pi(X; v_x)}\, \hat{p}_U(y|X) \tag{3.89}$$

where $\hat{p}_U(\cdot|X)$ is the Bayes estimator associated with the uniform prior on $\mathbb{R}^p$ given by (3.85). In addition, for any prior measure π, the Kullback-Leibler risk difference between $\hat{p}_U(\cdot|x)$ and the Bayesian predictive density $\hat{p}_\pi(\cdot|x)$ is given by

$$\mathscr{R}_{KL}(\mu, \hat{p}_U) - \mathscr{R}_{KL}(\mu, \hat{p}_\pi) = E_{\mu,v_w}\left[\log m_\pi(W; v_w)\right] - E_{\mu,v_x}\left[\log m_\pi(X; v_x)\right] \tag{3.90}$$

where $E_{\mu,v}$ denotes the expectation with respect to the normal $\mathscr{N}_p(\mu, vI_p)$ distribution.

Proof The marginal density of (X, Y) associated with π is equal to

$$\begin{aligned}
\hat{p}_\pi(x, y) &= \int_{\mathbb{R}^p} p(x|\mu, v_x)\, p(y|\mu, v_y)\, \pi(\mu)\, d\mu \\
&= \int_{\mathbb{R}^p} \frac{1}{(2\pi v_x)^{p/2}} \exp\left(-\frac{\|x-\mu\|^2}{2v_x}\right) \frac{1}{(2\pi v_y)^{p/2}} \exp\left(-\frac{\|y-\mu\|^2}{2v_y}\right) \pi(\mu)\, d\mu .
\end{aligned}$$

Applying (3.85) and (3.86) it follows that

$$\begin{aligned}
\hat{p}_\pi(x, y) &= \frac{1}{(2\pi)^p\,(v_x v_y)^{p/2}} \int_{\mathbb{R}^p} \exp\left(-\frac{\|y-x\|^2}{2(v_x+v_y)}\right) \exp\left(-\frac{\|\mu-w\|^2}{2v_w}\right) \pi(\mu)\, d\mu \\
&= \frac{(2\pi v_w)^{p/2}}{(2\pi)^p\,(v_x v_y)^{p/2}} \exp\left(-\frac{\|y-x\|^2}{2(v_x+v_y)}\right) m_\pi(w; v_w) \\
&= \hat{p}_U(y|x)\, m_\pi(w; v_w).
\end{aligned}$$

Since $\hat{p}_\pi(y|x) = \hat{p}_\pi(x, y)/m_\pi(x)$, (3.89) follows.

Hence, we can write the risk difference as

$$\begin{aligned}
&\mathscr{R}_{\mathrm{KL}}(\mu, \hat{p}_U) - \mathscr{R}_{\mathrm{KL}}(\mu, \hat{p}_\pi) \\
&= \int\int p(x|\mu, v_x)\, p(y|\mu, v_y) \log \frac{\hat{p}_\pi(y|x)}{\hat{p}_U(y|x)}\, dy\, dx \\
&= \int\int p(x|\mu, v_x)\, p(y|\mu, v_y) \log \frac{m_\pi(W(x, y); v_w)}{m_\pi(x; v_x)}\, dy\, dx \\
&= E^{X,Y} \log m_\pi(W(X, Y); v_w) - E^{X,Y} \log m_\pi(X; v_x) \\
&= E^{W} \log m_\pi(W|v_w) - E^{X} \log m_\pi(X|v_x).
\end{aligned}$$

□

Using this lemma, George et al. (2006) gave a simple proof of the result of Liang and Barron (2004) for the Gaussian setting. By taking the same sequence of priors $\{\pi_k\} = \mathscr{N}_p(0, kI_p)$, the difference of the Bayes risk equals (using constancy of the risk of $\hat{p}_U$)

$$\begin{aligned}
\mathscr{R}_{\mathrm{KL}}(\mu, \hat{p}_U) - r(\pi_k, \hat{p}_{\pi_k}) &= \int \pi_k(\mu) \left[E_{\mu,v_w} \log m_{\pi_k}(W, v_w) - E_{\mu,v_x} \log m_{\pi_k}(X, v_x)\right] d\mu \\
&= \int \pi_k(\mu) \Bigg[E_{\mu,v_w} \log \left\{(2\pi(v_w + k))^{-p/2} \exp\left(-\frac{\|W\|^2}{2(v_w + k)}\right)\right\} \\
&\quad - E_{\mu,v_x} \log \left\{(2\pi(v_x + k))^{-p/2} \exp\left(-\frac{\|X\|^2}{2(v_x + k)}\right)\right\} \Bigg] d\mu \\
&= \int \pi_k(\mu) \Bigg[- p/2 \log(2\pi(v_w + k)) - \frac{p v_w}{2(v_w + k)} \\
&\quad + p/2 \log(2\pi(v_x + k)) + \frac{p v_x}{2(v_x + k)} \Bigg] d\mu \\
&= -\frac{p}{2} \log \frac{v_w + k}{v_x + k} - \frac{p v_w}{2(v_w + k)} + \frac{p v_x}{2(v_x + k)}.
\end{aligned}$$

Hence, we see that $\lim_{k\to\infty} r(\pi_k, \hat{p}_U) - r(\pi_k, \hat{p}_{\pi_k}) = 0$ and so, $\hat{p}_U$ is minimax by Lemma 1.8. George et al. (2006) also show that the best predictive invariant density is dominated by any Bayesian predictive density relative to a superharmonic prior. This result parallels the result of Stein for the estimation of the mean under quadratic loss and the use differential operators discussed in Sect. 2.6. The following lemma from George et al. (2006) allows us to give sufficient conditions for domination. We use Stein's identity in the proof.

Lemma 3.18 *If $m_\pi(z; v_x)$ is finite for any z, then for any $v_w \le v \le v_x$ the marginal $m_\pi(z; v)$ is finite. In addition,*

$$\begin{aligned}\frac{\partial}{\partial v} E \log m_\pi(z; v) &= E_{\mu,v}\left[\frac{\Delta m_\pi(Z; v)}{m_\pi(Z; v)} - \frac{1}{2}\|\nabla \log m_\pi(Z; v)\|^2\right] \\ &= E_{\mu,v}\left[2\frac{\Delta\sqrt{m_\pi(Z; v)}}{\sqrt{m_\pi(Z; v)}}\right]. \end{aligned} \tag{3.91}$$

Proof For any $v_w \le v \le v_x$,

$$\begin{aligned} m_\pi(z; v) &= \int_{\mathbb{R}^p} \frac{1}{(2\pi v)^{p/2}} \exp\left(-\frac{\|z-\mu\|^2}{2v}\right)\pi(\mu)\, d\mu \\ &= \left(\frac{v_x}{v}\right)^{p/2} \int_{\mathbb{R}^p} \frac{1}{(2\pi v_x)^{p/2}} \exp\left(-\frac{v_x}{v}\frac{\|z-\mu\|^2}{2v_x}\right)\pi(\mu)\, d\mu \\ &\le \left(\frac{v_x}{v}\right)^{p/2} m_\pi(z; v_x) < \infty. \end{aligned}$$

Hence, the marginal m_π is finite. Setting $Z' = (Z-\mu)/\sqrt{v} \sim \mathcal{N}(0, I)$,

$$\begin{aligned} \frac{\partial}{\partial v} E_{\mu,v} \log m_\pi(Z; v) &= \frac{\partial}{\partial v}\int p(z|\mu, v)\log\left(m_\pi(z; v)\, dz\right) \\ &= \frac{\partial}{\partial v}\int p(z'|0, 1)\log\left(m_\pi(\sqrt{v}z' + \mu; v)\right) dz' \\ &= E_{Z'}\frac{(\partial/\partial v) m_\pi(\sqrt{v}Z' + \mu; v)}{m_\pi(\sqrt{v}Z' + \mu; v)} \end{aligned} \tag{3.92}$$

where

$$\begin{aligned} \frac{\partial}{\partial v} \quad m_\pi(\sqrt{v}z' + \mu; v) &= \frac{\partial}{\partial v}\int \frac{1}{(2\pi v)^{p/2}} \exp\left\{-\frac{\|\sqrt{v}z' + \mu - \mu'\|^2}{2v}\right\}\pi(\mu')\, d\mu' \\ = \frac{1}{(2\pi v)^{p/2}} &\int\left(-\frac{p}{2v} + \frac{\|z-\mu'\|^2}{2v^2} - \frac{\|z'\|^2}{2v} - \frac{2\langle z', \mu - \mu'\rangle}{2v^{3/2}}\right) p(z|\mu')\pi(\mu')\, d\mu' \\ = \frac{\partial}{\partial v} m_\pi(z; v) &- \int \frac{\langle z-\mu, z-\mu'\rangle}{2v^2} p(z|\mu')\pi(\mu')\, d\mu'. \end{aligned} \tag{3.93}$$

Note that

$$\nabla_z m_\pi(z, v) = \int \frac{-(z-\mu)}{v} p(z|\mu)\pi(\mu) d\mu \tag{3.94}$$

and

$$\Delta_z m_\pi(z, v) = \int \left[\frac{-p}{v} + \frac{\|z-\mu\|^2}{v^2} \right] p(z|\mu)\pi(\mu) d\mu$$
$$= 2\frac{\partial}{\partial v} m_\pi(z; v). \tag{3.95}$$

It follows that

$$E_{Z'} \frac{(\partial/\partial v) m_\pi(\sqrt{v}Z' + \mu; v)}{m_\pi(\sqrt{v}Z' + \mu; v)} = E_{\mu,v}\left(\frac{1}{2}\frac{\Delta m_\pi(Z; v)}{m_\pi(Z; v)} + \frac{\langle Z-\mu, \nabla \log m_\pi(Z; v)\rangle}{2v} \right).$$

Hence, using Stein's identity,

$$E_{\mu,v}\left[\frac{(Z-\mu)^{\mathrm{T}} \nabla \log m_\pi(Z; v)}{2v} \right] = E_{\mu,v}\left[\frac{1}{2} \Delta \log m_\pi(Z; v) \right]$$

$$= E_{\mu,v}\left[\frac{1}{2} \left(\frac{\Delta m_\pi(Z; v)}{m_\pi(Z; v)} - \|\nabla \log m_\pi(Z; v)\|^2 \right) \right],$$

which is the desired result. □

Lemmas 3.17 and 3.18 gives a result regarding minimaxity and domination from George et al. (2006). This result reveals parallels to those on minimax estimation of mean under quadratic loss in Sect. 3.1.1. Its proof is contained in the proof of Theorem 3.17.

Theorem 3.16 *Assume that $m_\pi(z; v_x)$ is finite for any z in $\mathbb{R}^p$. If $\Delta m_\pi \le 0$ for all $v_w \le v \le v_x$, then the Bayesian predictive density $\hat{p}_\pi(y|x)$ is minimax and dominates $\hat{p}_U$ (when π is not the uniform itself). If $\Delta \pi \le 0$, then the Bayesian predictive density $\hat{p}_\pi(y|x)$ is minimax and dominates $\hat{p}_U$ (when π is uniform).*

The next result from Brown et al. (2008) illuminates the link between the two problems of estimating the predictive density under the Kullback-Leibler loss and estimating the mean under quadratic loss. The result expresses this link in terms of risk differences.

Theorem 3.17 *Suppose the prior $\pi(\mu)$ is such that the marginal $m_\pi(z; v)$ is finite for any $z \in \mathbb{R}^p$. Then,*

$$\mathcal{R}_{KL}(\mu, \hat{p}_U) - \mathcal{R}_{KL}(\mu, \hat{p}_\pi) = \frac{1}{2}\int_{v_w}^{v_x} \frac{1}{v^2} \left(\mathcal{R}_Q^v(\mu, X) - \mathcal{R}_Q^v(\mu, \hat{\mu}_{\pi,v}) \right) dv. \tag{3.96}$$

Proof From (3.90) and (3.91) it follows

$$\mathscr{R}_{\text{KL}}(\mu, \hat{p}_U) - \mathscr{R}_{\text{KL}}(\mu, \hat{p}_\pi) = \int_{v_w}^{v_x} -\frac{\partial}{\partial v} E_{\mu,v}[\log m_\pi(Z; v)] \, dv$$
$$= \int_{v_w}^{v_x} E_{\mu,v}\left[2 \frac{\Delta\sqrt{m_\pi(Z; v)}}{\sqrt{m_\pi(Z; v)}}\right] dv. \quad (3.97)$$

On the other hand, Stein (1981) showed that

$$\mathscr{R}_Q^v(\mu, X) - \mathscr{R}_Q^v(\mu, \hat{\mu}_{\pi,v}) = -4v^2 E_{\mu,v} \frac{\Delta\sqrt{m_\pi(Z; v)}}{\sqrt{m_\pi(Z; v)}}. \quad (3.98)$$

Hence substituting (3.98) in the integral (3.97) gives (3.96). □

It is worth noting that using (3.88) and (3.96) leads to the following expression for the Kullback-Liebler risk of $\hat{p}_U$:

$$\frac{1}{2}\int_{v_w}^{v_x} \frac{1}{v^2}\left(\mathscr{R}_Q^v(\mu, X)\right) dv = \frac{1}{2}\int_{v_w}^{v_x} \frac{p}{v} \, dv$$
$$= \frac{p}{2} \log \frac{v_x}{v_w}$$
$$= \frac{p}{2} \log\left(1 + \frac{v_x}{v_y}\right).$$
$$= \mathscr{R}_{\text{KL}}(\mu, \hat{p}_U). \quad (3.99)$$

The area of predictive density estimation continues to develop. Recent research covers the case of restricted parameter (Fourdrinier et al. 2011), general α-divergence losses (Maruyama and Strawderman 2012; Boisbunon and Maruyama 2014), integrated $L1$ and $L2$ losses (Kubokawa et al. 2015, 2017). For a general review, see George and Xu (2010).

Chapter 4
Spherically Symmetric Distributions

4.1 Introduction

In the previous chapters, estimation problems were considered for the normal distribution setting. Stein (1956) showed that the usual estimator of a location vector could be improved upon quite generally for $p \geq 3$ and Brown (1966) substantially extended this conclusion to essentially arbitrary loss functions. Explicit results of the James-Stein type, however, have thus far been restricted to the case of the normal distribution. Recall the geometrical insight from Sect. 2.2.2, the development did not depend on the normality of X or even that θ is a location vector – this suggests that the improvement for Stein-type estimators may hold for more general distributions. Strawderman (1974a) first explored such an extension and considered estimation of the location parameter for scale mixtures of multivariate normal distributions. Other extensions of James-Stein type results to distributions other than scale mixtures of normal distributions are due to Berger (1975), Brandwein and Strawderman (1978), and Bock (1985). In this chapter, we will introduce the general class of spherically symmetric distributions; we will examine point estimation for variants of this general class in subsequent three chapters.

4.2 Spherically Symmetric Distributions

The normal distribution has been generalized in two important directions. First, as a special case of the exponential family and second, as a spherically symmetric distribution. In this chapter, we will consider the latter. There are a variety of equivalent definitions and characterizations of the class of spherically symmetric distributions; a comprehensive review is given in Fang et al. (1990). We now turn our interest to general orthogonally invariant distributions in $\mathbb{R}^n$ and a slightly more general notion of spherically symmetric distributions.

D. Fourdrinier et al., *Shrinkage Estimation*, Springer Series in Statistics,
https://doi.org/10.1007/978-3-030-02185-6_4

Definition 4.1 A random vector $X \in \mathbb{R}^n$ (equivalently the distribution of X) is spherically symmetric about $\theta \in \mathbb{R}^n$ if $X - \theta$ is orthogonally invariant. We denote this by $X \sim SS(\theta)$.

Note that Definition 4.1 states that $X \sim SS(\theta)$ if and only if $X = Z + \theta$ where $Z \sim SS(0)$. As an example, the uniform distribution $\mathscr{U}_{R,\theta}$ (cf. Definition 1.4) on the sphere $S_{R,\theta}$ of radius R and centered at θ is spherically symmetric about θ. Furthermore, if P is a spherically symmetric distribution about θ, then

$$P(HC + \theta) = P(C + \theta),$$

for any Borel set C of $\mathbb{R}^n$ and any orthogonal transformation H.

The following proposition is immediate from the definition.

Proposition 4.1 *If a random vector $X \in \mathbb{R}^n$ is spherically symmetric about $\theta \in \mathbb{R}^n$ then, for any orthogonal transformation H, HX is spherically symmetric about $H\theta$ ($X - \theta$ has the same distribution as $HX - H\theta$).*

The connection between spherical symmetry and uniform distributions on spheres is indicated in the following theorem.

Theorem 4.1 *A distribution P in $\mathbb{R}^n$ is spherically symmetric about $\theta \in \mathbb{R}^n$ if and only if there exists a distribution ρ in $\mathbb{R}_+$ such that $P(A) = \int_{\mathbb{R}_+} \mathscr{U}_{r,\theta}(A)\, d\rho(r)$ for any Borel set A of $\mathbb{R}^n$. Furthermore, if a random vector X has such a distribution P, then the radius $\|X - \theta\|$ has distribution ρ (called the radial distribution) and the conditional distribution of X given $\|X - \theta\| = r$ is the uniform distribution $\mathscr{U}_{r,\theta}$ on the sphere $S_{r,\theta}$ of radius r and centered at θ.*

Proof Sufficiency is immediate since the distribution $\mathscr{U}_{r,\theta}$ is spherically symmetric about θ for any $r \geq 0$.

It is clear that for the necessity it suffices to consider $\theta = 0$. Let X be distributed as P where P is $SS(0)$, $\nu(x) = \|x\|$, and ρ be the distribution of ν. Now, for any Borel sets A in $\mathbb{R}^n$ and B in $\mathbb{R}_+$ and for any orthogonal transformation H, we have (using basic properties of conditional distributions)

$$\begin{aligned}
\int_B P(H^{-1}(A) \mid \nu = r)\, d\rho(r) &= P(H^{-1}(A) \cap \nu^{-1}(B)) \\
&= P(H^{-1}(A \cap H(\nu^{-1}(B)))) \\
&= P(A \cap H(\nu^{-1}(B))) \\
&= P(A \cap \nu^{-1}(B)) \\
&= \int_B P(A \mid \nu = r)\, d\rho(r)
\end{aligned}$$

where we used the orthogonal invariance of the measure P and the function ν. Since the above equality holds for any B, then, almost everywhere with respect to ρ, we have

$$P(H^{-1}(A) \mid \nu = r) = P(A \mid \nu = r).$$

Equivalently, the conditional distribution given ν is orthogonally invariant on S_r. By unicity (see Lemma 1.1), it is the uniform distribution on S_r and the theorem follows. □

Corollary 4.1 *A random vector $X \in \mathbb{R}^n$ has a spherically symmetric distribution about $\theta \in \mathbb{R}^n$ if and only if X has the stochastic representation $X = \theta + R\,U$ where R ($R = \|X - \theta\|$) and U are independent, $R \geq 0$ and $U \sim \mathcal{U}$.*

Proof In the proof of Theorem 4.1, we essentially show that the distribution of $(X - \theta)/\|X - \theta\|$ is $\mathcal{U}$ independently of $\|X - \theta\|$. This is the necessity part of the corollary. The sufficiency part is direct. □

Also, the following corollary is immediate.

Corollary 4.2 *Let X be a random vector in R^n having a spherically symmetric distribution about $\theta \in \mathbb{R}^n$. Let h be a real valued function on R^n such that the expectation $E_\theta[h(X)]$ exists. Then*

$$E_\theta[h(X)] = E[E_{R,\theta}[h(X)]]\,,$$

where $E_{R,\theta}$ is the conditional expectation of X given $\|X - \theta\| = R$ (i.e. the expectation with respect to the uniform distribution $\mathcal{U}_{R,\theta}$ on the sphere $S_{R,\theta}$ of radius R and centered at θ) and E is the expectation with respect to the distribution of the radius $\|X - \theta\|$.

A more general class of distributions where $(X - \theta)/\|X - \theta\| \sim \mathcal{U}$ but not necessarily independently of $\|X - \theta\|$ is known as the isotropic distributions (see Philoche 1977). The class of spherically symmetric distributions with a density with respect to the Lebesgue measure is of particular interest. The form of this density and its connection with the radial distribution are the subject of the following theorem.

Theorem 4.2 *Let $X \in \mathbb{R}^n$ have a spherically symmetric distribution about $\theta \in \mathbb{R}^n$. Then the following two statements are equivalent.*

(1) *X has a density f with respect to the Lebesgue measure on $\mathbb{R}^n$.*
(2) *$\|X - \theta\|$ has a density h with respect to the Lebesgue measure on $\mathbb{R}_+$.*

Further, if (1) *or* (2) *holds, there exists a function g from $\mathbb{R}_+$ into $\mathbb{R}_+$ such that*

$$f(x) = g(\|x - \theta\|^2) \ a.e.$$

and

$$h(r) = \frac{2\pi^{n/2}}{\Gamma(n/2)} r^{n-1} g(r^2) \ a.e.$$

The function g is called the generating function and h the radial density.

Proof The fact that (1) implies (2) follows directly from the representation of X in polar coordinates. We can also argue that (2) implies (1) in a similar fashion using the independence of $\|X - \theta\|$, angles, and the fact that the angles have a density. The following argument shows this directly and, furthermore, gives the relationship between f, g, and h.

It is clear that it suffices to assume that $\theta = 0$. Suppose then that $R = \|X\|$ has a density h. According to Theorem 4.1, for any Borel set A of $\mathbb{R}^n$, we have

$$\begin{aligned}
P(X \in A) &= \int_0^\infty \int_{S_r} \mathbb{1}_A\,(y)\; d\mathcal{U}_r(y)\, h(r)\, dr \\
&= \int_0^\infty \int_{S_r} \mathbb{1}_A(y)\, \frac{d\sigma_r\,(y)}{\sigma_1(S_1) r^{n-1}}\; h(r)\; dr \quad \text{(by (1.4))} \\
&= \int_0^\infty \int_{S_r} \mathbb{1}_A(y)\, \frac{h(\|y\|)}{\sigma_1(S_1)\|y\|^{n-1}}\; d\sigma_r(y)\; dr \\
&= \int_{\mathbb{R}^n} \mathbb{1}_A(y) \frac{h(\|y\|)}{\sigma_1(S_1)\|y\|^{n-1}}\; dy \quad \text{(by Lemma 1.4)} \\
&= \int_A \frac{h(\|y\|)}{\sigma_1(S_1)\|y\|^{n-1}}\; dy.
\end{aligned}$$

This implies that the random vector X has density

$$f(x) = \frac{h(\|x\|)}{\sigma_1(S_1)\|x\|^{n-1}} = g(\|x\|^2)$$

with $h(r) = \sigma_1(S_1)\, r^{n-1}\, g(r^2)$, which is the announced formula for $h(r)$ since $\sigma_1(S_1) = 2\pi^{n/2}/\Gamma(n/2)$ by Corollary 1.1. □

We now turn our attention to the mean and the covariance matrix of a spherically symmetric distribution (when they exist).

Theorem 4.3 *Let $X \in \mathbb{R}^n$ be a random vector with a spherically symmetric distribution about $\theta \in \mathbb{R}^n$. Then, the mean of X exists if and only if the mean of $R = \|X - \theta\|$ exists, in which case $E[X] = \theta$. The covariance matrix of X exists if and only if $E[R^2]$ is finite, in which case*

$$cov(X) = \frac{E[R^2]}{n} I_n.$$

Proof Note that $X = Z + \theta$ where $Z \sim SS(0)$ and it suffices to consider the case $\theta = 0$. By the stochastic representation $X = R\,U$ in Corollary 4.1 with $R = \|X\|$ independent of U and $U \sim \mathcal{U}$, the expectation $E[X]$ exists if and only if the expectations $E[R]$ and $E[U]$ exist. However, since U is bounded, $E[U]$ exists and is equal to zero since $E[U] = E[-U]$ by orthogonal invariance.

Similarly, $E[\|X\|^2] = E[R^2]\,E[\|U\|^2] = E[R^2]$ and consequently the covariance matrix of X exists if and only if $E[R^2] < \infty$. Now

$$\text{cov}(RU) = E[R^2]\,E[UU^{\mathrm{T}}] = \frac{E[R^2]}{n} I_n .$$

Indeed $E[U_i^2] = E[U_j^2] = 1/n$ since U_i and U_j have the same distribution by orthogonal invariance and since $\sum_{i=1}^n U_i^2 = 1$. Furthermore, $E[U_i U_j] = 0$, for $i \neq j$, since $U_i U_j$ has the same distribution as $-U_i U_j$ by orthogonal invariance.

□

An interesting and useful subclass of spherically symmetric distributions consists of the spherically symmetric unimodal distributions. We only consider absolutely continuous distributions.

Definition 4.2 A random vector $X \in \mathbb{R}^n$ with density f is unimodal if the set $\{x \in \mathbb{R}^n \mid f(x) \geq a\}$ is convex for any $a \geq 0$.

Lemma 4.1 *Let $X \in \mathbb{R}^n$ be a spherically symmetric random vector about θ with generating function g. Then the distribution of X is unimodal if and only if g is nonincreasing.*

Proof Suppose first that the generating function g is nonincreasing. Take the left continuous version of g. For any $a \geq 0$, defining $g^{-1}(a) = \sup\{y \geq 0 \mid g(y) = a\}$ we have

$$\{x \in \mathbb{R}^n \mid g(\|x\|^2) \geq a\} = \{x \in \mathbb{R}^n \mid \|x\|^2 \leq g^{-1}(a)\}$$

which is a ball of radius $\sqrt{g^{-1}(a)}$ and convex. Conversely suppose that the set $\{x \in \mathbb{R}^n \mid g(\|x\|^2) \geq a\}$ is convex for any $a \geq 0$ and let $\|x\| \leq \|y\|$. Then, for $x^{\mathrm{T}} = y/\|y\|\,\|x\|$, we have $\|x^{\mathrm{T}}\| = \|x\|$ and $x^{\mathrm{T}} \in [-y, y]$ and hence, by the unimodality assumption, $g(\|x\|^2) = g(\|x^{\mathrm{T}}\|^2) \geq g(\|y\|^2)$. □

Theorem 4.1 showed that a spherically symmetric distribution is a mixture of uniform distributions on spheres. It is worth noting that, when the distribution is also unimodal, it is a mixture of uniform distributions on balls.

Theorem 4.4 *Let $X \in \mathbb{R}^n$ be a spherically symmetric random vector about $\theta \in \mathbb{R}^n$ with generating function g. Then the distribution of X is unimodal if and only if there exists a distribution ν in $\mathbb{R}_+$ with no point mass at 0 such that*

$$P[X \in A] = \int_{R_+} \mathcal{V}_{r,\theta}(A)\, d\nu(r) \tag{4.1}$$

for any Borel set A of $\mathbb{R}^n$*, where* $\mathscr{V}_{r,\theta}$ *is the uniform distribution on the ball* $B_{r,\theta} = \{x \in \mathbb{R}^n \mid \|x - \theta\| \leq r\}$.

Proof It is clear that it suffices to consider the case where $\theta = 0$. Suppose first that formula (4.1) is satisfied. Then expressing

$$\mathscr{V}_{r,0}(A) = \frac{1}{\lambda(B_r)} \int_{B_r} \mathbb{1}_A(x)\, dx$$

gives

$$\begin{aligned} P[X \in A] &= \int_{\mathbb{R}_+} \frac{1}{\lambda(B_r)} \int_{B_r} \mathbb{1}_A(x)\, dx\, d\nu\ (r) \\ &= \int_{\mathbb{R}_+} \frac{1}{\lambda(B_r)} \int_0^r \int_{S_u} \mathbb{1}_A(x)\, d\sigma_u(x)\, du\ d\nu(r) \\ &= \int_{\mathbb{R}_+} \int_{S_u} \mathbb{1}_A(x) \int_u^\infty \frac{1}{\lambda(B_r)}\, d\nu(r)\, d\sigma_u(x)\, du \end{aligned}$$

after applying Lemma 1.4 and Fubini's theorem. Then

$$\begin{aligned} P[X \in A] &= \int_0^\infty \int_{S_u} \mathbb{1}_A(x)\, g(\|x\|^2)\, d\sigma_u(x)\, du \\ &= \int_A g(\|x\|^2)\, dx \end{aligned}$$

again by Lemma 1.4 with the nonincreasing function

$$g(u^2) = \int_u^\infty \frac{1}{\lambda(B_r)}\, d\nu\ (r). \tag{4.2}$$

Hence according to Lemma 4.1, the distribution of X is unimodal.

Conversely, suppose that the distribution of X is unimodal. According to the above, this distribution will be a mixture of uniform distributions on balls if there exists a distribution ν on $\mathbb{R}_+$ with no point mass at 0 such that (4.2) holds. If such a distribution exists, (4.2) implies that ν can be expressed through a Stieltjes integral as

$$\nu(u) = \int_0^u \lambda(B_r)(-dg(r^2)).$$

It suffices therefore to show that ν is a distribution function on $\mathbb{R}_+$ with no point mass at 0. Note that, as g is nonincreasing, ν is the Stieltjes integral of a positive function with respect to a nondecreasing function and hence ν is nondecreasing. Since $\lambda(B_r) = \lambda(B_1)\, r^n = n\, \sigma_1(S_1)\, r^n$, an integration by parts gives

$$\nu(u) = \sigma_1(S_1) \int_0^u r^{n-1} g(r^2)\, dr - \lambda(B_1)^n\, g(u^2). \tag{4.3}$$

Note that the first term of the right hand side (4.3) is the distribution function of the radial distribution (see Theorem 4.2) and approaches 0 (respectively 1) when u approaches 0 (respectively ∞). Therefore, to complete the proof it suffices to show that

$$\lim_{u \to 0} u^n g(u^2) = \lim_{u \to \infty} u^n g(u^2) = 0\,.$$

Since

$$\int_0^\infty r^{n-1} g(r^2)\, dr < \infty,$$

we have

$$\lim_{r \to \infty} \int_{r/2}^r r^{n-1} g(u^2)\, du = 0.$$

By the monotonicity of g, we have

$$\int_{r/2}^r u^{n-1} g(u^2)\, du \geq (r^2) \int_{r/2}^r u^{n-1}\, du = g(r^2)\, r^n\, \frac{1}{n} \left(1 - \frac{1}{2^n}\right).$$

Hence, $\lim_{r \to \infty} r^n g(r^2) = 0$. The limit as r approaches 0 can be treated similarly and the result follows. □

It is possible to allow the possibility of a point mass at 0 for a spherically symmetric unimodal distribution, but we choose to restrict the class to absolutely continuous distributions. For a more general version of unimodality see Section 2.1 of Liese and Miescke (2008).

4.3 Elliptically Symmetric Distributions

By Definition 1.2, a random vector $X \in \mathbb{R}^n$ is orthogonally invariant if, for any orthogonal transformation H, HX has the same distribution as X. The notion of orthogonal transformation is relative to the classical scalar product $\langle x, y \rangle = \sum_{i=1}^n x_i y_i$. It is natural to investigate orthogonal invariance with respect to orthogonal transformations relative to a general scalar product $\langle x, y \rangle_\Gamma = x^{\mathrm{T}} \Gamma y =$

$\sum_{1\leq i,j\leq n} x_i \Gamma_{ij} y_j$ where Γ is a symmetric positive definite $n \times n$ matrix. We define a transformation H to be Γ-orthogonal if it preserves the scalar product in the sense that, for any $x \in \mathbb{R}^n$ and $y \in \mathbb{R}^n$, $\langle Hx, Hy\rangle_\Gamma = \langle x, y\rangle_\Gamma$ or, equivalently, if it preserves the associated norm $\|x\|_\Gamma = \sqrt{\langle x, x\rangle_\Gamma}$, that is, if $\|Hx\|_\Gamma = \|x\|_\Gamma$. Note that H is necessarily invertible since

$$\ker H = \{x \in \mathbb{R}^n \mid Hx = 0\} = \{x \in \mathbb{R}^n \mid \|Hx\|_\Gamma = 0\} = \{x \in \mathbb{R}^n \mid \|x\|_\Gamma = 0\} = \{0\}\,.$$

Then it can be seen that H is Γ-orthogonal if and only if $\langle Hx, y\rangle_\Gamma = \langle x, H^{-1}y\rangle_\Gamma$, for any $x \in \mathbb{R}^n$ and $y \in \mathbb{R}^n$ or, equivalently, if $H^{\mathrm{T}}\Gamma H = \Gamma$.

In this context, the Γ-sphere of radius $r \geq 0$ is defined as

$$S_r^\Gamma = \{x \in \mathbb{R}^n \mid x^{\mathrm{T}}\Gamma x = r^2\}\,.$$

Definition 4.3 A random vector $X \in \mathbb{R}^n$ (equivalently the distribution of X) is Γ-orthogonally invariant if, for any Γ-orthogonal transformation H, the distribution of $Y = HX$ is the same as that of X.

We can define a uniform measure on the ellipse S_r^Γ in a manner analogous to (1.3) and the resulting measure is indeed Γ-orthogonally invariant. It is not however the superficial measure mentioned at the end of Sect. 1.3, but is, in fact, a constant multiple of this measure where the constant of proportionality depends on Γ and reflects the shape of the ellipse. Whatever the constant of proportionality is, it allows the construction of a unique uniform distribution on S_r^Γ as in (1.4). The uniqueness follows from the fact that the Γ-orthogonal transformations form a compact group. We can then adapt the material from Sects. 1.3 and 4.2 to the case of a general positive definite matrix Γ. However, we present an alternative development.

The following discussion indicates a direct connection between the usual orthogonal invariance and Γ-orthogonal invariance. Suppose, for the moment, that $X \in \mathbb{R}^n$ has a spherically symmetric density given by $g(\|x\|^2)$. Let Σ be a positive definite matrix and A be a nonsingular matrix such that $AA^{\mathrm{T}} = \Gamma$. Standard change of variables gives the density of $Y = AX$ as $|\Sigma|^{-1/2}g(y^{\mathrm{T}}\Sigma^{-1}y)$. Let H be any Σ^{-1} orthogonal transformation and let $Z = HY$. The density of Z is $|\Sigma|^{-1/2}g(z^{\mathrm{T}}\Sigma^{-1}z)$ since H^{-1} is also Σ^{-1} -orthogonal and hence, $(H^{-1})^{\mathrm{T}}\Sigma^{-1}H^{-1} = \Sigma^{-1}$. This suggests that, in general, $Y = \Sigma^{\frac{1}{2}}X$ is Σ^{-1}-orthogonally invariant if and only if X is orthogonally invariant. The following result establishes this general fact.

Theorem 4.5 *Let Σ be a positive definite $n \times n$ matrix. A random vector $Y \in \mathbb{R}^n$ is Σ^{-1}-orthogonally invariant if and only if $Y = \Sigma^{1/2}X$ with X orthogonally invariant.*

Proof First note that, for any Σ^{-1}-orthogonal matrix H, $\Sigma^{-1/2}H\Sigma^{-1/2}$ is an I_n -orthogonal matrix since

$$
\begin{aligned}
(\Sigma^{-1/2} H \Sigma^{1/2})^{\mathrm{T}} (\Sigma^{-1/2} H \Sigma^{1/2}) &= \Sigma^{1/2} H^{\mathrm{T}} \Sigma^{-1} H \Sigma^{1/2} \\
&= \Sigma^{1/2} \Sigma^{-1} \Sigma^{1/2} \\
&= I_n .
\end{aligned}
$$

Then, if X is orthogonally invariant, for any Borel set C, of $\mathbb{R}^n$ we have

$$
\begin{aligned}
P[H \Sigma^{1/2} X \in C] &= P[\Sigma^{-1/2} H \Sigma^{1/2} X \in \Sigma^{-1/2} C] \\
&= P[X \in \Sigma^{-1/2} C] \\
&= P[\Sigma^{1/2} X \in C].
\end{aligned}
$$

Hence $Y = \Sigma^{1/2} X$ is Σ^{-1}-orthogonally invariant.

Similarly, for any orthogonal matrix G, $\Sigma^{1/2} G \Sigma^{-1/2}$ is a Σ^{-1}-orthogonal matrix. So, if $Y = \Sigma^{1/2} X$ is Σ^{-1}-orthogonally invariant, then X is orthogonally invariant. □

Note that, if X is orthogonally invariant and its covariance matrix exists, it is of the form $\sigma^2 I_n$ by Theorem 4.3. Therefore, if $Y = \Sigma^{1/2} X$, the covariance matrix of Y is $\sigma^2 \Sigma$, while, by Theorem 4.5, Y is Σ^{-1}-orthogonal invariant. In statistical models, it is often more natural to parametrize through a covariance matrix Σ than through its inverse (graphical models are the exception) and this motivates the following definition of elliptically symmetric distributions.

Definition 4.4 Let Σ be a positive definite $n \times n$ matrix. A random vector X (equivalently the distribution of X) is elliptically symmetric about $\theta \in \mathbb{R}^n$ if $X - \theta$ is Σ^{-1}-orthogonally invariant. We denote this by $X \sim ES(\theta, \Sigma)$.

Note that, if $X \sim SS(\theta)$, then $X \sim ES(\theta, I_n)$. If $Y \sim ES(\theta, \Sigma)$, then $\Sigma^{-1/2} Y \sim SS(\Sigma^{-1/2}\theta)$.

In the following, we briefly present some results for elliptically symmetric distributions that follow from Theorem 4.5 and are the analogues of those in Sects. 1.3 and 4.2. The proofs are left to the reader.

For the rest of this section, let Σ be a fixed positive definite $n \times n$ matrix and denote by $S_R^{\Sigma^{-1}} = \{x \in \mathbb{R}^n \mid x^{\mathrm{T}} \Sigma^{-1} x = R^2\}$ the Σ^{-1}- ellipse of radius R and by $\mathcal{U}_R^{\Sigma}$ the uniform distributions on $S_R^{\Sigma^{-1}}$.

Lemma 4.2

(1) *The uniform distribution $\mathcal{U}_R^{\Sigma}$ on $S_R^{\Sigma^{-1}}$ is the image under the transformation $Y = \Sigma^{\frac{1}{2}} X$ of the uniform distribution $\mathcal{U}_R$ on the sphere S_R, that is,*

$$
\mathcal{U}_R^{\Sigma}(\Omega) = \mathcal{U}_R(\Sigma^{-\frac{1}{2}} \Omega)
$$

for any Borel set Ω of $S_R^{\Sigma^{-1}}$.

(2) *If* X *is distributed as* $\mathcal{U}_R^{\Sigma}$, *then*

  (a) $\Sigma^{-1/2}X/(X^{\mathrm{T}}\Sigma^{-1}X)^{1/2}$ *is distributed as* $\mathcal{U}$ *and*
  (b) $X/(X^{\mathrm{T}}\Sigma^{-1}X)^{1/2}$ *is distributed as* $\mathcal{U}_1^{\Sigma}$.

Theorem 4.6 *A random vector* $X \in \mathbb{R}^n$ *is distributed as* $ES(\theta, \Sigma)$ *if and only if there exists a distribution* $\rho \in \mathbb{R}_+$ *such that*

$$P[X \in A] = \int_{\mathbb{R}_+} \mathcal{U}_{r,\theta}^{\Sigma}(A)\, d\rho\,(r)$$

for any Borel set A *on* $\mathbb{R}^n$, *where* $\mathcal{U}_{r,\theta}^{\Sigma}$ *is the uniform distribution* $\mathcal{U}_r^{\Sigma}$ *translated by* θ. *Equivalently* X *has the stochastic representation* $X = RU$ *where* $R = \|X - \theta\|_{\Sigma^{-1}} = ((x-\theta)^{\mathrm{T}}\Sigma^{-1}(x-\theta))^{1/2}$ *and* U *are independent,* $R \geq 0$ *and* $U \sim \mathcal{U}_1^{\Sigma}$. *For such* X, *the radius* R *has distribution* ρ *(called the radial distribution).*

Theorem 4.7 *Let* $X \in \mathbb{R}^n$ *be distributed as* $ES(\theta, \Sigma)$. *Then the following two statements are equivalent:*

(1) X *has a density* f *with respect to the Lebesgue measure on* $\mathbb{R}^n$; *and*
(2) $\|X - \theta\|_{\Sigma^{-1}}$ *has a density* h *with respect to Lebesgue measure on* $\mathbb{R}_+$.

Further, if (1) *or* (2) *holds, there exists a function* g *from* $\mathbb{R}_+$ *into* $\mathbb{R}_+$ *such that*

$$f(x) = g(\|x - \theta\|_{\Sigma^{-1}}^2)$$

and

$$h(r) = \frac{2\pi^{n/2}}{\Gamma(n/2)}|\Sigma|^{-1/2}r^{n-1}g(r^2).$$

Theorem 4.8 *Let* $X \in \mathbb{R}^n$ *be distributed as* $ES(\theta, \Sigma)$. *Then the mean of* X *exists if and only if the mean of* $R = \|X - \theta\|_{\Sigma^{-1}}$ *exists, in which case* $E[X] = \theta$. *The covariance matrix exists if and only if* $E[R^2]$ *is finite, in which case* $\mathrm{cov}(X) = E[R^2]\,\Sigma/n$.

Theorem 4.9 *Let* $X \in \mathbb{R}^n$ *be distributed as* $ES(\theta, \Sigma)$ *with generating function* g. *Then the distribution of* X *is unimodal if and only if* g *is nonincreasing. Equivalently there exists a distribution* $\nu \in \mathbb{R}_+$ *with no point mass at* 0 *such that*

$$P[X \in A] = \int_{\mathbb{R}_+} \mathcal{V}_{r,\theta}^{\Sigma}(A)\, d\nu(r)$$

for any Borel set A *of* $\mathbb{R}^n$, *where* $\mathcal{V}_{r,\theta}^{\Sigma}$ *is the uniform distribution on the ball (solid ellipse)*

$$B_{r,\theta}^{\Sigma} = \{x \in \mathbb{R}^n \mid \|x - \theta\|_{\Sigma^{-1}} \leq r\}.$$

4.4 Marginal and Conditional Distributions for Spherically Symmetric Distributions

In this section, we study marginal and conditional distributions of spherically symmetric distributions. We first consider the marginal distributions for a uniform distribution on S_R.

Theorem 4.10 *Let* $X = (X_1^{\mathrm{T}}, X_2^{\mathrm{T}})^{\mathrm{T}} \sim \mathscr{U}_R$ *in* $\mathbb{R}^n$ *where dim* $X_1 = p$ *and dim* $X_2 = n - p$. *Then, for* $1 \le p < n$, X_1 *has an absolutely continuous spherically symmetric distribution with generating function* g_R *given by*

$$g_R(\|x_1\|^2) = \frac{\Gamma(n/2)\, R^{2-n}}{\Gamma((n-p)/2)\, \pi^{p/2}} (R^2 - \|x_1\|^2)^{(n-p)/2-1}\, \mathbb{1}_{B_R}(x_1). \tag{4.4}$$

Proof The proof is based on the fact that $R\,Y/\|Y\| \sim \mathscr{U}_R$, for any random variable Y with a spherically symmetric distribution (see Lemma 1.2), in particular $\mathscr{N}_n(0, I_n)$, and on the fact that X_1 has an orthogonally invariant distribution in $\mathbb{R}^p$. To see this invariance, note that, for any $p \times p$ orthogonal matrix H_1 and any $(n-p) \times (n-p)$ orthogonal matrix H_2, the matrix

$$H = \begin{pmatrix} H_1 & 0 \\ 0 & H_2 \end{pmatrix},$$

is a block diagonal $n \times n$ orthogonal matrix. Hence

$$H \begin{pmatrix} X_1 \\ X_2 \end{pmatrix} = \begin{pmatrix} H_1 X_1 \\ H_2 X_2 \end{pmatrix} \tag{4.5}$$

is distributed as $(x_1^{\mathrm{T}}, x_2^{\mathrm{T}})^{\mathrm{T}}$ and it follows that $H_1 X_1 \sim X_1$ and so X_1 is orthogonally invariant.

Therefore, if $Y = (Y_1^{\mathrm{T}}, Y_2^{\mathrm{T}})^{\mathrm{T}} \sim \mathscr{N}_n(0, I_n)$, then $\|Y_1\|^2$ is independent of $\|Y_2\|^2$ and, according to standard results, $Z = \|Y_1\|^2/\|Y\|^2$ has a beta distribution, that is $Beta(p/2, (n-p)/2)$. It follows that $Z' = \|X_1\|^2/\|X\|^2 = \|X_1\|^2/R^2$ has the same distribution since both $X/\|X\|$ and $Y/\|Y\|$ have distribution $\mathscr{U}_R$.

Thus $\|X_1\|^2 = R^2 Z'$ has a $Beta(p/2, (n-p)/2)$ density scaled by R^2. By a change of variable, the density of $\|X_1\|$ is equal to

$$h_R(r) = \frac{2}{B(p/2, (n-p)/2)} \frac{r^{p-1}(R^2 - r^2)^{(n-p)/2-1}}{R^{n-2}}\, \mathbb{1}_{(0,R)}(r).$$

Hence, by Theorem 4.2, X_1 has the density given by (4.4). □

Corollary 4.3 *Let* $X = (X_1^{\mathrm{T}}, X_2^{\mathrm{T}})^{\mathrm{T}} \sim SS(\theta)$ *in* $\mathbb{R}^n$ *where dim* $X_1 = p$ *and dim* $X_2 = n - p$ *and where* $\theta = (\theta_1^{\mathrm{T}}, \theta_2^{\mathrm{T}})^{\mathrm{T}}$.

Then, for $1 \le p < n$, *the distribution of* X_1 *is an absolutely continuous spherically symmetric distribution* $SS(\theta_1)$ *on* $\mathbb{R}^p$ *with generating function given by* $\int g_R(\|X_1 - \theta_1\|^2)\, d\nu(R)$ *where* ν *is the radial distribution of* X *and* g_R *is given by* (4.4).

Unimodality properties of the densities of projections are given in the following result.

Corollary 4.4 *For the setup of Corollary* 4.3, *the density of* X_1 *is unimodal whenever* $n - p \ge 2$. *Furthermore, if* $p = n - 2$ *and* $X =\sim \mathcal{U}_{R,\theta}$, *then* X_1 *has the uniform distribution on* B_{R,θ_1} *in* $\mathbb{R}^{n-2}$.

In this book, we will have more need for the marginal distributions than the conditional distributions of spherically symmetric distributions. For results on conditional distributions, we refer the reader to Fang and Zhang (1990) and to Fang et al. (1990). We will however have use for the following result.

Theorem 4.11 *Let* $X = (X_1^{\mathrm{T}}, X_2^{\mathrm{T}})^{\mathrm{T}} \sim \mathcal{U}_{R,\theta}$ *in* $\mathbb{R}^n$ *where dim* $X_1 = p$ *and dim* $X_2 = n - p$ *and where* $\theta = (\theta_1^{\mathrm{T}}, \theta_2^{\mathrm{T}})^{\mathrm{T}}$. *Then the conditional distribution of* X_1 *given* X_2 *is the uniform distribution on the sphere in* $\mathbb{R}^p$ *of radius* $(R^2 - \|X_2 - \theta_2\|^2)^{1/2}$ *centered at* θ_1.

Proof First, it is clear that the support of the conditional distribution of X_1 given X_2 is the sphere in $\mathbb{R}^p$ of radius $(R^2 - \|X_2 - \theta_2\|^2)^{1/2}$ centered at θ_1. It suffices to show that the translated distribution centered at 0 is orthogonally invariant. To this end, note that, for any orthogonal transformation H on $\mathbb{R}^p$, the block diagonal transformation with blocks H and I_{n-p}, denoted by $\tilde{H}$, is orthogonal in $\mathbb{R}^n$. Then

$$\tilde{H}\big((X_1 - \theta_1)^{\mathrm{T}}, (X_2 - \theta_2)^{\mathrm{T}}\big)^{\mathrm{T}} \sim \big((X_1 - \theta_1)^{\mathrm{T}}, (X_2 - \theta_2)^{\mathrm{T}}\big)^{\mathrm{T}} \sim \mathcal{U}_{R,\theta}$$

that is,

$$\big((H(X_1 - \theta_1))^{\mathrm{T}}, (X_2 - \theta_2)^{\mathrm{T}}\big)^{\mathrm{T}} \sim \big((X_1 - \theta_1)^{\mathrm{T}}, (X_2 - \theta_2)^{\mathrm{T}}\big)^{\mathrm{T}} \sim \mathcal{U}_{R,\theta}\,.$$

Hence

$$H(X_1 - \theta_1)|(X_2 - \theta_2) \sim (X_1 - \theta_1)|(X_2 - \theta_2)\,,$$

and therefore, the distribution of X_1 given X_2 is orthogonally invariant, since θ_2 is fixed. The lemma follows. □

When properly interpreted, Corollaries 4.3 and 4.4 and Theorem 4.11 continue to hold for a general orthogonal projection π from $\mathbb{R}^n$ onto any subspace V of dimension p. See also Sect. 2.4.4 where the distribution is assumed to be normal.

4.5 The General Linear Model

This section is devoted to the general linear model, its canonical form and the issues of estimation, sufficiency and completeness.

4.5.1 *The Canonical Form of the General Linear Model*

Much of this book is devoted to some form of the following general problem. Let $(X^{\mathrm{T}}, U^{\mathrm{T}})^{\mathrm{T}}$ be a partitioned random vector in $\mathbb{R}^n$ with a spherically symmetric distribution around a vector partitioned as $(\theta^{\mathrm{T}}, 0^{\mathrm{T}})^{\mathrm{T}}$ where $\dim X = \dim\theta = p$ and $\dim U = \dim 0 = k$ with $p + k = n$. Such a distribution arises from a fixed orthogonally invariant random vector $(X_0^{\mathrm{T}}, U_0^{\mathrm{T}})^{\mathrm{T}}$ and a fixed scale parameter σ through the transformation

$$(X^{\mathrm{T}}, U^{\mathrm{T}})^{\mathrm{T}} = \sigma\,(X_0^{\mathrm{T}}, U_0^{\mathrm{T}})^{\mathrm{T}} + (\theta^{\mathrm{T}}, 0^{\mathrm{T}})^{\mathrm{T}}\,, \tag{4.6}$$

so that the distribution of $((X-\theta)^{\mathrm{T}}, U^{\mathrm{T}})^{\mathrm{T}}$ is orthogonally invariant. We also refer to θ as a location parameter.

We will assume that the covariance matrix of $(X^{\mathrm{T}}, U^{\mathrm{T}})^{\mathrm{T}}$ exists, which is equivalent to the finiteness of the expectation $E[R^2]$ where $R = (\|X-\theta\|^2 + \|U\|^2)^{1/2}$ is its radius (in this case, we have $\mathrm{cov}(X^{\mathrm{T}}, U^{\mathrm{T}})^{\mathrm{T}} = E[R^2]\, I_n/n)$. Then it will be convenient to assume that the radius $R_0 = (\|X_0\|^2 + \|U_0\|^2)^{1/2}$ of $(X_0^{\mathrm{T}}, U_0^{\mathrm{T}})^{\mathrm{T}}$ satisfies $E[R_0^2] = n$ since we have

$$\mathrm{cov}(X^{\mathrm{T}}, U^{\mathrm{T}})^{\mathrm{T}} = \sigma^2\,\mathrm{cov}(X_0^{\mathrm{T}}, U_0^{\mathrm{T}})^{\mathrm{T}} = \sigma^2\, I_n\,.$$

Note that when it is assumed that the distribution in (4.6) is absolutely continuous with respect to the Lebesgue measure on $\mathbb{R}^n$, the corresponding density may be represented as

$$\frac{1}{\sigma^n}\, g\!\left(\frac{\|z-\theta\|^2 + \|u\|^2}{\sigma^2}\right) \tag{4.7}$$

where g is the generating function.

This model also arises as the canonical form of the following seemingly more general model, the general linear model. For an $n \times p$ matrix V (often referred to as the design matrix and assumed here to be full rank p), suppose that an $n \times 1$ vector Y is observed such that

$$Y = V\beta + \varepsilon\,, \tag{4.8}$$

where β is a $p \times 1$ vector of (unknown) regression coefficients and ε is an $n \times 1$ vector with a spherically symmetric error distribution about 0. A common alternative representation of this model is $Y = \eta + \varepsilon$ where ε is as above and η is in the column space of V.

Using partitioned matrices, let $G = (G_1^{\mathrm{T}}\ G_2^{\mathrm{T}})^{\mathrm{T}}$ be an $n \times n$ orthogonal matrix partitioned such that the first p rows of G (*i.e.* the rows of G_1 considered as column vectors) span the column space of V. Now let

$$\begin{pmatrix} X \\ U \end{pmatrix} = G\,Y = \begin{pmatrix} G_1 \\ G_2 \end{pmatrix} V\,\beta + G\,\varepsilon = \begin{pmatrix} \theta \\ 0 \end{pmatrix} + G\,\varepsilon \tag{4.9}$$

with $\theta = G_1 V \beta$ and $G_2 V \beta = 0$ since the rows of G_2 are orthogonal to the columns of V. It follows from the definition that $(X^{\mathrm{T}}, U^{\mathrm{T}})^{\mathrm{T}}$ has a spherically symmetric distribution about $(\theta^{\mathrm{T}}, 0^{\mathrm{T}})^{\mathrm{T}}$. In this sense, the model given in the first paragraph is the canonical form of the above general linear model.

This model has been considered by various authors such as Cellier et al. (1989), Cellier and Fourdrinier (1995), Maruyama (2003b), Maruyama and Strawderman (2005), and Fourdrinier and Strawderman (2010). Also, Kubokawa and Srivastava in (2001) addressed the multivariate case where θ is a mean matrix (in this case where X and U are matrices as well).

4.5.2 Least Squares, Unbiased and Shrinkage Estimation

Consider the model in (4.9). Since the columns of G_1^{T} (the rows of G_1) and the columns of V span the same space, there exists a nonsingular $p \times p$ matrix A such that

$$V = G_1^{\mathrm{T}} A, \text{ which implies } A = G_1 V, \tag{4.10}$$

since $G_1 G_1^{\mathrm{T}} = I_p$. So

$$\theta = A\beta, \text{ that is, } \beta = A^{-1}\theta\,. \tag{4.11}$$

Noting that $V^{\mathrm{T}} V = A^{\mathrm{T}} G_1\, G_1^{\mathrm{T}} A = A^{\mathrm{T}} A$, it follows that the estimation of θ by $\hat{\theta}(X, U)$ under the loss

$$L(\theta, \hat{\theta}) = (\hat{\theta} - \theta)^{\mathrm{T}} (\hat{\theta} - \theta) = \|\hat{\theta} - \theta\|^2 \tag{4.12}$$

is equivalent to the estimation of β by

$$\hat{\beta}(Y) = A^{-1}\,\hat{\theta}(G_1 Y, G_2 Y) = (G_1 V)^{-1}\,\hat{\theta}(G_1 Y, G_2 Y) \tag{4.13}$$

under the loss

$$L^*(\beta, \hat{\beta}) = (\hat{\beta} - \beta)^{\mathrm{T}} A^{\mathrm{T}} A (\hat{\beta} - \beta) = (\hat{\beta} - \beta)^{\mathrm{T}} V^{\mathrm{T}} V (\hat{\beta} - \beta) \tag{4.14}$$

in the sense that the resulting risk functions are equal,

$$R^*(\beta, \hat{\beta}) = E[L^*(\beta, \hat{\beta}(Y))] = E[L(\theta, \hat{\theta})] = R(\theta, \hat{\theta}) \,.$$

Actually, the corresponding loss functions are equal. To see this, note that

$$\begin{aligned} L^*(\beta, \hat{\beta}(Y)) &= (\hat{\beta}(Y) - \beta)^{\mathrm{T}} A^{\mathrm{T}} A (\hat{\beta}(Y) - \beta) \\ &= (A(\hat{\beta}(Y) - \beta))^{\mathrm{T}} (A(\hat{\beta}(Y) - \beta)) \\ &= (\hat{\theta}(X, U) - \theta)^{\mathrm{T}} (\hat{\theta}(X, U) - \theta) \\ &= L(\theta, \hat{\theta}(X, U)) \,, \end{aligned}$$

where (4.13) and (4.11) were used for the third equality.

Note that the above equivalence between the estimation of θ, the mean vector of X, and the estimation of the regression coefficients β also holds for the respective invariant losses

$$L(\theta, \hat{\theta}, \sigma^2) = \frac{1}{\sigma^2} (\hat{\theta} - \theta)^{\mathrm{T}} (\hat{\theta} - \theta) = \frac{1}{\sigma^2} \|\hat{\theta} - \theta\|^2 \tag{4.15}$$

and

$$L^*(\beta, \hat{\beta}, \sigma^2) = \frac{1}{\sigma^2} (\hat{\beta} - \beta)^{\mathrm{T}} A^{\mathrm{T}} A (\hat{\beta} - \beta) = \frac{1}{\sigma^2} (\hat{\beta} - \beta)^{\mathrm{T}} V^{\mathrm{T}} V (\hat{\beta} - \beta) \,. \tag{4.16}$$

Additionally, the correspondence (4.13) can be reversed as

$$\hat{\theta}(X, U) = A \, \hat{\beta}(G_1^{\mathrm{T}} X + G_2^{\mathrm{T}} U) = G_1 X \, \hat{\beta}(G_1^{\mathrm{T}} X + G_2^{\mathrm{T}} U) \tag{4.17}$$

since, according to (4.9),

$$Y = G^{\mathrm{T}} \begin{pmatrix} X \\ U \end{pmatrix} = \begin{pmatrix} G_1 \\ G_2 \end{pmatrix}^{\mathrm{T}} \begin{pmatrix} X \\ U \end{pmatrix} = (G_1^{\mathrm{T}} \, G_2^{\mathrm{T}}) \begin{pmatrix} X \\ U \end{pmatrix} = G_1^{\mathrm{T}} X + G_2^{\mathrm{T}} U \,. \tag{4.18}$$

There is also a correspondence between the estimation of θ and the estimation of η in the following alternative representation of the general linear model. Here

$$\eta = G^{\mathrm{T}} \begin{pmatrix} \theta \\ 0 \end{pmatrix} = \begin{pmatrix} G_1 \\ G_2 \end{pmatrix}^{\mathrm{T}} \begin{pmatrix} \theta \\ 0 \end{pmatrix} = (G_1^{\mathrm{T}} \, G_2^{\mathrm{T}}) \begin{pmatrix} \theta \\ 0 \end{pmatrix} = G_1^{\mathrm{T}} \theta + G_2^{\mathrm{T}} 0 = G_1^{\mathrm{T}} \theta$$

and

$$G_1 \eta = G_1 \, G_1^{\mathrm{T}} \theta = \theta \, .$$

It follows that the estimation of $\theta \in \mathbb{R}^p$ by $\hat{\theta}(X, U)$ under the loss $\|\hat{\theta} - \theta\|^2$ (the loss (4.12)) is equivalent to the estimation of η in the column space of V under the loss $\|\hat{\eta} - \eta\|^2$ by

$$\hat{\eta}(Y) = G_1^{\mathrm{T}} \hat{\theta}(G_1 Y, G_2 Y) \tag{4.19}$$

in the sense that the risks functions are equal. The easy demonstration is left to the reader.

Consider the first correspondence expressed in (4.13) and (4.17) between estimators in Models (4.8) and (4.9). We will see that it can be made completely explicit for a wide class of estimators. First, note that the matrix G_1 can be easily obtained by the Gram-Schmidt orthonormalization process or by the QR decomposition of the design matrix X, where Q is an orthogonal matrix such that $Q^{\mathrm{T}} V = R$ and R is an $n \times p$ upper triangular matrix (so that $G_1 = Q_1^{\mathrm{T}}$ and $G_2 = Q_2^{\mathrm{T}}$). Second, a particular choice of A can be made that gives rise to a closed form of G_1.

To see this, let

$$A = (V^{\mathrm{T}} V)^{1/2} \tag{4.20}$$

(a square root of $V^{\mathrm{T}} V$, which is invertible since V has full rank) and set

$$G_1 = A \, (V^{\mathrm{T}} V)^{-1} V^{\mathrm{T}} = (V^{\mathrm{T}} V)^{-1/2} V^{\mathrm{T}} \, . \tag{4.21}$$

Then we have

$$G_1 \, V = A \, , \qquad V = G_1^{\mathrm{T}} A \, , \tag{4.22}$$

and

$$G_1 \, G_1^{\mathrm{T}} = (V^{\mathrm{T}} V)^{-1/2} V^{\mathrm{T}} V (V^{\mathrm{T}} V)^{-1/2} = I_p \, . \tag{4.23}$$

Hence, as in (4.10), (4.22) expresses that the columns of G_1^{T} (the rows of G_1) span the same space as the columns of V, noticing that (4.23) means that these vectors are orthogonal. Therefore, completing G_1^{T} through the Gram-Schmidt orthonormalization process, we obtain an orthogonal matrix $G = (G_1^{\mathrm{T}} G_2^{\mathrm{T}})^{\mathrm{T}}$, with G_1 in (4.21), such that

$$G \, V = \begin{pmatrix} (V^{\mathrm{T}} V)^{1/2} \\ 0 \end{pmatrix} . \tag{4.24}$$

The relationship linking A and G_1 in (4.21) is an alternative to (4.10) and is true in general; that is,

$$G_1 = A\,(V^{\mathrm{T}}V)^{-1}\,V^{\mathrm{T}} \text{ or equivalently } A = (V^{\mathrm{T}}G_1^{\mathrm{T}})^{-1}V^{\mathrm{T}}V\,.$$

Indeed, we have $V^{\mathrm{T}}V = A^{\mathrm{T}}A$ so that $(V^{\mathrm{T}}V)^{-1}(A^{\mathrm{T}}A) = I_p$. Hence, $(V^{\mathrm{T}}V)^{-1}A^{\mathrm{T}} = A^{-1}$, which implies $V\,(V^{\mathrm{T}}V)^{-1}A^{\mathrm{T}} = V\,A^{-1} = G_1^{\mathrm{T}}A\,A^{-1} = G_1^{\mathrm{T}}$, according to (4.10).

As a consequence, if $\hat{\beta}_{ls}$ is the least squares estimator of β, we have

$$\hat{\beta}_{ls}(Y) = (V^{\mathrm{T}}V)^{-1}\,V^{\mathrm{T}}\,Y \tag{4.25}$$

so that the corresponding estimator $\hat{\theta}_0$ of θ is the projection $\hat{\theta}_0(X, U) = X$ since

$$\hat{\theta}_0(X, U) = A\,\hat{\beta}_{ls}(Y) = A\,(V^{\mathrm{T}}V)^{-1}V^{\mathrm{T}}\,Y = G_1\,Y = X\,. \tag{4.26}$$

From this correspondence, the estimator $\hat{\theta}_0(X, U) = X$ of θ is often viewed as the standard estimator. Note that, with the choice of A in (4.20), we have the closed form

$$\hat{\beta}_{ls}(Y) = (V^{\mathrm{T}}V)^{-1/2}X\,. \tag{4.27}$$

Furthermore, the correspondence between $\hat{\theta}(X, U)$ and $\hat{\beta}_{ls}(Y)$ can be specified when $\hat{\theta}(X, U)$ depends on U only through $\|U\|^2$, in which case, with a slight abuse of notation, we write $\hat{\theta}(X, U) = \hat{\theta}(X, \|U\|^2)$. Indeed, first note that

$$\begin{aligned}
\|X\|^2 &= (A\,\hat{\beta}_{ls}(Y))^{\mathrm{T}}(A\,\hat{\beta}_{ls}(Y)) \\
&= (\hat{\beta}_{ls}(Y))^{\mathrm{T}}\,A^{\mathrm{T}}A\,(\hat{\beta}_{ls}(Y)) \\
&= (\hat{\beta}_{ls}(Y))^{\mathrm{T}}\,V^{\mathrm{T}}V\,(\hat{\beta}_{ls}(Y)) \\
&= (V\,\hat{\beta}_{ls}(Y))^{\mathrm{T}}V\,(\hat{\beta}_{ls}(Y)) \\
&= \|V\,\hat{\beta}_{ls}(Y)\|^2\,.
\end{aligned} \tag{4.28}$$

On the other hand, according to (4.9), we have $\|X\|^2 + \|U\|^2 = \|G\,Y\|^2 = \|Y\|^2$. Hence,

$$\begin{aligned}
\|U\|^2 &= \|Y\|^2 - \|X\|^2 \\
&= \|Y\|^2 - \|V\,\hat{\beta}_{ls}(Y)\|^2 \\
&= \|Y - V\,\hat{\beta}_{ls}(Y)\|^2
\end{aligned} \tag{4.29}$$

since $Y - V\,\hat{\beta}_{ls}(Y)$ is orthogonal to $V\,\hat{\beta}_{ls}(Y)$. Consequently, according to (4.13) and (4.10), Equations (4.29) and (4.26) give that the estimator $\hat{\beta}(Y)$ of β corresponding to the estimator $\hat{\theta}(X, \|U\|^2)$ of θ is

$$\hat{\beta}(Y) = (G_1\, V)^{-1}\, \hat{\theta}(G_1\, V\, \hat{\beta}_{ls}(Y), \|Y - V\, \hat{\beta}_{ls}(Y)\|^2)\,. \tag{4.30}$$

Note that, when ones chooses G_1 as in (4.21), $\hat{\beta}(Y)$ in (4.30) has the closed form

$$\begin{aligned}\hat{\beta}(Y) &= (V^{\mathrm{T}}\, V)^{-1/2}\, \hat{\theta}\Big((V^{\mathrm{T}}\, V)^{1/2}\, \hat{\beta}_{ls}(Y), \|Y - V\, \hat{\beta}_{ls}(Y)\|^2\Big) \\ &= (V^{\mathrm{T}}\, V)^{-1/2}\, \hat{\theta}\Big((V^{\mathrm{T}}\, V)^{-1/2} V^{\mathrm{T}} Y, \|Y - V\, \hat{\beta}_{ls}(Y)\|^2\Big)\,. \end{aligned} \tag{4.31}$$

In particular, we can see through (4.28), that the "robust" Stein-type estimators of θ,

$$\hat{\theta}_r(X, \|U\|^2) = \left(1 - a\, \frac{\|U\|^2}{\|X\|^2}\right) X \tag{4.32}$$

have as a correspondence the "robust" estimators of β

$$\begin{aligned}\hat{\beta}_r(Y) &= (G_1\, V)^{-1} \left(1 - a\, \frac{\|Y - V\, \hat{\beta}_{ls}(Y)\|^2}{\|V\, \hat{\beta}_{ls}(Y)\|^2}\right) G_1\, V\, \hat{\beta}_{ls}(Y) \\ &= \left(1 - a\, \frac{\|Y - V\, \hat{\beta}_{ls}(Y)\|^2}{\|V\, \hat{\beta}_{ls}(Y)\|^2}\right) \hat{\beta}_{ls}(Y) \end{aligned} \tag{4.33}$$

(note that the two $G_1\, V$ terms simplify). We use the term "robust" since, for appropriate values of the positive constant a, they dominate X whatever the spherically symmetric distribution, as we will see in Chap. 5 (see also Cellier et al. 1989; Cellier and Fourdrinier 1995).

According to the correspondence seen above between the risk functions of the estimators of θ and the estimators of β, using these estimators in (4.33) is then a good alternative to the least squares estimator: they dominate the least squares estimator of β simultaneously for all spherically symmetric error distributions with a finite second moment (see Fourdrinier and Strawderman (1996) for the use of these robust estimators when σ^2 is known and also Sect. 5.2).

4.5.3 Sufficiency in the General Linear Model

Suppose $(X^{\mathrm{T}}, U^{\mathrm{T}})^{\mathrm{T}}$ has a spherically symmetric distribution about $(\theta^{\mathrm{T}}, 0^{\mathrm{T}})^{\mathrm{T}}$ with $\dim X = \dim \theta = p > 0$ and $\dim U = \dim 0 = k > 0$. Furthermore, suppose that the distribution is absolutely continuous with respect to the Lebesgue measure on $\mathbb{R}^n$ for $n = p + k$. The corresponding density may be represented as in (4.7). We refer to θ as a location vector and to σ as a scale parameter. As seen in the previous section, such a distribution arises from a fixed orthogonally invariant random vector $(X_0^{\mathrm{T}}, U_0^{\mathrm{T}})^{\mathrm{T}}$ with generating function g through the transformation

$$\begin{pmatrix} X \\ U \end{pmatrix} = \sigma \begin{pmatrix} X_0 \\ U_0 \end{pmatrix} + \begin{pmatrix} \theta \\ 0 \end{pmatrix}.$$

Each of θ, σ^2 and $g(\cdot)$ may be known or unknown, but perhaps the most interesting case from a statistical standpoint is the following.

Suppose θ and σ^2 are unknown and $g(\cdot)$ is known. It follows immediately from the factorization theorem that $(X, \|U\|^2)$ is sufficient . It is intuitively clear that this statistic is also minimal sufficient since $\dim(X, \|U\|^2) = \dim(\theta, \sigma^2)$. Here is a proof of that fact.

Theorem 4.12 *Suppose that* $(X^{\mathrm{T}}, U^{\mathrm{T}})^{\mathrm{T}}$ *is distributed as* (4.7). *Then the statistic* $(X, \|U\|^2)$ *is minimal sufficient for* (θ, σ^2) *when g is known.*

Proof By Theorem 6.14 of Casella and Berger (2001), it suffices to show that if, for all $(\theta, \sigma^2) \in \mathbb{R}^p \times \mathbb{R}_+$,

$$\frac{g\left(\frac{\|x_1-\theta\|^2+\|u_1\|}{\sigma^2}\right)}{g\left(\frac{\|x_2-\theta\|^2+\|u_2\|^2}{\sigma^2}\right)} = c \tag{4.34}$$

where c is a constant then $x_1 = x_2$ and $\|u_1\|^2 = \|u_2\|^2$. Note that $0 < c < \infty$ since otherwise (4.7) cannot be a density.

Letting $\tau^2 = 1/\sigma^2$, (4.34) can be written, for all $\tau > 0$, as

$$g(\tau^2 v_1^2) = c g(\tau^2 v_2^2) \tag{4.35}$$

where $v_1^2 = \|x_1 - \theta\|^2 + \|u_1\|^2$ and $v_2^2 = \|x_2 - \theta\|^2 + \|u_2\|^2$ for each fixed $\theta \in \mathbb{R}^p$. First, we will show that $v_1^2 = v_2^2$. Note that

$$\begin{aligned} 1 &= \int_{\mathbb{R}^p \times \mathbb{R}^k} g(\|x\|^2 + \|u\|^2)\, dx\, du \\ &= K \int_0^\infty r^{p+k-1}\, g(r^2)\, dr \qquad \text{(by Theorem 4.2)} \\ &= K v^{p+k} \int_0^\infty \tau^{p+k-1}\, g(v^2\tau^2)\, d\tau \end{aligned} \tag{4.36}$$

for any $v > 0$. Then it follows from (4.35) and (4.36) that

$$\begin{aligned} 1 &= K v_1^{p+k} \int_0^\infty \tau^{p+k-1} g(v_1^2\tau^2)\, d\tau \\ &= cK v_1^{p+k} \int_0^\infty \tau^{p+k-1} g(v_2^2\tau^2)\, d\tau \\ &= c\frac{v_1^{p+k}}{v_2^{p+k}}. \end{aligned} \tag{4.37}$$

Let $F(b) = \int_0^b \tau^{p+k-1} g(\tau^2)\, d\tau$ and choose b such that F is strictly increasing at b. Suppose $v_1 > v_2$. Then, for any $v > 0$,

$$F(b) = v^{p+k} \int_0^{b/v} \tau^{p+k-1}\, g(v^2\tau^2)\, d\tau$$

and consequently

$$\begin{aligned}\int_0^{b/v_1} \tau^{p+k-1}\, g(v_1^2\tau^2) d\tau &= \frac{F(b)}{v_1^n} \\ &= c \int_0^{b/v_1} \tau^{p+k-1}\, g(v_2^2\tau^2) d\tau \\ &< c \int_0^{b/v_2} \tau^{p+h-1}\, g(v_2^2\tau^2) d\tau \\ &= c\, \frac{F(b)}{v_2^n}.\end{aligned}$$

It follows that $c\, v_1^{p+k}/v_2^{p+h} > 1$, which contradicts (4.37). A similar argument would give $c\, v_1^{p+k}/v_2^{p+h} < 1$ for $v_1 < v_2$ and $v_1 = v_2$. Now, setting $\theta = \frac{x_1+x_2}{2}$ in the expressions for v_1 and v_2 implies $\|u_1\|^2 = \|u_2^2\|$. It then follows that $\|x_1 - \theta\|^2 = \|x_2 - \theta\|^2$ for all $\theta \in \mathbb{R}^p$, which implies $x_1 = x_2$ by setting $\theta = x_2$ (or x_1). □

In the case where θ is unknown, σ^2 is known, and the distribution is multivariate normal, X is minimal sufficient (and complete). However, in the non-normal case, $(X, \|U\|^2)$ is typically minimally sufficient, and may or may not be complete, which is the subject of the next section.

4.5.4 Completeness for the General Linear Model

The section largely follows the development in Fourdrinier et al. (2014). In the case where both θ and σ^2 are unknown and g is known, the minimal sufficient statistic $(X, \|U\|^2)$ can be either complete or incomplete depending on g. If g corresponds to a normal distribution, the statistic is complete by standard results for exponential families. However, when the generating function is of the form $K\, g(t) \mathbb{1}_{(r_1,r_2)}(t)$ with $0 < r_1 < r_2 < \infty$ and K is the normalizing constant, $(X, \|U\|^2)$ is not complete. In fact incompleteness of $(X, \|U\|^2)$ follows from the fact that the minimal sufficient statistic, when θ is known, σ^2 is unknown and g is known, is incomplete.

Theorem 4.13

(1) *If* $X \sim f(x-\theta)$ *with* $\theta \in \mathbb{R}^p$ *where* f *has compact support, then* X *is not complete for* θ.
(2) *If* $X \sim 1/\sigma f(x/\sigma)$, *where* f *has support contained in an interval* $[a, b]$ *with* $0 < a < b < \infty$, *then* X *is not complete for* σ.

Before giving the proof of Theorem 4.13, note that if the generating function is of the form $K\, g(t)\mathbb{1}_{[r_1,r_2]}(t)$ for $0 < r_1 < r_2 < \infty$ and the value of θ is assumed to be known and equal to θ_0, then $T = \|X - \theta_0\|^2 + \|U\|^2$ is minimal sufficient and has density of the form $K/\sigma^{p+k}\, t^{(p+k)/2}\, g(t/\sigma^2)\mathbb{1}_{[r_1\sigma^2, r_2\sigma^2]}(T)$.

Therefore, T is not a complete statistic for σ^2 by Lemma 4.13 (2). It follows that there exists a function $h(\cdot)$ not equal to zero a.e. such that $E_\sigma[h(T)] = 0$ for all $\sigma > 0$. Since $E_{\sigma^2}[h(\beta\, T)] = E_{\beta\sigma^2}[h(T)]$, it follows that $E_{\sigma^2}[h(\beta\, T)] = 0$ for all $\sigma^2 > 0, \beta > 0$, and also that $M(t) = \int_0^1 E_{\sigma^2}[h(\beta\, t)]\, m(\beta)\, d\beta = 0$ for any function $m(\cdot)$ for which the integral exists. In particular, this holds when $m(\cdot)$ is the density of a Beta$(k/2, p/2)$ random variable (where finiteness of the integral is guaranteed since $E_{\sigma^2}[h(\beta\, T)]$ is continuous in β). Now, since $B = \|U\|^2/T$ has a Beta$(k/2, p/2)$ distribution, $\|U\|^2 = BT$, and $M(\sigma^2) = E_{\sigma^2}[h(B\, T) = E_{\sigma^2}[h(\|U\|^2)] \equiv 0$.

Since the distribution of $\|U\|^2$ does not depend on θ, it follows that when both θ and σ^2 are unknown, $E_{\theta,\sigma^2}[h(\|U\|^2)] \equiv 0$. Hence, $(X, \|U\|^2)$, while minimal sufficient, is not complete for the case of a generating function of the form $g(t)\mathbb{1}_{[r_1,r_2]}(T)$ with $0 < r_1 < r_2 < \infty$.

Note that whenever θ is unknown, σ^2 is known, and $(X, \|U\|^2)$ is minimal sufficient (so the distribution is not normal, since then X would be minimal sufficient) $\|U\|^2$ is ancillary and the minimal sufficient statistic is not complete.

Proof of Theorem 4.13 First, note that part (2) follows from part (1) by the standard technique of transforming a scale family to a location family by taking logs.

We will show the incompleteness of a location family in $\mathbb{R}$ when F has bounded support. We show first that, if F is a cdf with bounded support contained inside $[a, b]$, the characteristic function (c.f.) $\hat{f}$ is analytic in $\mathbb{C}$ (the entire complex plane) and is of order 1 (i.e., $|\hat{f}(\eta)|$ is $O\exp(|\eta|^{1+\varepsilon})$ for all $\epsilon > 0$ and is not $O(\exp(|\eta|^{1-\varepsilon})$ for any $\epsilon > 0$).

To see this, without loss of generality assume $0 < a < b < \infty$. Then

$$\begin{aligned} |\hat{f}(\eta)| &\le \int_a^b \exp(|\eta| X)\, dF(x) \\ &\le \exp(b|\eta|) \int_a^b dF(x) \\ &= \exp(b|\eta|) \\ &= O(\exp(|\eta|^{1+\varepsilon})). \end{aligned}$$

for all $\varepsilon > 0$. Also, if $\eta = -iv$ for $v > 0$, then

$$|\hat{f}(\eta)| = \int_a^b \exp(vx)\, dF(x) \geq \exp(av) \int_a^b dF(x) = \exp(av)\,.$$

However, $\exp(av)$ is not $O(\exp(v^{1-\varepsilon})$ for any $\varepsilon > 0$. Hence $\hat{f}(\eta)$ is of order 1.

In the step above, we used $0 < a < b < \infty$. Note that if either a and/or b is negative then the distribution of X is equal to the distribution of $z + \theta_0$ where θ_0 is negative and where the distribution of z satisfies the assumptions of the theorem. Hence $E\exp(i\eta x) = E\exp(i\eta z)e^{i\eta\theta_0}$, so $|E\exp(i\eta x)| \leq \exp(|\eta|b)\exp(|i\eta||\theta_0|)$ which is $O\exp(|\eta|^{1+\epsilon})$ for all $\epsilon > 0$.

Similarly, for $\eta = -iv$ (recall $\theta_0 < 0$),

$$\begin{aligned} |E\exp(i\eta x)| &= E\exp(tvz)\, \exp(-v\theta_0) \\ &\geq e^{v|\theta_0|}\exp(av) \\ &= \exp(v(a + |\theta_0|) \end{aligned}$$

and this is not $O(\exp^{v^{1-\varepsilon}})$ for any $\varepsilon > 0$. □

Note that $\hat{f}(\eta)$ exists in all of $\mathbb{C}$ since F has bounded support and is analytic by standard results in complex analysis (See e.g. Rudin 1966). To complete the proof of Theorem 4.13 we need the following lemma.

Lemma 4.3 *If $X \sim F(x)$ where the cdf F has bounded support in $\mathbb{R}$ and F is not degenerate, then the characteristic function $\hat{f}(\eta)$ has at least one zero in $\mathbb{C}$.*

Proof This follows almost directly from the Hadamard factorization theorem which implies that a function $\hat{f}(z)$ that is analytic in all of $\mathbb{C}$ and of order 1 is of the form $\hat{f}(z) = \exp(az + b)P(z)$. $P(z)$ is the so called canonical product formed from the zeros of $\hat{f}(z)$, where $P(0) = 1$ and $P(z) = 0$ for each such root. (See e.g., Titchmarsh (1932) for an extended discussion of the form of $P(z)$). Therefore, either $\hat{f}(z)$ has no zeros, in which case $\hat{f}(z) = \exp(az)$ (since $\hat{f}(0) = 1 = e^b \Rightarrow b = 0$) and $P(z) \equiv 1$, or $\hat{f}(z)$ has at least one zero. The case where $\hat{f}(z) = \exp(az)$ corresponds to the degenerate case where $\exp(az) = \hat{f}(z) = E\exp(izx)$ with $P[X = -ia] = 1$. Since F is assumed to not be degenerate, $\hat{f}(z)$ must have at least one zero by the uniqueness of the Fourier transform.

To finish the proof of Theorem 4.13 note that, by By Lemma 4.3, there exists an η_0 such that

$$\hat{f}(\eta_0) = \int_{-\infty}^{\infty} \exp(i\eta_0 x) f(x)\, dx = 0.$$

This implies that for any $\theta \in \mathbb{R}$,

$$
\begin{aligned}
0 &= \left(\int_{-\infty}^{\infty} \exp(i\eta_0 x) f(x)\, dx\right) \exp^{i\eta_0\theta} \\
&= \int_{-\infty}^{\infty} \exp(ix(\eta_0 + \theta)) f(x)\, dx \\
&= \int_{-\infty}^{\infty} \exp(i\eta_0 x) f(x - \theta)\, dx \\
&= E_\theta[\exp(i\eta_0 X) = E_\theta[\exp(i(a_0 + b_0 i))X] \\
&= E_\theta \exp(i\eta_0 X) \exp(-b_0 X)] \\
&= E_\theta[\exp(-b_0 X)\{\cos a_0 x + i \sin a_0 x\}].
\end{aligned}
$$

Hence, for any $\theta \in \mathbb{R}$, we have $E_\theta[\exp(-b_0 x)\cos(a_0 x)] \equiv 0$.

Additionally, $E_\theta[|\exp(-b_0 x)\cos(a_0 x)|] < \infty$ for all θ since $f(\cdot)$ has bounded support. The theorem then follows, since $h(X) = e^{-b_0 X} \cos a_0 X$ is an unbiased estimator of 0, which is not equal to 0 almost surely for each θ. This proves the result for $p = 1$. The extension from $\mathbb{R}$ to $\mathbb{R}^p$ is straightforward since the marginal distribution of each coordinate has compact support. □

4.6 Characterizations of the Normal Distribution

There is a large literature on characterizations of the normal distribution that has had a long history. A classical reference that covers a number of characterizations of the normal distribution is Kagan et al. (1973). We give only a small sample of these characterizations. The first result gives a characterization in terms of the normality of linear transformations.

Theorem 4.14 *Let $X \sim ES(\theta)$ in $\mathbb{R}^n$. If A is any fixed linear transformation of positive rank such that AX has a normal distribution then X has a normal distribution.*

Proof First note that it suffices to consider the case $\theta = 0$. Furthermore it suffices to prove the result for $X \sim SS(0)$ since an elliptically symmetric distribution is the image of a spherically symmetric distribution by a nonsingular transformation. Note also that, if $X \sim SS(0)$, its characteristic function $\varphi_X(t) = \Psi(t^{\mathrm{T}}t)$ since, for any orthogonal transformation H, the characteristic function φ_{HX} of HX satisfies

$$
\varphi_{HX}(t) = \varphi_X(H^{\mathrm{T}}t) = \varphi_X(t).
$$

Now the characteristic function φ_{AX} of AX equals

$$
\varphi_{AX}(t) = E[\exp\{it^{\mathrm{T}}AX\}] = E[\exp\{i(A^{\mathrm{T}}t)^{\mathrm{T}}X\}] = \Psi(t^{\mathrm{T}}AA^{\mathrm{T}}t). \tag{4.38}
$$

Also, by Theorem 4.3, $\text{Cov}(X) = E[R^2]/nI_n$. Hence $\text{Cov}(AX) = (E[R^2]/n)\, AA^{\text{T}}$ and the fact that AX is normal implies that $E[R^2] < \infty$ and that $\text{Cov}(AX) = \alpha\, AA^{\text{T}}$ for $\alpha \geq 0$. This implies that $\varphi_{AX}(t) = \exp\{-\alpha\, t^{\text{T}} AA^{\text{T}} t/2\}$. Therefore, by (4.38), $\Psi(z) = \exp\{-\alpha z/2\}$ and $\varphi_X(t) = \exp\{-\alpha t^{\text{T}} t/2\}$, so X is normal. □

Corollary 4.5 *Let $X \sim ES(\theta)$ in $\mathbb{R}^n$. If any orthogonal projection Π has a normal distribution (and, in particular, any marginal), then X has a normal distribution.*

The next theorem gives a characterization in terms of the independence of linear projections.

Theorem 4.15 *Let $X \sim ES(\theta)$ in $\mathbb{R}^n$. If A and B are any two fixed linear transformations of positive rank such that AX and BX are independent, then X has a normal distribution.*

Proof As in the proof of Theorem 4.14, we can assume that $X \sim SS(0)$. Then the characteristic function φ_X of X is $\varphi_X(t) = \Psi(t^{\text{T}} t)$. Hence, the characteristic functions φ_{AX} and φ_{BX} of AX and BX are $\varphi_{AX}(t_1) = \Psi(t_1^{\text{T}} AA^{\text{T}} t_1)$ and $\varphi_{BX}(t_2) = \Psi(t_2^{\text{T}} BB^{\text{T}} t_2)$, respectively. By the independence of AX and BX, we have

$$\Psi(t_1^{\text{T}} AA^{\text{T}} t_1 + t_2^{\text{T}} BB^{\text{T}} t_2) = \Psi(t_1^{\text{T}} AA^{\text{T}} t_1)\Psi(t_2^{\text{T}} BB^{\text{T}} t_2).$$

Since A and B are of positive rank this implies that, for any $u \geq 0$ and $v \geq 0$,

$$\Psi(u + v) = \Psi(u)\Psi(v).$$

This equation is known as Hamel's equation and its only continuous solution is $\Psi(u) = e^{\alpha u}$ for some $\alpha \in \mathbb{R}$ (see for instance Feller 1971, page 305). Hence, $\varphi_X(t) = e^{\alpha t^{\text{T}} t}$ for some $\alpha \leq 0$ since φ_X is a characteristic function. It follows that X has a normal distribution. □

Corollary 4.6 *Let $X \sim ES(\theta)$ in $\mathbb{R}^n$. If any two projections (in particular, any two marginals) are independent, then X has a normal distribution.*

Chapter 5
Estimation of a Mean Vector for Spherically Symmetric Distributions I: Known Scale

5.1 Introduction

In Chaps. 2 and 3 we studied estimators that improve over the "usual" estimator of the location vector for the case of a normal distribution. In this chapter, we extend the discussion to spherically symmetric distributions discussed in Chap. 4. Section 5.2 is devoted to a discussion of domination results for Baranchik type estimators while Sect. 5.3 examines more general estimators. Section 5.4 discusses Bayes minimax estimation. Finally, Sect. 5.5 discusses estimation with a concave loss.

We close this introductory section by extending the discussion of Sect. 2.2 on the empirical Bayes justification of the James-Stein estimator to the general multivariate (but not necessarily normal) case.

Suppose X has a p-variate distribution with density $f(\|x - \theta\|^2)$, unknown location vector θ and known scale matrix $\sigma^2 I_p$. The problem is to estimate θ under loss $L(\theta, \delta) = \|\delta - \theta\|^2$. Let the prior distribution on θ be given by $\pi(\theta) = f^{\star n}(\theta)$, the n-fold convolution of the density $f(\cdot)$ with itself. Note that the distribution of θ is the same as that of $\sum_{i=1}^n Y_i$ where the Y_i are iid with density $f(\cdot)$. Recall from Example 1.3 that the Bayes estimator of θ is given by

$$\delta_n(X) = \frac{n}{n+1} X = \left(1 - \frac{1}{n+1}\right) X .$$

Assume now that n is unknown. Since

$$E(X^{\mathrm{T}} X) = E\left(\sum_{i=0}^{n} Y_i^{\mathrm{T}} Y_i\right) = (n+1)\, E(Y_0^{\mathrm{T}} Y_0) = (n+1)\,(\operatorname{tr} \sigma^2 I) = (n+1)\, p\, \sigma^2 ,$$

D. Fourdrinier et al., *Shrinkage Estimation*, Springer Series in Statistics,
https://doi.org/10.1007/978-3-030-02185-6_5

an unbiased estimator of $n+1$ is $X^{\mathrm{T}}X/(p\sigma^2)$, and so $p\,\sigma^2/(X^{\mathrm{T}}X)$ is a reasonable estimator of $1/(n+1)$. Substituting $p\,\sigma^2/(X^{\mathrm{T}}X)$ for $1/(n+1)$ in the Bayes estimator, we have that

$$\delta^{EB}(X) = \left(1 - \frac{p\,\sigma^2}{X^{\mathrm{T}}X}\right)X$$

can be viewed as an empirical Bayes estimator of θ without any assumption on the form of the density (and in fact there is not even any need to assume there is a density). Hence this Stein-like estimator can be viewed as a reasonable alternative to X from an empirical Bayes perspective regardless of the form of the underlying distribution.

Note that Diaconis and Ylvisaker (1979) introduced the prior $f^{\star n}(\theta)$ as a reasonable conjugate prior for location families since it gives linear Bayes estimators. Strawderman (1992) gave the above empirical Bayes argument. In the normal case the sequence of priors corresponds to that in Sect. 2.2.3 with $\tau^2 = n\,\sigma^2$. The shrinkage factor $p\,\sigma^2$ in the present argument differs from $(p-2)\,\sigma^2$ in the normal case since in this general case we use a "plug-in" estimator of $1/(n+1)$ as opposed to the unbiased estimator (in the normal case) of $1/(\sigma^2+\tau^2)$.

5.2 Baranchik-Type Estimators

In this section, assuming that X has a spherically symmetric distribution with mean vector θ and that loss is $L(\theta, \delta) = \|\delta - \theta\|^2$, we consider estimators of the Baranchik-type, as (2.19) in the normal setting, for different families of densities. In Sect. 5.3, we consider results for general estimators of the form $X + g(X)$.

5.2.1 Variance Mixtures of Normal Distributions

We first consider spherically symmetric densities which are variance mixtures of normal distributions. Suppose

$$f(\|x-\theta\|^2) = \frac{1}{(2\pi)^{p/2}} \int_0^\infty \frac{1}{v^{p/2}} \exp\left\{-\frac{\|x-\theta\|^2}{2v}\right\} dG(v), \tag{5.1}$$

where $G(\cdot)$ is a probability distribution on $(0, \infty)$, i.e., a mixture of $\mathcal{N}_p(\theta, vI)$ distributions with mixing distribution $G(\cdot)$.

Our first result gives a domination result for Baranchik type estimators for such distributions. This result is analogous to Theorem 2.3 in the normal case.

Theorem 5.1 (Strawderman 1974b) *Let X have density of the form* (5.1) *and let*

$$\delta_{a,r}^{B}(X) = \left(1 - a\frac{r(\|X\|^2)}{\|X\|^2}\right)X,$$

where the function $r(\cdot)$ is absolutely continuous. Assume the expectations $E[V]$ and $E[V^{-1}]$ are finite where V has distribution G. Then $\delta_{a,r}^{B}(X)$ is minimax for the loss $L(\theta, \delta) = \|\delta - \theta\|^2$ provided

(1) $0 \le a \le 2(p-2)/E[V^{-1}]$,
(2) $0 \le r(t) \le 1$ *for any* $t \ge 0$,
(3) $r(t)$ *is nondecreasing in* t, *and*
(4) $r(t)/t$ *is nonincreasing in* t.

Furthermore, $\delta_{a,r}^{B}(X)$ dominates X provided the inequalities in (1) *or* (2) *(on a set of positive measure) are strict or $r'(t)$ is strictly increasing on a set of positive measure.*

Proof The proof proceeds by calculating the conditional risk given $V = v$, noting that the distribution of $X|V = v$ is normal $N(\theta, vI_p)$. First note that $E[V] < \infty$ is equivalent to $E_0[\|X\|^2] < \infty$ so that the risk of X is finite. Similarly, it can be seen that $E[V^{-1}] < \infty$ if and only if $E_0[\|X\|^{-2}] < \infty$. Then, thanks to (2), we have $E_0[r^2(\|X\|^2)\|X\|^{-2}] < \infty$. Actually, we will see below that, for any θ, $E_\theta[\|X\|^{-2}] \le E_0[\|X\|^{-2}]$, and hence, $E_\theta[r^2(\|X\|^2)\|X\|^{-2}] < \infty$ which guarantees that the risk of $\delta_{a,r}^{B}(X)$ is finite. Note that, conditionally on V, $\|X\|^2/V$ has a noncentral chi-square distribution with p degrees of freedom and noncentrality parameter $\|\theta\|^2/V$. Hence, since the family of noncentral chi-square distributions have monotone (increasing) likelihood ratios in the noncentrality parameter (and therefore are stochastically increasing), $\|X\|^2/V$ is (conditionally) stochastically decreasing in V and increasing in $\|\theta\|^2$.

Hence,

$$E_\theta\left[\frac{1}{\|X\|^2/V}\right] \le E_0\left[\frac{1}{\|X\|^2/V}\right]$$

and, as a result,

$$\begin{aligned} E_\theta\left[\frac{1}{\|X\|^2}\right] &= E\left[E_\theta\left[\frac{1}{\|X\|^2}\middle| V\right]\right] \\ &= E\left[\frac{1}{V}E_\theta\left[\frac{1}{\|X\|^2/V}\middle| V\right]\right] \\ &\le E\left[\frac{1}{V}E_0\left[\frac{1}{\|X\|^2/V}\right]\right] \\ &= E_0\left[\frac{1}{\|X\|^2}\right]. \end{aligned}$$

This sufficies to establish finiteness of the risk of $\delta^B_{a,r}(X)$. We now deal with the main part of the theorem. Using Corollary 2.1 and Theorem 2.3, we have

$$\begin{aligned}
R(\theta, \delta^B_{a,r}) &= E\{E[\|\delta^B_{a,r}(X) - \theta\|^2 \,|V]\} \\
&= E\left\{E\left[\|X-\theta\|^2 + V^2\left(\frac{a^2 r^2(\|X\|^2)}{V^2\|X\|^2} - \frac{2a(p-2)}{V}\frac{r(\|X\|^2)}{\|X\|^2}\right)\right.\right. \\
&\qquad \left.\left. -4\,aVr'(\|X\|^2)\middle|V\right]\right\} \\
&\leq R(\theta, X) + E\left\{aE\left[\frac{r(\|X\|^2)}{\|X\|^2/V}\middle|V\right]\left(\frac{a}{V} - 2(p-2)\right)\right\}, \qquad (5.2)
\end{aligned}$$

since $r^2(\|X\|^2) \leq r(\|X\|^2)$ and $r'(\|X\|^2) \geq 0$. Now, as a consequence of the above monotone likelihood property, $\|X\|^2/V$ is stochastically decreasing in V. It follows that the conditional expectation in (5.2) is nondecreasing in V since, if $v_1 < v_2$, we have

$$\begin{aligned}
E\left[\frac{r(\|X\|^2)}{\|X\|^2/V}\middle|V = v_1\right] &= E\left[\frac{r\left(v_1\frac{\|X\|^2}{V}\right)}{\|X\|^2/V}\middle|V = v_1\right] \\
&\leq E\left[\frac{r\left(v_2\frac{\|X\|^2}{V}\right)}{\|X\|^2/V}\middle|V = v_1\right] \\
&\leq E\left[\frac{r\left(v_2\frac{\|X\|^2}{V}\right)}{\|X\|^2/V}\middle|V = v_2\right] \\
&= E\left[\frac{r(\|X\|^2)}{\|X\|^2/V}\middle|V = v_2\right].
\end{aligned}$$

The first inequality follows since $r(\|X\|^2)$ is nondecreasing while the second since $r(t)/t$ is nonincreasing and $\|X\|^2/V$ is stochastically decreasing in V. Finally, using the fact that $aV^{-1} - 2(p-2)$ is decreasing in V, and the fact that $E[g(Y)h(Y)] \leq E[g(Y)]E[h(Y)]$ if g and h are monotone in opposite directions, it follows that

$$\begin{aligned}
R(\theta, \delta^B_{a,r}) &\leq R(\theta, X) + aE\left[\frac{Vr(\|X\|^2)}{\|X\|^2}\right]E\left[\frac{a}{V} - 2(p-2)\right] \\
&\leq R(\theta, X) \qquad (5.3)
\end{aligned}$$

by assumption (a). Hence $\delta^B_{a,r}(X)$ is minimax, since X is minimax.

The dominance result follows since the inequality in (5.2) is strict if there is strict inequality in (2) or if $r'(\cdot)$ is strictly positive on a set of positive measure and the inequality in (5.3) is strict if the inequalities in (1) are strict. □

Example 5.1 (The multivariate Student-t distribution) The multivariate Student-t distribution: If V has an inverse Gamma $(v/2, v/2)$ distribution (that is, $V \sim v/\chi_v^2$), then the distribution of X is a multivariate Student-t distribution with ν degrees of freedom. Since $E[V] = E[v/\chi_v^2] = v/(v-2)$ for $v > 2$ and $E[V^{-1}] = E[\chi_v^2/v] = 1$, the conditions of Theorem 5.1 requires $0 \leq a \leq 2(p-2)$ and $v > 2$.

Example 5.2 (Examples of the function $r(t)$) The James-Stein estimator has $r(t) \equiv 1$ and hence satisfies conditions (2), (3) and (4) of Theorem 5.1. Also $r(t) = t/(t+b)$ satisfies these conditions. Similarly, the positive-part James-Stein estimator $\big(1 - a/X^{\mathrm{T}}X\big)_+ X$ is such that

$$r(t) = \begin{cases} t/a & \text{for } 0 \leq t \leq a \\ 1 & \text{for } t \geq a \end{cases}$$

and

$$\frac{r(t)}{t} = \begin{cases} 1/a & \text{for } 0 \leq t \leq a \\ 1/t & \text{for } t \geq a \end{cases}$$

hence also satisfies the conditions (2), (3) and (4) of Theorem 5.1.

It is worth noting, and easy to see, that if the sampling distribution is $N(\theta, I_p)$ and the prior distribution is any variance mixture of normal distributions as in (3.4), in the Baranchik representation of the Bayes estimator (see Corollary 3.1), the function $r(t)/t$ is always nonincreasing. This fact leads to the following observation on the (sampling distribution) robustness of Bayes minimax estimators for a normal sampling distribution. If $\delta^\pi(X) = \big(1 - a\, r(\|X\|^2)/\|X\|^2\big)X$ is a Bayes minimax estimator with respect to a scale mixture of normal priors for a $N(\theta, I_p)$ sampling distribution, and if $r(t)$ is nondecreasing, this Bayes minimax estimator remains minimax for a multivariate-t sampling distribution in Example 5.1 as long as the degrees of freedom is greater than two.

It is also interesting to note that, in general, there will be no uniformly optimal choice of the shrinkage constant "a" in the James-Stein estimator if the mixing distribution $G(\cdot)$ is nondegenerate. The optimal choice will typically depend on $\|\theta\|^2$. This is in contrast to the normal sampling distribution case, where $G(\cdot)$ is degenerate, and where the optimal choice is $a = (p-2)\sigma^2$.

5.2.2 *Densities with Tails Flatter Than the Normal*

In this section we consider the subclass of spherically symmetric densities $f(\|x - \theta\|^2)$ such that, for any $t \geq 0$ for which $f(t) > 0$,

$$\frac{F(t)}{f(t)} \geq c > 0 \tag{5.4}$$

for some fixed positive c, where

$$F(t) = \frac{1}{2}\int_t^\infty f(u)du. \tag{5.5}$$

This class was introduced in Berger (1975) (without the constant 1/2 multiplier).

This class of densities contains a large subclass of variance mixtures of normal densities but also many others. The following lemma gives some conditions which guarantee inclusion or exclusion from the class satisfying (5.4) and (5.5).

Lemma 5.1 *Suppose X has density $f(\|x-\theta\|^2)$.*

(1) *(Mixture of normals). If, for some distribution G on $(0,\infty)$,*

$$f(\|x-\theta\|^2) = \left(\frac{1}{\sqrt{2\pi}}\right)^p \int_0^\infty v^{-p/2}\exp\left\{-\frac{\|x-\theta\|^2}{2v}\right\} dG(v)$$

where $E[V^{-p/2}]$ is finite, E denoting the expectation with respect to G, then $f(\cdot)$ is in the class (5.4) with $c = E[V^{-p/2+1}]/E[V^{-p/2}]$ for $p \geq 3$.

(2) *If $f(t) = h(t)e^{-at}$ with $h(t)$ nondecreasing, then $f(\cdot)$ is in the class (5.4).*

(3) *If $f(t) = e^{-atg(t)}$ where $g(t)$ is nondecreasing and $\lim_{t\to\infty} g(t) = \infty$, then $f(t)$ is not in the class (5.4).*

Proof (1) Applying the definition of F in (5.5) we have

$$\begin{aligned} F(t) &= \frac{1}{2}\int_t^\infty f(u)du \\ &= \frac{1}{2(\sqrt{2\pi})^p}\int_t^\infty \int_0^\infty v^{-p/2}\exp\{-u/2v\}\, dG(v)du \\ &= \frac{1}{(\sqrt{2\pi})^p}\int_0^\infty v^{-p/2+1}\exp\{-t/2v\}\, dG(v). \end{aligned}$$

Hence the ratio in (5.4) equals

$$\begin{aligned} \frac{F(t)}{f(t)} &= \frac{\int_0^\infty v^{-p/2+1}\exp\{-t/2v\}\, dG(v)}{\int_0^\infty v^{-p/2}\exp\{-t/2v\}\, dG(v)} \\ &\geq \frac{\int_0^\infty v^{-p/2+1}\, dG(v)}{\int_0^\infty v^{-p/2} dG(v)} \\ &= \frac{E[V^{-p/2+1}]}{E[V^{-p/2}]}. \end{aligned} \tag{5.6}$$

The inequality follows since the family of densities proportional to the function $v \mapsto v^{-p/2}\exp\{-t/2v\}$ has monotone (increasing) likelihood ratio in the parameter t. Note that if $p \geq 3$, $E[V^{-p/2}] < \infty$ implies $E[V^{-p/2+1}] < \infty$. This completes the proof of (1).

(2) In this case it follows

$$\begin{aligned}\frac{F(t)}{f(t)} &= \frac{\frac{1}{2}\int_t^{\infty} h(u)e^{-au}du}{h(t)e^{-at}} \\ &\geq \frac{1}{2}\int_t^{\infty} e^{-a(u-t)}du \\ &= \frac{1}{2a}.\end{aligned}$$

Hence (5.4) is satisfied with $c = 1/2a$, which proves (2).

(3) In this case it follows

$$\begin{aligned}2\lim_{t\to\infty}\frac{F(t)}{f(t)} &= \lim_{t\to\infty}\frac{\int_t^{\infty}\exp\{-aug(u)\}du}{\exp\{-atg(t)\}} \\ &= \lim_{t\to\infty}\int_t^{\infty}\exp\{-aug(u)+atg(t)\}du \\ &= \lim_{t\to\infty}\int_0^{\infty}\exp\{-a(u+t)g(u+t)+atg(t)\}du \\ &\leq \lim_{t\to\infty}\int_0^{\infty}\exp\{-aug(t)\}du \\ &= \lim_{t\to\infty}\frac{1}{ag(t)} \\ &= 0.\end{aligned}$$

Hence $f(t)$ is not in the class (5.4), which shows (c). □

Part (2) of the lemma shows that densities with tails flatter than the normal (and including the normal) are in the class (5.4), while densities with tails "sufficiently lighter" than the normal are not included. Also the condition in part (3) is stronger than necessary in that it suffices that the condition hold only for all t larger than some positive K. See Berger (1975) for further details and discussion.

Example 5.3 Some specific examples in the class (5.4) include (see Berger 1975 for more details)

(1) $f(t) = K/\cosh t \qquad (c \approx 1/2)$

(2) $f(t) = Kt(1+t^2)^{-m}$ with $m > p/4 \qquad (c = m/2)$

(3) $f(t) = Ke^{-\alpha t-\beta}/(1+e^{-\alpha t-\beta})^2 \qquad (c = \alpha/2)$

(4) $f(t) = Kt^n e^{-t/2}$ for $n \geq 0 \qquad (c = 1)$.

The latter two distributions are known as the logistic type and Kotz , respectively.

The following lemma plays the role of Stein's lemma (Theorem 2.1) for the family of spherically symmetric densities.

Lemma 5.2 *Let X have density $f(\|x-\theta\|^2)$ and let $g(X)$ be a weakly differentiable function such that $E_\theta[|(X-\theta)^{\mathrm{T}}g(X)|] < \infty$. Then*

$$E_\theta[(X-\theta)^{\mathrm{T}}g(X)] = E_\theta\left[\frac{F(\|X-\theta\|^2)}{f(\|X-\theta\|^2)}\operatorname{div} g(X)\right]$$
$$= C\,E^*_\theta\Big[\operatorname{div} g(X)\Big]$$

where $F(t)$ is defined as in (5.5) and E^*_θ denotes expectation with respect to the density

$$x \mapsto \frac{1}{C}\,F(\|x-\theta\|^2)$$

and where it is assumed that

$$C = \int_{\mathbb{R}^p} F(\|x-\theta\|^2)\,dx < \infty\,.$$

Proof Note that the existence of the expectations in Lemma 5.2 will be guaranteed for any function $g(x)$ such that $E_\theta[\|g(x)\|^2] < \infty$ as soon as $E_0[\|X\|^2] < \infty$. The proof will follow along the lines of Sect. 2.4 making use of Stokes' theorem. It follows that

$$E[(X-\theta)^{\mathrm{T}}g(X)]$$
$$= \int_{R^p} (x-\theta)^{\mathrm{T}}g(x)f(\|x-\theta\|^2)\,dx$$
$$= \int_0^\infty \int_{S_{R,\theta}} (x-\theta)^{\mathrm{T}}g(x)\,f(\|x-\theta\|^2)\,d\sigma_{R,\theta}(x)\,dR \quad \text{(by Lemma 1.4)}$$

$$= \int_0^\infty \int_{S_{R,\theta}} \left(\frac{x-\theta}{\|x-\theta\|}\right)^{\mathrm{T}} d\sigma_{R,\theta}(x)\,R\,f(R^2)\,dR$$
$$= \int_0^\infty \int_{B_{R,\theta}} \operatorname{div} g(x)\,dx R\,f(R^2)\,dR \qquad \text{(Stokes' theorem)}$$
$$= \int_{R^p} \operatorname{div} g(x) \int_{\|x-\theta\|}^\infty Rf(R^2)\,dR\,dx \qquad \text{(Fubini's theorem)}$$
$$= \int_{R^p} \operatorname{div} g(x)\,F(\|x-\theta\|^2)\,dx$$

$$= E_\theta\left[\operatorname{div} g(x)\frac{F(\|x-\theta\|^2)}{f(\|x-\theta\|^2)}\right]$$

$$= C\, E_\theta^*\left[\operatorname{div} g(X)\right]$$

□

Now, with the important analog of Stein's lemma in hand, we can extend some of the minimaxity results from the Gaussian setting to the case of spherically symmetric distributions. The following result gives conditions for minimaxity of estimators of the Baranchik type.

Theorem 5.2 *Let X have density $f(\|x-\theta\|^2)$ which satisfies* (5.4) *for some $0 < c < \infty$. Assume also that $E_0[\|X\|^2] < \infty$ and $E_0[\|X\|^{-2}] < \infty$. Let*

$$\delta^B_{a,r}(X) = \left(1 - \frac{a\, r(\|X\|^2)}{\|X\|^2}\right) X$$

where $r(\cdot)$ is absolutely continuous. Then $\delta^B_{a,r}(X)$ is minimax for $p \geq 3$ provided

(1) $0 < a \leq 2c\,(p-2)$,
(2) $0 \leq r(t) \leq 1$, *and*
(3) $r(\cdot)$ *is nondecreasing.*

Furthermore $\delta^B_{a,r}(X)$ dominates X provided both inequalities are strict in (1) *or in* (2) *on a set of positive measure or if $r'(\cdot)$ is strictly positive on a set of positive measure.*

Proof We note that the conditions ensure finiteness of the risk so that Lemma 5.2 is applicable. Hence we have

$$R(\theta, \delta^B_{a,r}) = E_\theta\left[\|X-\theta\|^2 + \frac{a^2 r^2(\|X\|^2)}{\|X\|^2} - 2\,\frac{a\, r(\|X\|^2) X^{\mathrm{T}}(X-\theta)}{\|X\|^2}\right]$$

$$= R(\theta, X) + a\, E_\theta\left[\frac{a r^2(\|X\|^2)}{\|X\|^2} - 2 \operatorname{div}\left(\frac{r(\|X\|^2) X}{\|X\|^2}\right)\frac{F(\|X-\theta\|^2)}{f(\|X-\theta\|^2)}\right]$$

by Lemma 5.2. Therefore the risk difference between $\delta^B_{a,r}(X)$ and X equals

$$\Delta_\theta = a\, E_\theta\left[\frac{a r^2(\|X\|^2)}{\|X\|^2} - \left(\frac{2(p-2) r(\|X\|^2)}{\|X\|^2} + 4\, r'(\|X\|^2)\right)\frac{F(\|X-\theta\|^2)}{f(\|X-\theta\|^2)}\right]$$

$$\leq a\, E_\theta\left[\frac{r(\|X\|^2)}{\|X\|^2}\left(a - 2(p-2)\,\frac{F(\|X-\theta\|^2)}{f(\|X-\theta\|^2)}\right)\right]$$

$$\leq a\, E_\theta\left[\frac{r(\|X\|^2)}{\|X\|^2}(a - 2(p-2)\, c)\right]$$

$$\leq 0.$$

The domination part follows as in Theorem 5.1. □

Theorem 5.2 applies for certain densities for which Theorem 5.1 is not applicable and additionally lifts the restriction that $r(t)/t$ is nonincreasing. However, if the density is a mixture of normals, and both theorems apply, the shrinkage constant "a" given by Theorem 5.1 (with $a = 2(p-2)/E[V^{-1}]$) is strictly larger than that for Theorem 5.2 (with $a = 2(p-2)c$) whenever the mixing distribution $G(\cdot)$ is not degenerate. To see this note that

$$\frac{1}{E[V^{-1}]} > c = \frac{E[V^{-p/2+1}]}{E[V^{-p/2}]}$$

or equivalently

$$E[V^{-p/2}] > E[V^{-1}]E[V^{-p/2+1}]$$

whenever the positive random variable V is non-degenerate. Note also that $E[V^{-1}] < \infty$ whenever $E[V^{-p/2}] < \infty$ and $p \geq 3$.

Example 5.4 (The multivarite Student-t distribution, continued) Suppose X has a p-variate Student-t distribution with ν degrees of freedom as in Example 5.1, so that V has an inverse $Gamma(\nu/2, \nu/2)$ distribution. In this case

$$E[V^{-p/2}] = \frac{2^{p/2}\Gamma\left(\frac{p+\nu}{2}\right)}{\nu^{p/2}\Gamma\left(\frac{\nu}{2}\right)}$$

which is finite for all $\nu > 0$ and $p > 0$.

The bound on the shrinkage constant, "a", in Theorem 5.1 is $2(p-2)$ as shown in Example 5.1, while the bound on "a", in Theorem 5.2, as indicated above, is given by

$$2(p-2)\frac{E[V^{-p/2+1}]}{E[V^{-p/2}]} = 2(p-2)\left(\frac{\nu}{\nu+p-2}\right) < 2(p-2).$$

Hence, for large p, the bound on the shrinkage factor "a" can be substantially less for Theorem 5.2 than for Theorem 5.1 in the case of a multivariate-t sampling distribution. Note that, for fixed p, as ν tends to infinity the smaller bound tends to the larger one (and the Student-t distribution tends to the normal).

Example 5.5 (Examples 5.3 *continued)* All of the distributions in Example 5.3 satisfy the assumptions of Theorem 5.2 (under suitable moment conditions for the second density). It is interesting to note that for the Kotz distribution , the value of c ($= 1$), as in (5.4), doesn't depend on the parameter $n > 0$. Hence the bound on the shrinkage factor "a" is $2(p-2)$ and is also independent of n, indicating a certain distributional robustness of the minimaxity property of Baranchik type estimators with $a < 2(p-2)$.

With additional assumptions on the function $F(t)/f(t)$ in (5.4) (i.e. it is either monotone increasing or monotone decreasing), theorems analogous to Theorem 5.2 can be developed which further improve the bounds on the shrinkage factor "a". These typically may involve additional assumptions on the function $r(\cdot)$. We will see examples of this type in the next section.

5.3 More General Minimax Estimators

We now consider minimaxity of general estimators of the form $X + a\, g(X)$. The initial results rely on Lemma 5.2. The first result follows immediately from this lemma and gives an expression for the risk.

Corollary 5.1 *Let X have a density $f(\|x-\theta\|^2)$ such that $E_0[\|X\|^2] < \infty$ and let $g(X)$ be weakly differentiable and be such that $E_\theta[\|g(X)\|^2] < \infty$.*

Then, for loss $L(\theta, \delta) = \|\delta - \theta\|^2$, the risk of $X + a\, g(X)$ can be expressed as

$$R(\theta, X + a\, g(X)) = R(\theta, X) + E_\theta\big[a^2\, \|g(X)\|^2 + 2\, a\, Q(\|X-\theta\|^2)\ \mathrm{div}\, g(X)\big] \tag{5.7}$$

where

$$Q(\|X-\theta\|^2) = \frac{F(\|X-\theta\|^2)}{f(\|X-\theta\|^2)} \tag{5.8}$$

and where $F(\|X-\theta\|^2)$ is defined in (5.5).

An immediate consequence of Corollary 5.1 when the density of f satisfies (5.4), i.e. $Q(t) \ge c > 0$ for some constant c, is the following.

Corollary 5.2 *Under the assumptions of Corollary* 5.1, *assume that, for some $c > 0$, we have $Q(t) \ge c$ for any $t \ge 0$. Then $X + g(X)$ is minimax and dominates X provided, for any $x \in \mathbb{R}^p$,*

$$\|g(x)\|^2 + 2\, c\ \mathrm{div}\ g(x) \le 0$$

with strict inequality on a set of positive measure.

The following two theorems establish minimaxity results under the assumption that $Q(t)$ is monotone.

Theorem 5.3 (Brandwein et al. 1993) *Suppose X has density $f(\|x-\theta\|^2)$ such that $E_0[\|X\|^2] < \infty$ and that $Q(t)$ in (5.8) is nonincreasing. Suppose there exists a nonpositive function $h(U)$ such that $E_{R,\theta}[h(U)]$ is nondecreasing where $U \sim U_{R,\theta}$ (the uniform distribution on the sphere of radius R centered at θ) and such that $E_\theta[|h(x)|] < \infty$. Furthermore suppose that $g(X)$ is weakly differentiable and also satisfies*

(1) $\operatorname{div} g(X) \le h(X)$,
(2) $\|g(X)\|^2 + 2\,h(X) \le 0$, *and*
(3) $0 \le a \le E_0(\|X\|^2)/p$.

Then $\delta(X) = X + ag(X)$ *is minimax. Also* $\delta(X)$ *dominates* X *provided* $g(\cdot)$ *is nonzero with positive probability and strict inequality holds with positive probability in* (1) *or* (2)*, or both inequalities are strict in* (3).

Proof Note that $g(x)$ satisfies the conditions of Corollary 5.1. Then we have

$$\begin{aligned} R(\theta,\delta) &= R(\theta,X) + a\,E[a\,\|g(X)\|^2 + 2\,Q(\|X-\theta\|^2)\operatorname{div} g(X)] \\ &= R(\theta,X) + a\,E[E_{R,\theta}[a\,\|g(X)\|^2 + 2\,Q\,(R^2)\operatorname{div} g(X)]] \end{aligned}$$

where $E_{R,\theta}$ is as above and E denotes the expectation with respect to the radial distribution. Now, using (1) and (2), we have

$$\begin{aligned} R(\theta,\delta) &\le R(\theta,X) + a\,E[E_{R,\theta}[-2\,a\,h(X) + 2\,Q(R^2)\,h(X)]] \\ &= R(\theta,X) + 2\,a\,E[(a - Q(R^2))\,E_{R,\theta}[-h(X)]] \\ &\le R(\theta,X) + 2\,a\,E[a - Q(R^2)]\,E_\theta[-h(X)] \end{aligned}$$

by the monotonicity assumptions on $E_{R,\theta}[h(\cdot)]$ and $Q(t)$ as well as the covariance inequality.

Hence, since $-h(X) \ge 0$, we have $R(\theta,\delta) \le R(\theta,X)$, provided $0 \le a \le E[Q(R^2)]$. Now $E[Q(R^2)] = E_0[\|X\|^2]/p$ by Lemma 5.3 below, hence δ is minimax. The domination result follows since the additional conditions imply strict inequality between the risks. □

Lemma 5.3 *For any* $k > -p$ *such that* $E[R^{k+2}] < \infty$,

$$E[R^k\,Q(R^2)] = \frac{1}{p+k}\,E[R^{k+2}].$$

In particular, we have

$$E[Q(R^2)] = \frac{1}{p}\,E[R^2] = \frac{1}{p}\,E_0[\|X\|^2]$$

and, for $p \ge 3$,

$$E\left[\frac{Q(R^2)}{R^2}\right] = \frac{1}{p-2}.$$

Proof Recall that the radial density $\varphi(r)$ of $R = \|X-\theta\|$ can be expressed as $\varphi(r) = \sigma(S)r^{p-1}f(r^2)$ where $\sigma(S)$ is the area of the unit sphere S in R^p. By (5.8) and (5.5), we have

$$
\begin{aligned}
E[R^k Q(R^2)] &= \frac{1}{2}\int_{R^p} \|x\|^k \int_{\|x\|^2}^{\infty} f(t)\,dt\,dx \\
&= \frac{1}{2}\int_0^{\infty}\int_{B_{\sqrt{t}}} \|x\|^k dx\; f(t)\,dt \quad \text{by Fubini's theorem} \\
&= \frac{1}{2}\int_0^{\infty}\int_0^{\sqrt{t}} \sigma(S)\, r^{k+p-1} dr\; f(t)\,dt \quad \text{by Lemma 1.4} \\
&= \frac{1}{2}\int_0^{\infty} \sigma(S)\,\frac{t^{(k+p)/2}}{k+p}\, f(t)\,dt \\
&= \frac{1}{k+p}\int_0^{\infty} r^{k+2}\varphi(r)\,dr \quad \text{by the change of variable } t = r^2 \\
&= \frac{1}{k+p}\, E[R^{k+2}].
\end{aligned}
$$

Note that positivity of integrands and $E[R^{k+2}] < \infty$ implies $E[R^k Q(R^2)] < \infty$.

□

The next theorem reverses the monotonicity assumption on $Q(\cdot)$ and changes the condition on the function $h(X)$ which, in turn, bounds the divergence of $g(X)$.

Theorem 5.4 (Brandwein et al. 1993) *Suppose X has a density $f(\|x-\theta\|^2)$ such that $E_0[\|X\|^2] < \infty$ and $E_0[1/\|X\|^2] < \infty$ and such that $Q(t)$ in (5.8) is nondecreasing. Suppose there exists a nonpositive function $h(X)$ such that $E_{R,\theta}\left[R^2 h(U)\right]$ is nonincreasing where $U \sim U_{R,\theta}$ and such that $E_\theta[-h(X)] < \infty$.*

Furthermore suppose that $g(X)$ is weakly differentiable and also satisfies

(1) $\operatorname{div} g(X) \le h(X)$,
(2) $\|g(X)\|^2 + 2\,h(X) \le 0$, *and*
(3) $0 \le a \le \frac{1}{(p-2)E_0(1/\|X\|^2)}$.

Then $\delta(X) = X + a\,g(X)$ is minimax. Also $\delta(X)$ dominates X provided $g(\cdot)$ is nonzero with positive probability and strict inequality holds with positive probability in (1) or (2), or both inequalities are strict in (3).

Proof As in the proof of Theorem 5.3, we have

$$
\begin{aligned}
R(\theta,\delta) &\le R(\theta,X) + 2\,a\,E[(a - Q(R^2))\,E_{R,\theta}[-h(X)]] \\
&= R(\theta,X) + 2\,a\,E\left[\left(\frac{a}{R^2} - \frac{Q(R^2)}{R^2}\right) E_{R,\theta}[-R^2\,h(X)]\right] \\
&\le R(\theta,X) + 2\,a\,E\left[\frac{a}{R^2} - \frac{Q(R^2)}{R^2}\right] E_{R_0,\theta}[-R_0^2\,h(X)]
\end{aligned}
$$

where R_0 is a point such that $a - Q(R_0^2) = 0$, provided such a point exists. Here we have used the version of the covariance inequality that states

$$Ef(X)g(X) \le Ef(X)g(X_0)$$

provided that $g(X)$ is nondecreasing (respectively, nonincreasing) and $f(X)$ changes sign once from $+$ to $-$ (respectively, $-$ to $+$) at X_0. But such a point R_0 does exist provided

$$E\left[\frac{a}{R^2} - \frac{Q(R^2)}{R^2}\right] \le 0$$

since $Q(R^2)$ is nondecreasing.

It follows that $R(\theta, \delta) \le R(\theta, X)$ provided that $aE[\frac{1}{R^2}] \le E[\frac{Q(R^2)}{R^2}]$. However $E[\frac{Q(R^2)}{R^2}] = \frac{1}{p-2}$ by Lemma 5.3 and hence the result follows as in Theorem 5.3. □

Note that the bound on "a" in both of these theorems is strictly larger than the bound in Theorem 5.2 provided $Q(R^2)$ is not constant. This is so since the bound in Theorem 5.2 is based on $c = \inf Q(R^2)$ while, in these results, the bound is equal to a (possibly weighted) average of $Q(R^2)$.

We indicate the utility of these two results by applying them to the James-Stein estimator.

Corollary 5.3 *Let $X \sim f(\|x-\theta\|^2)$ for $p \ge 4$ and let $\delta_b^{JS}(X) = (1 - b/\|X\|^2)X$. Assume also that $E_0[\|X\|^2] < \infty$ and $E_0[1/\|X\|^2] < \infty$. Then $\delta_b^{JS}(X)$ is minimax and dominates X provided either*

(1) *$Q(R^2)$ is nonincreasing and*

$$0 < b < 2(p-2)\frac{E_0\|X\|^2}{p}, \quad or$$

(2) *$Q(R^2)$ is nondecreasing and*

$$0 < b < \frac{2}{E_0(1/\|X\|^2)}.$$

Proof We apply Theorems 5.3 and 5.5 with $g(X) = -[2\,(p-2)/\|X\|^2]X$, $\operatorname{div} g(X) = -2\,(p-2)^2/\|X\|^2 = h(X)$. It follows from Lemma A.5 in Appendix A.10 that when $p \ge 4$, $E_{\theta,R}[h(U)]$ is nondecreasing in R and $E_{\theta,R}[R^2 h(U)]$ is nonincreasing in R. Hence, if $Q(R^2)$ is nonincreasing, Theorem 5.3 implies that

$$\delta_a(X) = X - \frac{2\,(p-2)\,a}{\|X\|^2}X = \delta^{JS}_{2\,(p-2)\,a}(X)$$

is minimax and dominates X provided $0 < a < E_0[\|X\|^2]/p$ or equivalently $0 < 2\,(p-2)\,a < 2\,(p-2)\,E_0(\|X\|^2)/p$ which is (1) with $b = 2\,(p-2)\,a$. Similarly, applying Theorem 5.5 when $Q(R^2)$ is nondecreasing, we find that $\delta_a(X)$ is minimax and dominates X if

$$0 < a < \frac{1}{(p-2)E_0(1/\|X\|^2)}$$

which is (2). □

Example 5.6 (Densities with increasing and decreasing $Q(R^2)$) Note first that variance mixtures of normal distributions have increasing $Q(R^2)$ since, by (5.6) and (5.8), $Q(R^2)$ may be viewed as the expected value of V with respect to a family of distributions with monotone increasing likelihood ratio in $t = R^2$. Note also that the bound for the shrinkage constant "a" in a James-Stein estimator is the same in Corollary 5.3 as it is in Theorem 5.1 for mixtures of normals.

We also note that, if we consider $f(t)$ to be proportional to a density of a positive random variable, then $2\,Q(t)$ is the reciprocal of the hazard rate. There is a large literature on increasing and decreasing hazard rates (see, for example, Barlow and Proschan 1981).

We note that the monotonicity of $Q(t)$ may be determined in many cases by studying the log-convexity or the log-concavity of $f(t)$. In particular, if $\ln f(t)$ is convex (concave), then $Q(t)$ is nondecreasing (nonincreasing). To see this, note that

$$Q(t) = \frac{1}{2}\frac{\int_t^\infty f(u)\,du}{f(t)} = \frac{1}{2}\int_0^\infty \frac{f(s+t)}{f(t)}\,ds$$

and hence $Q(t)$ will be nondecreasing (nonincreasing) if $\frac{f(s+t)}{f(t)}$ is nondecreasing (nonincreasing) in t for each $s > 0$. But, assuming for simplicity that f is differentiable, for any $t \geq 0$ such that $f(t) > 0$,

$$\begin{aligned}
\frac{d}{dt}\left(\frac{f(s+t)}{f(t)}\right) &= \frac{f(t)f'(s+t) - f(s+t)f'(t)}{f^2(t)} \\
&= \frac{f(s+t)}{f(t)}\left[\frac{f'(s+t)}{f(s+t)} - \frac{f'(t)}{f(t)}\right] \\
&= \frac{f(s+t)}{f(t)}\left[\frac{d}{dt}\ln f(s+t) - d/dt \ln f(t)\right].
\end{aligned}$$

This is positive or negative when $\ln f(s+t)$ is convex or concave in t, respectively. For example if X has a Kotz distribution with parameter n, $f(t) \propto t^n e^{-t/2}$. Then $\ln f(t) = K + n\ln t - \frac{t}{2}$ which is concave if $n \geq 0$ and convex if $n \leq 0$. Hence $Q(t)$ is decreasing if $n > 0$ and increasing if $n < 0$. Of course the log-convexity (log-concavity) of $f(t)$ is not a necessary condition for the nondecreasing (nonincreasing) monotonicity of $Q(t)$. Thus, it is easy to check that $f(t) \propto \exp(-t^2)\exp[-1/2\int_0^t \exp(-u^2)\,du]$ leads to $Q(t) = \exp(t^2)$, which is increasing. But $\log f(t)$ is not convex.

An important class of distributions is covered by the following corollary.

Corollary 5.4 *Let $X \sim f(\|x-\theta\|^2)$ for $p \geq 4$ with $f(t) \propto \exp(-\beta t^\alpha)$ where $\alpha > 0$ and $\beta > 0$. Then $\delta_b^{JS}(X) = (1 - b/\|X\|^2)X$ is minimax and dominates X provided either*

(1) $\alpha \leq 1$ *and* $0 < b < \frac{2}{\beta^{1/\alpha}} \frac{p-2}{p} \frac{\Gamma((p+2)/2\alpha)}{\Gamma(p/2\alpha)}$ *or*

(1) $\alpha > 1$ *and* $0 < b < \frac{2}{\beta^{1/\alpha}} \frac{\Gamma(p/2\alpha)}{\Gamma((p-2)/2\alpha)}$.

Proof By the above discussion, $Q(R^2)$ is nonincreasing (nondecreasing) for $\alpha \geq 1$ ($\alpha \leq 1$). Then the result follows from Corollary 5.3 and the fact that

$$E_0[\|X\|^k] = \frac{1}{\beta^{k/2\alpha}} \frac{\Gamma(\frac{p+k}{2\alpha})}{\Gamma(\frac{p}{2\alpha})}$$

for $k > -p$. □

The final theorem of this section gives conditions for minimaxity of estimators of the form $X + a\, g(X)$ for general spherically symmetric distributions. Note that no density is needed for this result which relies on the radial distribution defined in Theorem 4.1.

We first need the following lemma which will play the role of the Stein lemma in the proof of the domination and minimaxity results.

Lemma 5.4 *Let X have a spherically symmetric distribution around θ, and let $g(X)$ be a weakly differentiable function such that $E_\theta[\,|(X-\theta)^{\mathrm{T}} g(X)|\,] < \infty$. Then*

$$E_\theta[(X-\theta)^{\mathrm{T}} g(X)] = \frac{1}{p} E\left[R^2 \int_{B_{R,\theta}} \operatorname{div} g(X)\, d\mathscr{V}_{R,\theta}(X)\right]$$

where E denotes the expectation with respect to the radial distribution and where $\mathscr{V}_{R,\theta}(\cdot)$ is the uniform distribution on $B_{R,\theta}$, the ball of radius R centered at θ.

Proof Let ρ be the radial distribution and according to Theorem 4.1, we have

$$\begin{aligned}
E[(X-\theta)^{\mathrm{T}} g(X)] &= \int_{\mathbb{R}_+} \int_{S_{R,\theta}} (x-\theta)^{\mathrm{T}} g(x)\, d\mathscr{U}_{R,\theta}(x)\, d\rho(R) \\
&= \int_{\mathbb{R}_+} \frac{R}{\sigma_{R,\theta}(S_{R,\theta})} \int_{S_{R,\theta}} \frac{(x-\theta)^{\mathrm{T}}}{\|x-\theta\|} g(x)\, d\sigma_{R,\theta}(x)\, d\rho(R) \\
&= \int_{\mathbb{R}_+} \frac{R}{\sigma_{R,\theta}(S_{R,\theta})} \int_{B_{R,\theta}} \operatorname{div} g(x)\, dx\, d\rho(R) \;\text{ by Stokes' theorem} \\
&= \frac{1}{p} \int_{\mathbb{R}_+} \int_{B_{R,\theta}} \operatorname{div} g(x)\, d\mathscr{V}_{R,\theta}(x)\, R^2 d\rho(R)
\end{aligned}$$

since the volume of $B_{R,\theta}$ equals $\lambda(B_{R,\theta}) = R\sigma_{R,\theta}(S_{R,\theta})/p$. □

Theorem 5.5 (Brandwein and Strawderman 1991a) *Let X have a spherically symmetric distribution around θ, and suppose $E_0[\|X\|^2] < \infty$ and $E_0[1/\|X\|^2] < \infty$. Suppose there exists a nonpositive function $h(\cdot)$ such that $h(X)$ is subharmonic and $E_{R,\theta}[R^2\, h(U)]$ is nonincreasing where $U \sim \mathcal{U}_{R,\theta}$ and such that $E_\theta[|h(x)|] < \infty$. Furthermore suppose that $g(X)$ is weakly differentiable and also satisfies*

(1) div $g(X) \le h(X)$,
(2) $\|g(X)\|^2 + 2\,h(X) \le 0$, *and*
(3) $0 \le a \le \frac{1}{p E_0(1/\|X\|^2)}$.

Then $\delta(X) = X + a\, g(X)$ is minimax. Also $\delta(X)$ dominates X provided $g(\cdot)$ is non-zero with positive probability and strict inequality holds with positive probability in (1) *or* (2), *or both inequalities are strict in* (3).

Proof Using Lemma 5.4 and Conditions (1) and (2), we have

$$\begin{aligned} R(\theta,\delta) &= R(\theta,X) + a\,E_\theta\big[a\,\|g(X)\|^2 + 2\,(X-\theta)^{\mathrm{T}} g(X)\big] \\ &\le R(\theta,X) + 2a\,E_\theta\big[-a\,h(X) + (X-\theta)^{\mathrm{T}} g(X)\big] \\ &= R(\theta,X) + 2a\Big\{E_\theta\big[-a\,h(X)\big] + \frac{1}{p}E\Big[R^2 \int_{B_{R,\theta}} \operatorname{div} g(X)\, d\mathcal{V}_{R,\theta}(X)\Big]\Big\} \\ &\le R(\theta,X) + 2a\Big\{E_\theta\big[-a\,h(X)\big] + \frac{1}{p}E\Big[R^2 \int_{B_{R,\theta}} h(X)\, d\mathcal{V}_{R,\theta}(X)\Big]\Big\}. \end{aligned}$$

By subharmonicity of h (see Appendix A.8 and Sections 1.3 and 2.5 in du Plessis 1970),

$$\int_{B_{R,\theta}} h(X) d\mathcal{V}_{R,\theta}(X) \le \int_{S_{R,\theta}} h(X) d\mathcal{U}_{R,\theta}(X).$$

Hence,

$$\begin{aligned} R(\theta,\delta) &\le R(\theta,X) + 2a\Big\{E_\theta\big[-a\,h(X)\big] + \frac{1}{p}E\Big[R^2 \int_{S_{R,\theta}} h(X)\, d\mathcal{U}_{R,\theta}(X)\Big]\Big\} \\ &= R(\theta,X) + 2a\,E\Big[\Big(\frac{a}{R^2} - \frac{1}{p}\Big)\cdot\Big(-R^2 \int_{S_{R,\theta}} h(X) d\mathcal{U}_{R,\theta}(X)\Big)\Big] \\ &= R(\theta,X) + 2a\,E\Big[\Big(\frac{a}{R^2} - \frac{1}{p}\Big)\big(-E_{R,\theta}[R^2 h(X)]\big)\Big] \\ &\le R(\theta,X) + 2a\,E\Big[\Big(\frac{a}{R^2} - \frac{1}{p}\Big)\Big] E\big[-E_{R,\theta}[R^2 h(X)]\big]. \end{aligned}$$

The last inequality follows from the monotonicity of $E_{R,\theta}[R^2 h(X)]$ and the covariance inequality. Hence $R(\theta, \delta) \le R(\theta, X)$ when $E\big[a/R^2 - 1/p\big] \le 0$ which is equivalent to (3). The domination part follows as before. □

We note that the shrinkage constant in the above result $1/\{pE_0[1/\|X\|^2]\}$ is somewhat smaller than the constant in Theorem 5.4 ($a = 1/\{(p-2)E_0[1/\|X\|^2]\}$), but Theorem 5.5 has essentially no restrictions on the distribution of X aside from moment conditions (which coincide in Theorems 5.4 and 5.5). In particular we do not even assume that a density exists! However there is an additional assumption of subharmonicity of h.

The following useful corollary gives minimaxity for James-Stein estimators in dimension $p \ge 4$ for all spherically symmetric distributions with finite $E_0[\|X\|^2]$ and $E_0[1/\|X\|^2]$.

Corollary 5.5 *Let X have a spherically symmetric distribution with $p \ge 4$, and suppose $E_0[\|X\|^2] < \infty$ and $E_0[1/\|X\|^2] < \infty$. Then*

$$\delta_a^{JS}(X) = \left(1 - \frac{a}{\|X\|^2}\right) X$$

is minimax and dominates X provided

$$0 < a < \frac{1}{pE_0(1/\|X\|^2)}.$$

Proof Here $g(X) = -X/\|X\|^2$ and is weakly differentiable for $p \ge 3$. Then $\operatorname{div} g(X) = -(p-2)/\|X\|^2$ and $\|g(X)\|^2 = 1/\|X\|^2$ so that Conditions (1) and (2) of Theorem 5.5 are satisfied with $h(X) = -\alpha/\|X\|^2$ where $0 \le \alpha \le p-2$. Now the subharmonicity of $h(X)$ and its monotonicity condition hold since it is shown in the appendix that, for $p \ge 4$, $1/\|X\|^2$ is super-harmonic (so that $E_{R,\theta}[1/\|X\|^2]$ is nonincreasing in R) and that $R^2 E_{R,\theta}[1/\|U\|^2]$ is nondecreasing in R.

Furthermore, it is worth noting that $E_{R,\theta}[1/\|U\|^2]$ is nonincreasing in $\|\theta\|$ (see Lemma A.5 and remark that follows). Hence, for any $\theta \in \mathbb{R}^p$, we have $E_\theta[-h(X)] < \infty$ since

$$E_{R,\theta}[1/\|X\|^2] \le E_{R,0}[1/\|X^2\|]$$

so that

$$E_\theta[1/\|X\|^2] \le E_0[1/\|X\|^2] < \infty\,,$$

by assumption. □

Example 5.7 (Nonspherical minimax estimators) In Sect. 2.4.4, we considered estimators which shrink toward a subspace. Theorem 5.5 allows us to show that estimators of this type are minimax for general spherically symmetric distributions.

To be specific, suppose V is a $s < p$ dimensional linear subspace and let

$$\delta_a(X) = P_V X + \left(1 - \frac{a}{\|X - P_V X\|^2}\right)(X - P_V X).$$

As in the proof of Theorem 2.6, it can be shown that the risk of $\delta_a(X)$ equals

$$R(\theta, \delta_a(X)) = E_{\nu_1}[\|Y_1 - \nu_1\|^2] + E_{\nu_2}\left[\left\|\left(1 - \frac{a}{\|Y_2\|^2}\right)Y_2 - \nu_2\right\|^2\right], \tag{5.9}$$

where Y_1, Y_2, ν_1 and ν_2 are as in Theorem 2.6.

In the present case, Y_2 has a spherically symmetric distribution about ν_2 of dimension $p - s$. Hence, by Theorem 5.5,

$$\begin{aligned} E(\theta, \delta_a(X)) &\leq E_{\nu_1}[\|Y_1 - \nu_1\|^2] + E_{\nu_2}[\|Y_2 - \nu_2\|^2] \\ &= E_\theta \|X - \theta\|^2 \\ &= R(\theta, X), \end{aligned}$$

provided $p - s \geq 4$ and

$$0 < a < \frac{1}{(p - s)\, E_0[1/\|X - P_V X\|^2]}.$$

5.4 Bayes Estimators

In this section, we consider (generalized) Bayes estimators of the location vector $\theta \in \mathbb{R}^p$ of a spherically symmetric distribution. More specifically let X be a random vector in $\mathbb{R}^p$ with density $f(\|x - \theta\|^2)$ and let $\pi(\theta)$ be a prior density. Under quadratic loss $\|\delta - \theta\|^2$, the (generalized) Bayes estimator of θ is the posterior mean given by

$$\delta_\pi(X) = X + \frac{1}{m(X)} \int_{\mathbb{R}^p} (\theta - X)\, f(\|X - \theta\|^2)\, \pi(\theta)\, d\theta \tag{5.10}$$

where $m(x)$ is the marginal

$$m(x) = \int_{\mathbb{R}^p} f(\|x - \theta\|^2) \pi(\theta)\, d\theta. \tag{5.11}$$

Recall from Sect. 3.1.1 that, in the normal case (that is, $f(t) \propto \exp(-t/2\sigma^2)$ with σ^2 known) the superharmonicity of $\sqrt{m(x)}$ is a sufficient condition for minimaxity of $\delta_\pi(X)$. This superharmonicity is implied by that of $m(x)$ and in turn by that of $\pi(\theta)$. While in the nonnormal case minimaxity has been studied by many authors (for example, see Strawderman (1974b); Berger (1975); Brandwein and Strawderman (1978, 1991a)) relatively few results on minimaxity of Bayes estimators are known. The primary technique to establish minimaxity is through a Baranchik representation of the form $(1 - a\,r(\|X\|^2)/\|X\|^2)X$. The minimaxity conditions are essentially those developed in Theorems 5.3 and 5.4 and most of the derivations are in the context of variance mixtures of normals. See Strawderman (1974b), Maruyama (2003a) and Fourdrinier et al. (2008) for more discussion and results on Bayes estimation in this setting.

The main difficulty in using Theorem 5.1 with mixtures of normals densities for the sampling distribution is to prove the monotonicity (and boundedness) properties of the function $r(\cdot)$. Maruyama (2003a) and Fourdrinier et al. (2008) consider priors which are mixtures of normals as well. Their main condition for obtaining minimaxity of the corresponding Bayes estimator is that the mixing density g of the sampling distribution has monotone nondecreasing likelihood ratio when considered as a scale parameter family. In Fourdrinier et al. (2008), explicit use is made of that monotone likelihood ratio property for the mixing (possibly generalized) density h of the prior distribution.

The main result of Fourdrinier et al. (2008) is the following. Consult that paper for the somewhat technical proof.

Theorem 5.6 *Let X be a random vector in $\mathbb{R}^p$ ($p \geq 3$) distributed as a variance mixture of multivariate normal distributions with density*

$$f(x) = \int_0^\infty \frac{1}{(2\pi v)^{p/2}} \exp\left(-\frac{1}{2} \frac{\|x-\theta\|^2}{v} \right) g(v)\, dv \tag{5.12}$$

where g is the density of a known nonnegative random variable V. Let π be a (generalized) prior with density of the form

$$\pi(\theta) = \int_0^\infty \frac{1}{(2\pi t)^{p/2}} \exp\left(-\frac{1}{2} \frac{\|\theta\|^2}{t} \right) h(t)\, dt \tag{5.13}$$

where h is a function from $\mathbb{R}_+$ into $\mathbb{R}_+$ such that this integral exists.

Assume that the mixing density g is such that

$$E[V] = \int_0^\infty v\, g(v)\, dv < \infty \text{ and } E[V^{-p/2}] = \int_0^\infty v^{-p/2}\, g(v)\, dv < \infty. \tag{5.14}$$

Assume also that the mixing function h of the (possibly improper) prior density π is absolutely continuous and satisfies

$$\lim_{t\to\infty} \frac{h(t)}{t^\beta} = c \tag{5.15}$$

for some $\beta < p/2 - 1$ *and some* $0 < c < \infty$*. Assume, finally, that* h *and* g *have monotone increasing likelihood ratio when considered as a scale parameter family.*

Then, if there exist $K > 0$*,* $t_0 > 0$ *and* $\alpha < 1$ *such that*

$$h(t) \le K\, t^{-\alpha} \quad \text{for } 0 < t < t_0, \tag{5.16}$$

the (generalized or proper) Bayes estimator δ_h *with respect to the prior distribution corresponding to the mixing function* h *is minimax provided that* β *satisfies*

$$-(p-2)\left[\frac{E[V^{-p/2+1}]}{E[V]E[V^{-p/2}]} - \frac{1}{2}\right] \le \beta. \tag{5.17}$$

For priors with mixing distribution h satisfying (5.16) and (5.17) an argument as in Maruyama (2003a) using Brown (1979) and a Tauberian theorem suggests that the resulting generalized Bayes estimator is admissible if $\beta \le 0$. Maruyama and Takemura (2008) have verified this under additional conditions which imply, in the setting of Theorem 5.6, that $E_\theta[\|X\|^3] < \infty$.

As an illustration assume that the sampling distribution is a p-variate Student-t with n_0 degrees of freedom which corresponds to the inverse gamma mixing density $(n_0/2, n_0/2)$, that is, to $g(v) \propto v^{-(n_0+2)/2}\exp(-n_0/2v)$. Let the prior be a Student-t distribution with n degrees of freedom, that is, with mixing density $h(t) \propto t^{-(n+2)/2}\exp(-n/2t)$. It is clear that Conditions (5.14) and (5.15) are satisfied with $n_0 \ge 7$. It is also clear that Condition (5.16) holds for any $\alpha < 1$. Finally a simple calculation shows that

$$\frac{E[V^{-p/2+1}]}{E[V]E[V^{-p/2}]} = \frac{n_0 - 2}{p + n_0 - 2}$$

so that Condition (5.17) reduces to

$$n \le (p-2)\left[\frac{2(n_0 - 2)}{p + n_0 - 2} - 1\right] - 2.$$

Note that, as $n > 0$, this condition holds if and only if $p \ge 5$ and

$$n_0 \ge 3 + p\frac{p}{p-4}.$$

Other examples (including generalized priors) can be found in Fourdrinier et al. (2008).

In the following, we consider broader classes of spherically symmetric distributions which are not restricted to variance mixtures of normals. Minimaxity of generalized Bayes estimators is obtained for unimodal spherically symmetric superharmonic priors $\pi(\|\theta\|^2)$ under the additional assumption that the Laplacian of $\pi(\|\theta\|^2)$ is a nondecreasing function of $\|\theta\|^2$. The results presented below are derived in Fourdrinier and Strawderman (2008a). An interesting feature is that their approach does not rely on the Baranchik representation used in Maruyama (2003a) and Fourdrinier et al. (2008). Note, however, that the superharmonicity property of the priors implies that the corresponding Bayes estimators cannot be proper (see Theorem 3.2).

First note that, for any prior $\pi(\theta)$, the Bayes estimator in (5.10) can be written as

$$\delta_\pi(X) = X + \frac{\nabla M(X)}{m(X)} \tag{5.18}$$

where, for any $X \in \mathbb{R}^p$,

$$M(x) = \int_{\mathbb{R}^p} F(\|x-\theta\|^2)\, \pi(\theta)\, d\theta$$

with F given in (5.5). Thus $\delta_\pi(X)$ has the general form $\delta_\pi(X) = X + g(X)$ (with $g(X) = \nabla M(X)/m(X)$). If the density $f(\|x-\theta\|^2)$ is as in Sect. 5.2.1, that is, such $F(t)/f(t) \geq c > 0$ for some fixed positive constant c, then Corollary 5.2 applies and $\delta_\pi(X) = X + g(X) = X + \nabla M(X)/m(X)$ is minimax provided, for any $x \in \mathbb{R}^p$,

$$2\,c\ \operatorname{div} g(x) + \|g(x)\|^2 \leq 0.$$

In particular, it follows that if

$$2\,c\,\frac{\Delta M(x)}{m(x)} - 2\,c\,\frac{\nabla M(x)\cdot\nabla m(x)}{m^2(x)} + \frac{\|\nabla M(x)\|^2}{m^2(x)} \leq 0 \tag{5.19}$$

and

$$E_\theta\left[\left\|\frac{\nabla M(X)}{m(X)}\right\|^2\right] < \infty,$$

δ_π is minimax.

For a spherically symmetric prior $\pi(\|\theta\|^2)$, the main result of Fourdrinier and Strawderman (2008a) is the following.

Theorem 5.7 *Assume that X has a spherically symmetric distribution in $\mathbb{R}^p$ with density $f(\|x-\theta\|^2)$. Assume that $\theta \in \mathbb{R}^p$ has a superharmonic prior $\pi(\|\theta\|^2)$ such that $\pi(\|\theta\|^2)$ is nonincreasing and $\Delta\pi(\|\theta\|^2)$ is nondecreasing in $\|\theta\|^2$. Assume also that*

$$E_\theta\left[\left\|\frac{\nabla M(X)}{m(X)}\right\|^2\right] < \infty.$$

Then the Bayes estimator δ_π is minimax under quadratic loss provided that $f(t)$ is log-convex, $c = \frac{F(0)}{f(0)} > 0$ and

$$\int_0^\infty f(t) t^{p/2} dt \le 4c \int_0^\infty -f'(t) t^{p/2} dt < \infty. \tag{5.20}$$

To prove Theorem 5.7 we need some preliminary lemmas whose proofs are given in Appendix A.9. Note first that it follows from the spherical symmetry of π that, for any $x \in \mathbb{R}^p$, $m(x)$ and $M(x)$ are functions of $t = \|x\|^2$. Then, setting

$$m(x) = m(t) \quad \text{and} \quad M(x) = M(t),$$

we have

$$\nabla m(x) = 2m'(t)\, x \quad \text{and} \quad \nabla M(x) = 2\, M'(t)\, x. \tag{5.21}$$

Lemma 5.5 *Assume that $\pi'(t) \le 0$, for any $t \ge 0$. Then we have $M'(t) \le 0$, for any $t \ge 0$.*

Lemma 5.6 *For any $x \in \mathbb{R}^p$,*

$$x \cdot \nabla m(x) = -2 \int_0^\infty H(u,t)\, u^{p/2}\, f'(u)\, du$$

and

$$x \cdot \nabla M(x) = \int_0^\infty H(u,t)\, u^{p/2}\, f(u)\, du$$

where, for $u \ge 0$ and for $t \ge 0$,

$$H(u,t) = \lambda(B) \int_{B_{\sqrt{u},x}} x \cdot \theta\, \pi'(\|\theta\|^2)\, d\mathcal{V}_{\sqrt{u},x}(\theta) \tag{5.22}$$

and $\mathcal{V}_{\sqrt{u},x}$ is the uniform distribution on the ball $B_{\sqrt{u},x}$ of radius $\sqrt{u}$ centered at x and $\lambda(B)$ is the volume of the unit ball.

Lemma 5.7 *For any $t \ge 0$, the function $H(u,t)$ in (5.22) is nondecreasing in u provided that $\Delta\pi(\|\theta\|^2)$ is nondecreasing in $\|\theta\|^2$.*

Lemma 5.8 *Let $h(\|\theta - x\|^2)$ be a unimodal density and let $\psi(\theta)$ be a symmetric function. Then*

$$\int_{\mathbb{R}^p} x \cdot \theta\, \psi(\theta)\, h(\|\theta - x\|^2)\, d\theta \ge 0$$

as soon as ψ is nonnegative.

Proof (Proof of Theorem 5.7*)* By the superharmonicity of $\pi(\|\theta\|^2)$, we have $\Delta M(x) \leq 0$ for all $x \in \mathbb{R}^p$ so that by (5.19), it suffices to prove that

$$-2\,c\,\nabla M(x)\cdot\nabla m(x) + \|\nabla M(x)\|^2 \leq 0 \tag{5.23}$$

for all $x \in \mathbb{R}^p$. Since m and M are spherically symmetric, by (5.21), (5.23) reduces to $-2cM'(t)m'(t) + \big(M'(t)\big)^2 \leq 0$ where $t = \|x\|^2$. Since $M'(t) \leq 0$ by Lemma 5.5, (5.23) reduces to $-2cm'(t) + M'(t) \geq 0$ or, by (5.21), to $-2\,c\,x\cdot\nabla m(x) + x\cdot\nabla M(x) \geq 0$ or, by Lemma 5.6, to

$$4cE\left[H(u,t)\frac{f'(u)}{f(u)}\right] + E[h(u,t)] \geq 0, \tag{5.24}$$

where E denotes the expectation with respect to the density proportional to $u^{p/2} f(u)$. Since, by assumption, $\Delta\pi(\|\theta\|^2)$ is nondecreasing in $\|\theta\|^2$, $H(u,t)$ is nondecreasing in u by Lemma 5.7. Furthermore $f'(u)/f(u)$ is nondecreasing by log-convexity of f so that (5.16) is satisfied as soon as

$$4\,c\,E[H(u,t)]\,E\left[\frac{f'(u)}{f(u)}\right] + E[H(u,t)] \geq 0. \tag{5.25}$$

Finally, as $\pi'(\|\theta\|^2) \leq 0$ by assumption, Lemma 5.2 guarantees that $H(u,t) \leq 0$ (note that $V_{\sqrt{u},x}$ has a unimodal density) and hence (5.25) reduces to

$$4cE\left[\frac{f'(u)}{f(u)}\right] + 1 \leq 0$$

which is equivalent to (5.20). □

Several examples of priors and sampling distributions which satisfy the assumptions of Theorem 5.7 are given in Fourdrinier and Strawderman (2008a). We briefly summarize these.

Example 5.8 (Priors related to the fundamental harmonic prior) Let $\pi(\|\theta\|^2) = \left(\dfrac{1}{A + \|\theta\|^2}\right)^c$ with $A \geq 0$ and $0 \leq c \leq \frac{p}{2} - 1$.

Example 5.9 (Mixtures of priors) Let $(\pi_\alpha)_{\alpha\in A}$ be a family of priors such that the assumptions of Theorem 5.7 are satisfied for any $\alpha \in A$. Then any mixture of the form $\displaystyle\int_A \pi_\alpha(\|\theta\|^2)\,dH(\alpha)$ where H is a probability measure on A satisfies these assumptions as well. For instance, Example 5.8 with $c = 1$, $p \geq 4$, $A = \alpha$ and the gamma density $\alpha \longmapsto \dfrac{\beta^{1-v}}{\Gamma(1-v)}\alpha^{-v}e^{-\beta\alpha}$ with $\beta > 0$ and $0 < v < 1$ leads to the prior

$$\|\theta\|^{-2-v}\,e^{\beta\|\theta\|^2}\,\Gamma(v,\beta\|\theta\|^2),$$

where

$$\Gamma(v, y) = \int_y^\infty e^{-x} x^{v-1}\, dx$$

is the complement of the incomplete gamma function.

Example 5.10 (Variance mixtures of normals) Let

$$\pi(\|\theta\|^2) = \int_o^\infty \left(\frac{u}{2\pi}\right)^{p/2} \exp\left(\frac{-u\|\theta\|^2}{2}\right) h(u)du$$

a mixture of normals with respect to the inverse of the variance . As soon as, for any $u > 0$,

$$\frac{uh'(u)}{h(u)} \leq -2,$$

the prior $\pi(\|\theta\|^2)$ satisfies the assumptions of Theorem 5.7. Note that the priors in Example 5.10 arise as such a mixture with $h(u) \propto \alpha u^{k-p/2-1} \exp(-A/2u)$.

Other examples can be given and a constructive approach is proposed in Fourdrinier and Strawderman (2008a).

We now give examples of sampling distributions which satisfy the assumptions of Theorem 5.7.

Example 5.11 (Variance mixtures of normals) Let

$$f(t) = (2\pi)^{-p/2} \int_0^\infty v^{-p/2} \exp\left(-\frac{t}{2v}\right) h(v)\, dv$$

where h is a mixing density and let V be a nonnegative random variable with density proportional to $f(t)$. If $E[V^{-p/2}] < \infty$ and $E[V]\, E[V^{-p/2}]/E[V^{-p/2+1}] < 2$ then the sampling density f satisfies the assumptions of Theorem 5.7.

Example 5.12 (Densities proportional to $e^{-\alpha t^\beta}$) Let

$$f(t) = K\, e^{-\alpha t^\beta}$$

where $\alpha > 0$, $\frac{1}{2} < \beta \leq 1$ and K is the normalizing constant. Then the sampling density f satisfies the assumptions of Theorem 5.7 as soon as β is in a neighborhood of the form $]1 - \epsilon, 1]$ with $\epsilon > 0$. However, note that these are not satisfied when $\beta = 1/2$.

Fourdrinier and Strawderman (2008a) give other examples with densities proportional to $e^{-\alpha t + \beta\varphi(t)}$ where φ is a convex function.

5.5 Shrinkage Estimators for Concave Loss

In this section we consider improved shrinkage estimators for loss functions that are concave functions of squared error loss. The basic results are due to Brandwein and Strawderman (1980, 1991b) and we largely follow the method of proof in the later paper. The general nature of the main result is that (under mild conditions) if an estimator can be shown to dominate X under squared error loss then the same estimator, with a suitably altered shrinkage constant, will dominate X for a loss which is a concave function of squared error loss.

Let X have a spherically symmetric distribution around θ, and let $g(X)$ be a weakly differentiable function. The estimators considered are of the form

$$\delta(X) = X + ag(X). \tag{5.26}$$

The loss functions are of the form

$$L(\theta, \delta) = \ell(||\delta - \theta||^2), \tag{5.27}$$

where $\ell(\cdot)$ is a differentiable nonnegative, nondecreasing concave function (so that, in particular $\ell'(\cdot) \geq 0$).

One basic tool needed for the main result is Theorem 5.5, and the other is the basic property of the concave function $\ell(\cdot)$ that $\ell(t + a) \leq \ell(t) + a\ell'(t)$.

The following result shows that shrinkage estimators that improve on X for squared error loss also improve on X for concave loss provided the shrinkage constant is adjusted properly.

Theorem 5.8 (Brandwein and Strawderman 1991a) *Let X have a spherically symmetric distribution around θ, let $g(X)$ be a weakly differentiable function, and let the loss be given by* (5.27).

Suppose there exists a subharmonic function $h(\cdot)$ such that $E_{\theta,R}[R^2 h(U)]$ is nonincreasing where $U \sim \mathcal{U}_{R,\theta}$. Furthermore suppose that the function $g(\cdot)$ satisfies $E^*_\theta[||g(X)||^2] < \infty$ and also satisfies

(1) div $g(x) \leq h(x)$, *for any $x \in \mathbb{R}^p$,*
(2) $\|g(x)\|^2 + 2h(x) \leq 0$, *for any $x \in \mathbb{R}^p$, and*
(3) $0 \leq a \leq \frac{1}{pE^*_0(1/\|X\|^2)}$,

where E^*_θ refers to the expectation with respect to the distribution whose Radon-Nikodyn derivative with respect to the distribution of X is proportional to $\ell'(||X - \theta||^2)$.

Then $\delta(X) = X + ag(X)$ is minimax. Also $\delta(X)$ dominates X provided $g(\cdot)$ is non-zero with positive probability and strict inequality holds with positive probability in (1) *or* (2), *or both inequalities are strict in* (3).

Proof Note, by concavity of $\ell(\cdot)$ and the usual identity

$$\begin{aligned} R(\theta, \delta) &= E_\theta[\ell(||\delta(X) - \theta||^2)] \\ &\le E_\theta[\ell(||X - \theta||^2)] \\ &\qquad + E_\theta[\ell'(||X - \theta||^2)(a^2||g(X)||^2 + 2a(X - \theta)'g(X))]. \end{aligned}$$

Hence, the difference in risk, $R(\theta, \delta) - R(\theta, X)$ is bounded by

$$\begin{aligned} R(\theta, \delta) - R(\theta, X) &\le E_\theta[\ell'(||X - \theta||^2)(a^2||g(X)||^2 + 2a(X - \theta)'g(X))] \\ &= E^*_\theta[(a^2||g(X)||^2 + 2a(X - \theta)'g(X))] \\ &\le 0, \end{aligned}$$

by Theorem 5.5 applied to the distribution corresponding to E^*_θ. □

Chapter 6
Estimation of a Mean Vector for Spherically Symmetric Distributions II: With a Residual

6.1 The General Linear Model Case with Residual Vector

In this chapter, we consider the canonical form of the general linear model introduced in Sect. 4.5 when a residual vector U is available. Recall that (X, U) is a random vector around $(\theta, \mathbf{0})$ (such that dim X = dim $\theta = p$ and dim U = dim $\mathbf{0} = k$) with a spherically symmetric distribution, that is, $(X, U) \sim SS_{p+k}(\theta, \mathbf{0})$. Estimation of θ under quadratic loss $\|\delta - \theta\|^2$ parallels the normal situation presented in Sects. 2.3 and 2.4 where $X \sim \mathcal{N}_p(\theta, \sigma^2 I_p)$ (with σ^2 known) and the estimators of θ are of the form $\delta(X) = X + \sigma^2 g(X)$. In the case where σ^2 is unknown (see Sect. 2.4.3), the corresponding estimators are

$$\delta(X) = X + \frac{S}{k+2}\, g(X)$$

where $S \sim \sigma^2 \chi_k^2$ independent of X. Note that, when $(X^{\mathrm{T}}, U^{\mathrm{T}})^{\mathrm{T}} \sim \mathcal{N}((\theta^{\mathrm{T}}, 0^{\mathrm{T}})^{\mathrm{T}}, \sigma^2 I_{p+k})$, $S = \|U\|^2$. This most basic case of the general linear model suggests considering improved shrinkage estimators of the form

$$\delta(X) = X + \frac{\|U\|^2}{k+2}\, g(X) \tag{6.1}$$

for some function g from $\mathbb{R}^p$ into $\mathbb{R}^p$. In this section,

$$\sigma^2 = \mathrm{Var}(X_i) = \mathrm{Var}(U_i) = \frac{1}{p} E_\theta[\|X - \theta\|^2] = \frac{1}{k} E_\theta[\|U\|^2] = \frac{1}{p+k} E[R^2],$$

where $R = (\|X - \theta\|^2 + \|U\|^2)^{1/2}$, can be considered as known or unknown. When σ^2 is unknown, $\|U\|^2/k$ is an unbiased estimator of σ^2. Also, when σ^2 is unknown, it is perhaps preferable to use the invariant loss $\|\delta - \theta\|^2/\sigma^2$ since the estimator X

D. Fourdrinier et al., *Shrinkage Estimation*, Springer Series in Statistics,
https://doi.org/10.1007/978-3-030-02185-6_6

has constant risk p and is minimax for this loss provided the variance of X is finite, while the minimax risk for the loss $\|\delta - \theta\|^2$ is infinite. Note that domination of an estimator under one of these losses implies domination under the other.

When σ^2 is known, estimators of the form $\delta(X) = X + \sigma^2 g(X)$ can be used and we will contrast these estimators with estimators (6.1) in the next section. One advantage of the estimators in (6.1) is that they share a striking robustness property, namely that, if $\|g(X)\|^2 + 2\,\mathrm{div}\, g(X) \leq 0$, then $X + g(X)\,\|U\|^2/(k+2)$ dominates X for any spherically symmetric distribution of (X, U). In particular, the form of the density may not be known and indeed there is no need that a density exists. The proof of this robustness property is given below and follows closely that of Cellier and Fourdrinier (1995).

Assuming the risk of X is finite (i.e., $E_\theta[\|X - \theta\|^2] = E_0[\|X\|^2] < \infty$) the risk of $\delta(X)$ is finite if and only if $E_\theta[\|U\|^4 \|g(X)\|^2] < \infty$ and the difference in risk between $\delta(X)$ and X is

$$\begin{aligned}\Delta(\theta) &= R(\theta, \delta) - R(\theta, X) \\ &= E_\theta\left[2\,(X-\theta)^{\mathrm{T}} g(X)\,\frac{\|U\|^2}{k+2} + \|g(X)\|^2 \frac{\|U\|^4}{(k+2)^2}\right]. \end{aligned} \tag{6.2}$$

The cross product term, that is, the first term in the right-hand side of (6.2) will be analyzed as in the normal case. The following is the key adaptation of Stein's identity.

Lemma 6.1 *(Stein type lemma for the general linear model: Cellier and Fourdrinier 1995) Assume that* $(X, U) \sim SS(\theta, \mathbf{0})$ *where* $\dim X = \dim\theta = p$ *and* $\dim U = \dim \mathbf{0} = k$. *Then, for any weakly differentiable function* g *from* $\mathbb{R}^p$ *into* $\mathbb{R}^p$ *such that*

$$E_\theta\big[|(X-\theta)^{\mathrm{T}} g(X)|\big] < \infty,$$

we have

$$E_\theta\big[(X-\theta)^{\mathrm{T}} g(X)\,\|U\|^2\big] = E_\theta\left[\mathrm{div}\, g(X)\,\frac{\|U\|^4}{k+2}\right]. \tag{6.3}$$

Proof We will show that, conditionally on the radius $R = \|X - \theta\|^2 + \|U\|^2$, (6.3) holds. First, conditionally on R, the left-hand side of (6.3) is expressed as (see Corollary 4.2)

$$\begin{aligned} E_{R,\theta}\big[(X-\theta)^{\mathrm{T}} g(X)\,\|U\|^2\big] &= \int_{S_{R,\theta}} (x-\theta)^{\mathrm{T}} g(x)\,\|u\|^2\, d\mathcal{U}_{R,\theta}(x,u) \\ &= \int_{S_{R,\theta}} (x-\theta)^{\mathrm{T}} g(x)\,(R^2 - \|x-\theta\|^2)\, d\mathcal{U}_{R,\theta}(x,u) \\ &= \int_{B_{R,\theta}} (x-\theta)^{\mathrm{T}} g(x)\, C_R^{p,k} (R^2 - \|x-\theta\|^2)^{k/2}\, dx \end{aligned} \tag{6.4}$$

since, according to (4.4), X given R has density

$$\psi_{R,\theta}(x) = C_R^{p,k} (R^2 - \|x - \theta\|^2)^{k/2-1} \, \mathbb{1}_{B_{R,\theta}}(x)$$

with

$$C_R^{p,k} = \frac{\Gamma((p+k)/2)}{\Gamma(k/2)} \frac{R^{2-(p+k)}}{\pi^{p/2}}.$$

Now, note that

$$(R^2 - \|x - \theta\|^2)^{k/2} (x - \theta) = \nabla\gamma(x)$$

where

$$\gamma(x) = \frac{-(R^2 - \|x - \theta\|^2)^{k/2+1}}{k+2}.$$

Hence, using the classical identity

$$\big(\nabla\gamma(x)\big)^{\mathrm{T}} g(x) = \operatorname{div}\big(\gamma(x)\, g(x)\big) - \gamma(x)\operatorname{div} g(x),$$

it follows from (6.4) that

$$E_{R,\theta}\big[(X-\theta)^{\mathrm{T}} g(X)\, \|U\|^2\big] = A + B \tag{6.5}$$

where

$$A = C_R^{p,k} \int_{B_{R,\theta}} \operatorname{div}\big(\gamma(x)\, g(x)\big) dx \tag{6.6}$$

and

$$B = C_R^{p,k} \int_{B_{R,\theta}} -\gamma(x) \operatorname{div} g(x) dx. \tag{6.7}$$

Applying Stokes' theorem to the integral in (6.6) gives

$$A = C_{S_R^{p,k}} \int_{S_{R,\theta}} \gamma(x)\, g(x) \frac{x-\theta}{\|x-\theta\|} \, d\sigma_{R,\theta}(x) = 0 \tag{6.8}$$

since, for any $x \in S_{R,\theta}$, $\gamma(x) = 0$. The B term in (6.7) can be expressed as

$$B = \int_{B_{R,\theta}} \operatorname{div} g(x) \frac{(R^2 - \|x-\theta\|^2)^2}{k+2} \psi_{R,\theta}(x)\, dx = E_{R,\theta}\left[\operatorname{div} g(X) \frac{\|U\|^4}{k+2}\right]$$

and, finally, the lemma follows from (6.4), (6.5) and (6.8). □

As a consequence of Lemma 6.1, we can derive a sufficient condition of domination of $\delta(X) = X + \|U\|^2/(k+2)g(X)$ over the usual estimate X.

Theorem 6.1 *Let* $(X, U) \sim SS_{p+k}(\theta, 0)$ *and the loss be given by* $\|\delta - \theta\|^2$. *Assume that* $E_\theta[\|X\|^2] < \infty$ *and* $E_\theta[\|U\|^4 \, \|g(X)\|^2] < \infty$. *Then an unbiased estimator of the risk difference* $\Delta(\theta)$ *in* (6.2) *between* $\delta(X) = X + g(X)\,\|U\|^2/(k+2)$ *and* X *is*

$$[2\,\mathrm{div}\, g(X) + \|g(X)\|^2]\,\frac{\|U\|^4}{(k+2)^2}. \tag{6.9}$$

A sufficient condition for domination of $\delta(X)$ *over* X *is that, for any* $x \in \mathbb{R}^p$,

$$2\;\mathrm{div}\, g(x) + \|g(x)\|^2 \le 0 \tag{6.10}$$

with strict inequality on a set a positive measure on $\mathbb{R}^p$.

Proof The proof of (6.9) follows immediately from (6.3) and (6.2). The domination condition (6.10) is a direct consequence of (6.9). □

Remark 6.1 The addition of the residual term U in the estimate yields an interesting and strong robustness property. Note that the hypotheses in Theorem (6.1) are independent of the radial distribution and are consequently valid for any spherically symmetric distribution. This is in contrast with the results of Sect. 6.2 which require conditions on the radial distribution.

Differential expressions that lead to risk domination results, such as in Theorem 6.1, have been extended to spherical and ellipitical location models by several authors (see, for example, Cellier et al. 1989, Chou and Strawderman 1990, Brandwein and Strawderman 1980, Brandwein and Strawderman 1991a, Cellier and Fourdrinier 1995, Fourdrinier et al. 2003, Fourdrinier et al. 2006, Kubokawa 1991, Maruyama 2003a, and Fourdrinier and Strawderman 2008a,b). A notable aspect of many of the papers, in the presence of a residual vector U, is the development of robust estimators in the sense that they are minimax for a wide class of spherically symmetric distributions (see particularly, for example, Cellier et al. 1989, Cellier and Fourdrinier 1995, and Fourdrinier et al. 2006).

The improved estimators in Sect. 5.3,without residual vector, require two critical hypotheses. The first is the superharmonicity condition on an auxillary function h such that $\|g\|^2/2 \le -h \le -\mathrm{div}\, g$. Secondly these estimators require the assumption that the function $R \to R^2 E_{R,\theta}[h]$ is nonincreasing. In contrast, the conditions for improvement of the improved estimator with the residual term included share the same set of hypotheses as the general Stein type estimators in the normal case (see Sect. 2.3). As a result, estimators which dominate X (through the differential inequality) in the normal case dominate X simultaneously for all spherically symmetric distributions (subject to the finiteness of the risk). At this point, we will focus on the so-called robust James-Stein estimators rather than discussing general examples as in Sect. 2.3.

Consider

$$\delta^a_{RJS}(X) = \left(1 - \frac{a}{\|X\|^2}\frac{\|U\|^2}{k+2}\right) X$$

where a is a positive constant which is of the form (6.1) with $g(X) = -aX/\|X\|^2$. Note this is the shrinkage in the basic James-Stein estimator in (2.13) with $\sigma^2 = 1$. Using the divergence calculation of this $g(X)$ from (2.16), the unbiased estimator of the risk difference implied by (6.9) is,

$$\left(a^2 - 2\,a(p-2)\right) \frac{1}{\|X\|^2} \frac{\|U\|^4}{(k+2)^2},$$

and so it follows that domination occurs for $0 < a < 2(p-2)$, and the optimal constant a (i.e., with minimum risk) is $a = p - 2$. Note that this optimal a is independent of the sampling distribution and yields improvement on X for any spherically symmetric distribution. Hence the best a also has a nice robust optimality property.

An alternative approach to the results of this section can be based on the approach used in Lemma 5.2 where a density is assumed, that is, $(X, U) \sim f(\|x-\theta\|^2 + \|U\|^2)$. This second approach has been used by many authors in this and more general settings. For spherically symmetric distributions with a density it is essentially related to the above method. A statement of this connection is given at the end of this section. The proof is provided in the Appendix. Thus a straightforward adaptation of the proof of Lemma 5.2 leads to

$$\begin{aligned} E_\theta\left[(X-\theta)^{\mathrm{T}} g(X)\,\|U\|^2\right] &= E_\theta\left[\frac{F(\|X-\theta\|^2 + \|U\|^2)}{f(\|X-\theta\|^2 + \|U\|^2)}\,\mathrm{div}_X g(X)\,\frac{\|U\|^2}{k+2}\right] \\ &= C\,E^*_\theta\left[\mathrm{div}_X g(X)\,\frac{\|U\|^2}{k+2}\right] \end{aligned} \tag{6.11}$$

where C and E^*_θ are defined in Lemma 5.2. Similarly

$$\begin{aligned} E_\theta\left[\|g(X)\|^2 \frac{\|U\|^4}{(k+2)^2}\right] &= E_\theta\left[U^{\mathrm{T}}\left(U\,\frac{\|U\|^2}{(k+2)^2}\,\|g(X)\|^2\right)\right] \\ &= E_\theta\left[\frac{F(\|X-\theta\|^2 + \|U\|^2)}{f(\|X-\theta\|^2 + \|U\|^2)}\,\mathrm{div}_U\left(U\|U\|^2\right)\|g(X)\|^2\right] \\ &= E_\theta\left[\frac{F(\|X-\theta\|^2 + \|U\|^2)}{f(\|X-\theta\|^2 + \|U\|^2)}\,\frac{\|U\|^2}{k+2}\,\|g(X)\|^2\right] \\ &= C\,E^*_\theta\left[\frac{\|U\|^2}{k+2}\,\|g(X)\|^2\right]. \end{aligned} \tag{6.12}$$

Hence the difference in risk between $X+g(X)\,\|U\|^2/(k+2)$ and X can be written as

$$C\,E_\theta^*\left[\left(2\operatorname{div} g(X)+\|g(X)\|^2\right)\frac{\|U\|^2}{k+2}\right]. \tag{6.13}$$

Note that the normalizing constant

$$C=\int_{\mathbb{R}^p\times\mathbb{R}^k} F(\|x-\theta\|^2+\|u\|^2)\,dx\,du\,. \tag{6.14}$$

can be expressed, through a straightforward application of the Fubini theorem, as

$$C=\frac{1}{p+k}\int_0^\infty r^2\,h(r)\,dr \tag{6.15}$$

where $h(r)$ is the radial density. Thus C is the common variance of each coordinate of (X,U). Therefore it follows from (6.13) that condition (6.10) is sufficient for the minimaxity of the estimator $X+g(X)\,\|U\|^2/(k+2)$, provided we treat the density $f(\cdot)$ as fixed and known, which implies implicitly that σ^2 is known. Alternatively, if

$$(X,U)\sim\frac{1}{\sigma^{p+k}}\,f\left(\frac{\|x-\theta\|^2+\|u\|^2}{\sigma^2}\right)$$

where σ^2 is unknown, and the loss is $\|\delta-\theta\|^2/\sigma^2$, then X is minimax simultaneously for all such families where $E_\theta[\|X\|^2]<\infty$. Hence (6.10) implies simultaneous minimaxity for the entire class as well.

6.1.1 More General Estimators

In this section, we give results for a more general class of estimators of θ of the form $\delta=\delta(X,\|U\|^2)$. The loss will be invariant squared error loss, i.e.

$$\eta\,\|\delta-\theta\|^2\,, \tag{6.16}$$

where $\eta=1/\sigma^2$, so that the risk is

$$R(\theta,\eta,\delta)=E_{\theta,\eta}\left[\eta\,\|\delta(X,U)-\theta\|^2\right], \tag{6.17}$$

where $E_{\theta,\eta}$ denotes the expectation with respect to the density (6.33) with $\eta=1/\sigma^2$. For the rest of this section, we assume

$$E_{\theta,\eta}\left[\|X-\theta\|^2\right]<\infty, \tag{6.18}$$

which guarantees that the standard estimator X has finite risk and is minimax. As $\delta(X, \|U\|^2)$ can be written as $\delta(X, \|U\|^2) = X + g(X, \|U\|^2)$, the finiteness of its risk is guaranteed by

$$E_{\theta,\eta}\left[\|g(X, \|U\|^2)\|^2\right] < \infty . \tag{6.19}$$

A version of the following lemma can be found in Fourdrinier et al. (2003). Its proof follows closely the pattern of (6.11) and (6.12).

Lemma 6.2 *Assume that the function $g(x, \|u\|^2)$ is weakly differentiable from $\mathbb{R}^{p+k}$ into $\mathbb{R}^p$. Then*

$$\eta\, E_{\theta,\eta}\left[(X-\theta)^{\mathrm{T}} g(X, \|U\|^2)\right] = C\, E^*_{\theta,\eta}\left[\mathrm{div}_X g(X, \|U\|^2)\right], \tag{6.20}$$

where $E^*_{\theta,\eta}$ is the expectation with respect to the density

$$\frac{\eta^{p+k}}{C} F\Big(\eta\left(\|x-\theta\|^2 + \|u\|^2\right)\Big), \tag{6.21}$$

provided either of the above expectations exists.

Similarly, for any weakly differentiable function h from $\mathbb{R}^{p+k}$ into $\mathbb{R}^p$,

$$\eta E_{\theta,\eta}\left[U^{\mathrm{T}} h(X, U)\right] = C\, E^*_{\theta,\eta}\left[\mathrm{div}_U h(X, U)\right] , \tag{6.22}$$

provided either of these expectations exists.

Thanks to Lemma 6.2, an expression of the risk difference between $\delta(X, \|U\|^2)$ and X is given in the following proposition.

Proposition 6.1 *Assume that $E_{\theta,\eta}\left[\|g(X, U)\|^2\right] < \infty$. The risk difference between $\delta(X, \|U\|^2) = X + g(X, \|U\|^2)$ and X equals*

$$\mathscr{R}(\theta, \eta, \delta) - \mathscr{R}(\theta, \eta, X) = C\, E^*_{\theta,\eta}\left[\mathscr{O} g(X, \|U\|^2)\right],$$

where

$$\mathscr{O} g(X, \|U\|^2) = 2\, \mathrm{div}_X g(X, \|U\|^2) + \frac{k-2}{\|U\|^2} \|g(X, \|U\|^2)\|^2 + 2 \frac{\partial}{\partial S} \|g(X, S)\|^2 \bigg|_{S=\|U\|^2} . \tag{6.23}$$

Proof A straightforward calculation of the risk difference gives

$$\Delta(\theta,\eta) = \eta\, E_{\theta,\eta}\left[2\,(X-\theta)^{\mathrm{T}} g(X,\|U\|^2) + \|g(X,\|U\|^2)\|^2\right]$$
$$= \eta\, E_{\theta,\eta}\left[2\,(X-\theta)^{\mathrm{T}} g(X,\|U\|^2) + U^{\mathrm{T}}\frac{U}{\|U\|^2}\|g(X,\|U\|^2)\|^2\right].$$

Using Lemma 6.2 on each term in the brackets, we obtain

$$\Delta(\theta,\eta) = C\, E^*_{\theta,\eta}\left[2\,\mathrm{div}_X g(X,\|U\|^2) + \mathrm{div}\left(\frac{U}{\|U\|^2}\,\|g(X,\|U\|^2)\|^2\right)\right]$$
$$= C\, E^*_{\theta,\eta}\Bigg[2\,\mathrm{div}_X g(X,\|U\|^2) + \frac{k-2}{\|U\|^2}\,\|g(X,\|U\|^2)\|^2$$
$$+\frac{U^{\mathrm{T}}}{\|U\|^2}\,\nabla_U \|g(X,\|U\|^2)\|^2\Bigg]$$

by the divergence formula. Finally expressing the gradient gives

$$\Delta(\theta,\eta) = C\, E^*_{\theta,\eta}\left[2\,\mathrm{div}_X g(X,\|U\|^2) + \mathrm{div}\left(\frac{U}{\|U\|^2}\,\|g(X,\|U\|^2)\|^2\right)\right]$$
$$= C\, E^*_{\theta,\eta}\Bigg[2\,\mathrm{div}_X g(X,\|U\|^2) + \frac{k-2}{\|U\|^2}\,\|g(X,\|U\|^2)\|^2$$
$$+2\,\frac{\partial}{\partial S}\|g(X,S)\|^2\bigg|_{S=\|U\|^2}\Bigg].$$

□

This result will be used in Sect. 6.3 to develop generalized Bayes minimax estimators. An easy corollary applicable to Baranchik type estimators of the form

$$\left(1 - a\,r\left(\frac{\|X\|^2}{S}\right)\frac{S}{\|X\|^2}\right)X \tag{6.24}$$

is the following. The proof is left to the reader.

Corollary 6.1 *The estimator* (6.24) *dominates* X *simultaneously for all spherically symmetric distributions* $SS_{p+k}(\theta, 0)$ *for which* $E^*_{\theta,\eta}[\|X\|^2] < \infty$ *under loss* (6.16) *provided*

(a) $0 < a \le 2\,(p-2)$,
(b) $0 \le r(\cdot) \le 1$, *and*
(c) $r(\cdot)$ *is nondecreasing.*

6.1.2 A Link Between Expectations with Respect to E^*_{θ,σ^2} and E_{θ,σ^2}

We mentioned above that the two approaches to the results of this section are connected. Here is a lemma, whose proof is postponed to Appendix A.6, which makes explicit this connection thanks to a link between expectations with respect to E^*_{θ,σ^2} and E_{θ,σ^2}.

Lemma 6.3 (Fourdrinier and Strawderman 2015) *For any function γ defined on $\mathbb{R}^p \times \mathbb{R}_+$ and for any $\theta \in \mathbb{R}^p$, we have*

$$\sigma^2 \, C \, E^*_{\theta,\sigma^2}\Big[\gamma\big(X, \|U\|^2\big)\Big] = E_{\theta,\sigma^2}\left[\frac{1}{2}\,\frac{1}{\|U\|^{k-2}} \int_0^{\|U\|^2} \gamma(X,s)\, s^{k/2-1}\, ds\right], \tag{6.25}$$

provided these expectations exist, where C is defined in (6.14).

6.2 A Paradox Concerning Shrinkage Estimators

In this section, we contrast the result of the previous section and Sect. 5.2. We continue our study of the problem of estimating the mean vector θ of a spherically symmetric distribution when the scale σ^2 is known but when a residual vector U is available.

In Sect. 5.2, we studied the important class of improved estimators, the James-Stein estimators $\delta^a_{JS}(X) = \big(1 - a\,\sigma^2/\|X\|^2\big)X$. The previous section provided an alternative class of robust James-Stein estimators, that is, $\delta^a_{RJS}(X,U) = \big(1 - a/\|X\|^2\;\|U\|^2/(k+2)\big)X$. In this section, we show that there often exist situations where $\delta^{p-2}_{RJS}(X,U)$ dominates $\delta^a_{JS}(X)$ simultaneously for all a and hence that the use of the residual vector U to estimate σ^2 may be superior to using its known value. This phenomenon seems paradoxical in the sense that the risk behavior of an estimator may be improved by substituting an estimate for a known quantity. This phenomenon adds to the attractiveness of the robust James-Stein class by demonstrating not only domination of the usual estimator X simultaneously for all spherically symmetric distributions, but also domination of the usual James-Stein estimators in many cases. A similar paradox was found in the context of goodness of fit testing by Wells (1990). The results of this section are Fourdrinier and Strawderman (1996) and Fourdrinier et al. (2004).

Note that the paradox cannot occur in the case of a normal distribution since by the Rao-Blackwell theorem, when σ^2 is known in the normal case, X is a complete sufficient statistic so that the conditional expectation of $\delta^a_{RJS}(X,U)$ given X reduces to $\delta^{ak/(k+2)}_{JS}(X)$ which dominates $\delta^a_{RJS}(X,U)$. Note also that, if the paradox holds for one value of σ^2 for a particular family, it holds for all values of σ^2 by the scale equivariance of $\delta^a_{RJS}(X, U)$ and, therefore, holds for any scale mixture. Hence, as the normal distribution arises as a mixture of uniform distributions on spheres, and also as a mixture of uniform distributions on balls, the paradox cannot occur for these distributions as well.

For ease of presentation, it is convenient to define the general estimator $\delta^a_\alpha(X, U) = \big(1 - a\|U\|^{2\alpha}/\|X\|^2\big)X$ for $\alpha = 0$ or 1 and to assume $\sigma^2 = 1$. Note that, for $\alpha = 0$, $\delta^a_0 = \delta^a_{JS}$ and, for $\alpha = 1$, $\delta^a_1 = \delta^{a/(k+2)}_{RJS}$. As in Sect. 6.1, we assume the finiteness of the risk of X (i.e., $E_0[\|X\|^2] < \infty$) and it is clear that the finiteness of the risk of $\delta^a_\alpha(X, U)$ is guaranteed as soon as $E_\theta\big[\|U\|^{2\alpha}/\|X\|^2\big] < \infty$. Under that condition, the following proposition yields the risk of δ^a_α.

Proposition 6.2 *Let the loss be* $\|\delta - \theta\|^2$*. The risk of* δ^a_α *equals*

$$R(\delta^a_\alpha, \theta) = E_0[\|X\|^2] + a^2\, E_\theta\left[\frac{\|U\|^{4\alpha}}{\|X\|^2}\right] - 2\,a\frac{p-2}{k+2\alpha}\; E_\theta\left[\frac{\|U\|^{2(\alpha+1)}}{\|X\|^2}\right].$$

Proof The risk calculation is a straightforward extension of the one in Lemma 6.1, with $g(x, s) = s^\alpha x/\|x\|^2$. □

It is easy to deduce from Lemma 6.2 that, for any $\theta \in \mathbb{R}^p$, the constant a for which the risk of δ^a_α is minimum is

$$a(\theta) = \frac{p-2}{k+2\alpha}\,\frac{E_\theta\big[\frac{\|U\|^{2(\alpha+1)}}{\|X\|^2}\big]}{E_\theta\big[\frac{\|U\|^{4\alpha}}{\|X\|^2}\big]}.$$

The corresponding risk is

$$R\big(\delta^{a(\theta)}_\alpha, \theta\big) = E_0\big[\|X\|^2\big] - \left(\frac{p-2}{k+2\alpha}\right)^2 \frac{\Big(E_\theta\big[\frac{\|U\|^{2(\alpha+1)}}{\|X\|^2}\big]\Big)^2}{E_\theta\big[\frac{\|U\|^{4\alpha}}{\|X\|^2}\big]}. \tag{6.26}$$

We already noticed in Sect. 6.1 that, for $\alpha = 1$, the optimal a does not depend on θ and equals $\frac{p-2}{k+2}$, which can also be easily seen from the above expression. For $\alpha = 0$, the optimal a depends on θ and equals

$$a(\theta) = \frac{p-2}{k}\,\frac{E_\theta\big[\frac{\|U\|^2}{\|X\|^2}\big]}{E_\theta\big[\frac{1}{\|X\|^2}\big]}. \tag{6.27}$$

Then the paradox will occur if, for any $a \geq 0$, $R\big(\delta_1^{(p-2)/(k+2)}, \theta\big) < R\big(\delta^a_0, \theta\big)$ and will certainly occur if $R\big(\delta_1^{(p-2)/(k+2)}, \theta\big) < R\big(\delta_0^{a(\theta)}, \theta\big)$ with $a(\theta)$ as in (6.27). By (6.26), this is equivalent to

$$\left(\frac{p-2}{k}\right)^2 \frac{\left(E_\theta\left[\frac{\|U\|^2}{\|X\|^2}\right]\right)^2}{E_\theta\left[\frac{1}{\|X\|^2}\right]} < \left(\frac{p-2}{k+2}\right)^2 E_\theta\left[\frac{\|U\|^4}{\|X\|^2}\right],$$

that is, to

$$\frac{\left(E_\theta\left[\frac{\|U\|^2}{\|X\|^2}\right]\right)^2}{E_\theta\left[\frac{\|U\|^4}{\|X\|^2}\right]E_\theta\left[\frac{1}{\|X\|^2}\right]} < \left(\frac{k}{k+2}\right)^2. \tag{6.28}$$

Expression (6.28) is a general condition for the paradox to occur. Fourdrinier and Strawderman (1996) developed a series of bounds for the quantities in the left-hand side of (6.28). However the resulting sufficient condition was complex and could be verified in a limited number of cases, the primary example being the Student Student-t distribution case. Subsequently Fourdrinier et al. (2004) developed an effective approach to deal with the expectations in (6.28) for the case of mixtures of normals.

Assume that (X, U) has a scale mixture of normals distribution with the representation

$$(X, U)|\,(Z = z) \sim \mathcal{N}_{p+k}\big((\theta, 0), z\, I_{p+k}\big) \tag{6.29}$$

where Z is a positive random variable. For model (6.29), expressions of the expectations in (6.28) are given by the following lemma.

Lemma 6.4 *Assume that (X, U) is a scale mixture of normals as in (6.29) and that $p \geq 3$. Let $q > -k/2$ and assume that $E[Z^{q-1}] < \infty$. Then we have*

$$E_\theta\left[\frac{\|U\|^{2q}}{\|X\|^2}\right] = 2^q\,\frac{\Gamma(k/2+q)}{\Gamma(k/2)}\;E\left[Z^{q-1} f_p\left(\frac{\|\theta\|^2}{Z}\right)\right]$$

where $f_p(\gamma) = E[Y^{-1}]$ for a random variable Y having a noncentral chi-square distribution with p degrees of freedom and noncentrality parameter γ.

Proof Note that X and U are independent conditional on Z and $(\|U\|^2/Z)|\,Z \sim \chi^2_k(0)$ and $(\|X\|^2/Z)|\,Z \sim \chi^2_p\big(\|\theta\|^2/Z\big)$. Hence we can write

$$\begin{aligned}
E_\theta\left[\frac{\|U\|^{2q}}{\|X\|^2}\,\middle|\, Z\right] &= E[\|U\|^{2q}\,|\,Z]E_\theta\left[\frac{1}{\|X\|^2}\,\middle|\, Z\right] \\
&= Z^{q-1}E\left[\left(\frac{\|U\|^2}{Z}\right)^q \,\middle|\, Z\right]E_\theta\left[\frac{Z}{\|X\|^2}\,\middle|\, Z\right] \\
&= Z^{q-1}2^q\,\frac{\Gamma(k/2+q)}{\Gamma(k/2)}\,f_p\left(\frac{\|\theta\|^2}{Z}\right)
\end{aligned}$$

since $q > -k/2$. Now use the fact that f_p is bounded if $p \geq 3$ and $E[Z^{q-1}] < \infty$ and uncondition to complete the proof. □

It follows directly from Lemma 6.4 for $q = 0, 1, 2$ that (6.28) is equivalent to

$$H_Z(\lambda) = \frac{\left(E\left[f_p(\lambda^2/Z)\right]\right)^2}{E\left[Zf_p(\lambda^2/Z)\right]E\left[Z^{-1}f_p(\lambda^2/Z)\right]} < \frac{k}{k+2} \tag{6.30}$$

for all $\lambda = \|\theta\| \geq 0$.

Alternatively note that

$$H_Z(\lambda) = \left(E_\lambda[W]E_\lambda[W^{-1}]\right)^{-1} \tag{6.31}$$

where W is a positive random variable with density

$$h_\lambda(w) = c(\lambda) f_p(\lambda^2 w) g(w)$$

where g is the density of $V = Z^{-1}$ and $c(\lambda)$ is a normalizing constant. Then (6.28) can also be expressed as

$$E_\lambda[W]E_\lambda[W^{-1}] > 1 + \frac{2}{k} \tag{6.32}$$

for all $\lambda \geq 0$.

The following main result shows that the paradox occurs for any nondegenerate mixture of normals when the dimension of the residual vector U is sufficiently large.

Theorem 6.2 *Assume that (X, U) is a scale mixture of normals as in (6.29), with Z nondegenerate, $E[Z] < \infty$ and $E[Z^{-1}] < \infty$. Then, for any $p \geq 3$, there exists a positive integer k_0 such that, for any integer $k \geq k_0$, the optimal robust James-Stein estimator $\delta^{(p-2)}_{RJS}$ $(= \delta_1^{(p-2)/(k+2)})$ simultaneously dominates all James-Stein estimators δ^a_{JS} $(= \delta^a_0)$.*

Proof Setting $\bar{H} = \sup_{\lambda \geq 0} H_Z(\lambda)$, Condition (6.30) reduces to $k > 2\frac{\bar{H}}{1-\bar{H}}$. From (6.31) we know (by covariance inequality) that $H_Z(\lambda) \leq 1$ with equality if and only if W is degenerate, that is, if and only if Z is degenerate, which corresponds to the normal case. Then $\bar{H} \leq 1$ and we only need to show that $\bar{H} < 1$ since H_Z is continuous, and hence $\bar{H}$ does not depend on k.

Now it can be shown (see Lemma 3 in Fourdrinier et al. 2004) that

$$\begin{aligned}
\lim_{\lambda \to \infty} H_Z(\lambda) &= \left(\lim_{\lambda \to \infty} E_\lambda[W] \lim_{\lambda \to \infty} E_\lambda[W^{-1}]\right)^{-1} \\
&= \left(\frac{1}{E[Z]} \cdot \frac{E[Z^2]}{E[Z]}\right)^{-1} \\
&= \frac{(E[Z])^2}{E[Z^2]} \\
&< 1,
\end{aligned}$$

for $p \geq 3$ and nondegenerate Z. Since $H_Z(\lambda) < 1$ for all λ and $\lim_{\lambda\to\infty} H_Z(\lambda) < 1$, this implies $\bar{H} < 1$. □

The necessity of nondegeneracy of Z is explicit in the proof of Theorem 6.2. Therefore the paradox occurs only in the case of nondegenerate mixtures of normals and not in the normal case, as previously noted.

Outside the class of mixtures of normals little is known. In the case where the radial distribution is concentrated on two points, Fourdrinier and Strawderman (1996) show that the paradox can occur for suitable weights. Showing the existence of the paradox in other families of spherically symmetric distributions is an open question.

6.3 Bayes Estimators

Let (X, U) be a random vector in $\mathbb{R}^p \times \mathbb{R}^k$ with density

$$\frac{1}{\sigma^{p+k}} f\left(\frac{\|x-\theta\|^2 + \|u\|^2}{\sigma^2}\right), \tag{6.33}$$

where $\theta \in \mathbb{R}^p$ and $\sigma \in \mathbb{R}_+\backslash\{0\}$ are unknown. We assume throughout that $p \geq 3$.

We consider generalized Bayes estimators of θ for priors of the form

$$\pi(\|\theta\|^2)\,\eta^b\,, \tag{6.34}$$

where $\eta = 1/\sigma^2$, under the invariant quadratic loss in (6.16).

We first show that, under weak moment conditions, such generalized Bayes estimators are robust in the sense that they do not depend on the underlying density f. Furthermore, we exhibit a large class of superharmonic priors π for which these generalized Bayes estimators dominate the usual minimax estimator X for the entire class of densities (6.33). Hence this subclass of estimators has the extended robustness property of being simultaneously generalized Bayes and minimax for the entire class of spherically symmetric distributions.

Note that, paralleling Sect. 4.5, the above model arises as the canonical form of the general linear model $Y = V\beta + \varepsilon$ where V is a $(p+k) \times p$ design matrix, β is a $p \times 1$ vector of unknown regression coefficients, and ε is an $(p+k) \times 1$ error vector with spherically symmetric density $f(\|\varepsilon\|^2/\sigma^2)/\sigma^{p+k}$.

In the following, for a real valued function $g(x, \|u\|^2)$, we denote by $\nabla_x g(x, u)$ and $\Delta_x g(x, \|u\|^2)$ the gradient and the Laplacian of $g(x, \|u\|^2)$ with respect to the variable x. Analogous notations hold with respect to the variable u. When $g(x, \|u\|^2)$ is a vector valued function, $\text{div}_x g(x, \|u\|^2)$ is the divergence with respect to x (here $\dim g(x, \|u\|^2) = \dim\, x$).

As previously noted, Stein (1981) shows that, when the density in (6.33) is normal with known scale, the generalized Bayes estimator corresponding to a prior

$\pi(\theta)$, for which the square root of the marginal density $m(x)$ is superharmonic, is minimax under the loss (6.16). Fundamental to this result is the development of an unbiased estimator of risk based on a differential expression involving $m(x)$ which has become a basic tool in proving minimaxity.

Another line of research pertinent to this section is the development of Bayes and generalized Bayes minimax estimators. In the case of a normal distribution with known scale, see Sect. 3.1, When the scale is unknown, see Sect. 3.4. For variance mixture of normals and, more generally, for spherically symmetric distributions with no residual, see Sect. 5.4.

Maruyama (2003b) showed that, for spherically symmetric distributions with a residual vector U and unknown scale parameter, the generalized Bayes estimator with respect to a prior on θ and η proportioned to $\eta^b \, \|\theta\|^{-a}$ is independent of the density f and is minimax under conditions on a and b and under weak moment conditions (see also Maruyama and Takemura 2008 and Maruyama and Strawderman 2005, 2009).

The goal of this section is to extend the phenomenon in Maruyama (2003b) to a broader class of priors of the form $\pi(\|\theta\|^2)\,\eta^b$ with $\pi(\|\theta\|^2)$ superharmonic. In particular, in Sect. 6.3.1, we show that the generalized Bayes estimators do not depend on the density f under weak moment conditions and, in Sect. 6.3.2, we prove that these generalized Bayes estimators are minimax provided the prior $\pi(\|\theta\|^2)$ is superharmonic and its Laplacian $\Delta\pi(\|\theta\|^2)$ is a nondecreasing function of $\|\theta\|^2$, under conditions on b, p and k.

In the case of a known scale parameter, Fourdrinier and Strawderman (2008a) studied the same class of priors $\pi(\|\theta\|^2)$ and proved minimaxity of generalized Bayes estimators for a large subclass of unimodal densities. We rely strongly on the techniques of that paper, as presented in Sect. 5.4.

6.3.1 *Form of the Bayes Estimators*

In Sect. 3.2 generalized Bayes estimators for the normal setting with an unknown variance were discussed. In this subsection we extend the normal case to the spherical setting with a residual vector, that is when the sampling distribution is of the form of (6.33). In the normal setting the generalized Bayes estimators in (3.25) were of the form $X - \frac{r(F)}{F} X$ where $F = ||X||^2/\|U\|^2$. In the more general setting of this subsection the shrinkage function is not a function of only F but is a more general function of both X and $\|U\|^2$ as in (3.17).

The results of this subsection and the next closely follow the developments in Fourdrinier and Strawderman (2010). We will see that for the sampling distribution in (6.33) and priors of the form (6.34), the generalized Bayes estimators do not depend on the density (6.33); more precisely their expressions depend only on π and b provided that

$$\int_0^\infty f(\tau)\, \tau^{(p+k)/2+b+1}\, d\tau < \infty, \tag{6.35}$$

which is equivalent to

$$E_{0,1}\Big[(\|X\|^2 + \|U\|^2)^{2(b+2)}\Big] < \infty .$$

Proposition 6.3 *For a prior of the form* (6.34) *and loss* (6.16), *the generalized Bayes estimator* $\delta(X, \|U\|^2) = X + g(X, \|U\|^2)$ *is such that, for any* $(x, u) \in \mathbb{R}^p \times \mathbb{R}^k$,

$$g(x, \|u\|^2) = \frac{\int_{\mathbb{R}^p} \frac{\theta - x}{(\|x-\theta\|^2+\|u\|^2)^{(p+k)/2+b+2}}\, \pi(\|\theta\|^2)\, d\theta}{\int_{\mathbb{R}^p} \frac{1}{(\|x-\theta\|^2+\|u\|^2)^{(p+k)/2+b+2}}\, \pi(\|\theta\|^2)\, d\theta}, \tag{6.36}$$

provided (6.35) *holds and* (6.36) *exists and hence* $\delta(X, \|U\|^2)$ *does not depend on* $f(\cdot)$.

Note that $g(x, \|u\|^2)$ arises as

$$\frac{\nabla_x M(x, \|u\|^2)}{m(x, \|u\|^2)},$$

where $m(x, \|u\|^2)$ is the marginal associated to π and the density

$$\varphi\left(\|x-\theta\|^2 + \|u\|^2\right) \propto \frac{1}{(\|x-\theta\|^2 + \|u\|^2)^{(p+k)/2+b+2}}, \tag{6.37}$$

and M is the marginal associated to ϕ with

$$\phi(t) = \frac{1}{2}\int_t^\infty \varphi(v)\, dv . \tag{6.38}$$

Therefore, for each fixed u, $\delta(X, u) = X + g(X, u)$ with $g(X, u)$ in (6.36) can be interpreted as the Bayes estimator of θ under the density φ and the prior π for fixed scale parameter $\|u\|$ under the loss$\|\delta - \theta\|^2$. This observation will be important in the next subsection since it will allow us to use results in Sect. 5.4 (Fourdrinier and Strawderman 2008a) which are developed for the case of known scale parameter.

Finally, note that existence of (6.36) will be guaranteed by the stronger finiteness risk condition developed in the proof of Theorem 6.3. More generally, it suffices that π be locally integrable and have tails that do not grow too fast at infinity. In particular, superharmonic priors are locally integrable and have bounded tails.

Proof of Proposition 6.3. The Bayes estimator under loss (6.16) is

$$\delta(X, \|U\|^2) = \frac{E[\eta\,\theta|X, U]}{E[\eta|X, U]} = X + g(X, \|U\|^2),$$

with, for any $(x, u) \in \mathbb{R}^p \times \mathbb{R}^k$,

$$\begin{aligned} g(x, \|u\|^2) &= \frac{E[\eta\,(\theta - x) \mid x, u]}{E[\eta|x, u]} \\ &= \frac{\int_0^\infty \int_{\mathbb{R}^p} \eta\,(\theta - x)\,\eta^{(p+k)/2}\, f(\eta\,(\|x-\theta\|^2 + \|u\|^2))\,\pi(\|\theta\|^2)\,\eta^b\, d\theta\, d\eta}{\int_0^\infty \int_{\mathbb{R}^p} \eta^{(p+k)/2+1}\, f(\eta\,(\|x-\theta\|^2 + \|u\|^2))\,\pi(\|\theta\|^2)\eta^b\, d\theta\, d\eta} \\ &= \frac{\int_{\mathbb{R}^p} \left(\int_0^\infty \eta^{(p+k)/2+b+1}\, f(\eta\,(\|x-\theta\|^2 + \|u\|^2))\, d\eta\right)(\theta - x)\,\pi(\|\theta\|^2)\, d\theta}{\int_{\mathbb{R}^p} \left(\int_0^\infty \eta^{(p+k)/2+b+1}\, f(\eta\,(\|x-\theta\|^2 + \|u\|^2))\, d\,\eta\right)\pi(\|\theta\|^2)\, d\theta}, \end{aligned}$$

by Fubini's theorem. Now, through the change of variable $\tau = \eta\,(\|x-\theta\|^2 + \|u\|^2)$ in the innermost integrals, we obtain

$$\begin{aligned} g(x, \|u\|^2) &= \frac{\int_{\mathbb{R}^p} \int_0^\infty \tau^{(p+k)/2+b+1}\, f(\tau)\, d\tau\, \frac{(\theta-x)\,\pi(\|\theta\|^2)}{(\|x-\theta\|^2+\|u\|^2)^{(p+k)/2+b+2}}\, d\theta}{\int_{\mathbb{R}^p} \int_0^\infty \tau^{(p+k)/2+b+1}\, f(\tau)\, d\tau\, \frac{\pi(\|\theta\|^2)}{(\|x-\theta\|^2+\|u\|^2)^{(p+k)/2+b+2}}\, d\theta} \\ &= \frac{\int_{\mathbb{R}^p} \frac{(\theta-x)\,\pi(\|\theta\|^2)}{(\|x-\theta\|^2+\|u\|^2)^{(p+k)/2+b+2}}\, d\theta}{\int_{\mathbb{R}^p} \frac{\pi(\|\theta\|^2)}{(\|x-\theta\|^2+\|u\|^2)^{(p+k)/2+b+2}}\, d\theta} \end{aligned}$$

thanks to (6.35). □

6.3.2 *Minimaxity of Generalized Bayes Estimators*

According to the expression of $g(X, \|U\|^2)$ in (6.36), we give an expression of the differential operator $\mathcal{O}g(X, \|U\|^2)$ in (6.23). The proof of Proposition 6.4 follows from straightforward calculations.

Proposition 6.4 *For* $g(X, \|U\|^2) = \frac{\nabla_X M(X, \|U\|^2)}{m(X, \|U\|^2)}$, (6.23) *can be expressed as*

$$\begin{aligned} \mathcal{O}g(X, \|U\|^2) = 2\,\frac{\Delta_X M(X, \|U\|^2)}{m(X, \|U\|^2)} &- 2\,\frac{\nabla_X m(X, \|U\|^2)^{\mathrm{T}} \nabla_X M(X, \|U\|^2)}{m^2(X, \|U\|^2)} \qquad (6.39) \\ &+ \frac{k-2}{\|U\|^2} \left\| \frac{\nabla_X M(X, \|U\|^2)}{m(X, \|U\|^2)} \right\|^2 + 2\,\frac{\partial}{\partial s} \left\| \frac{\nabla_X M(X, s)}{m(X, s)} \right\|^2 \Bigg|_{s=\|U\|^2}, \end{aligned}$$

where, for any $(x, u) \in \mathbb{R}^p \times \mathbb{R}^k$,

$$m(x, \|u\|^2) = \int_{\mathbb{R}^p} \varphi(\|x - \theta\|^2 + \|u\|^2)\, \pi(\|\theta\|^2)\, d\theta, \tag{6.40}$$

and

$$M(x, \|u\|^2) = \int_{\mathbb{R}^p} \phi(\|x - \theta\|^2 + \|u\|^2)\, \pi(\|\theta\|^2)\, d\theta \tag{6.41}$$

with φ *and* ϕ *given by* (6.37) *and* (6.38).

In Sect. 5.4, we studied Bayes minimax estimation of a location vector in the case of spherically symmetric distributions with known scale parameter. For a subclass of spherically symmetric densities, we proved minimaxity of generalized Bayes estimators for spherically symmetric priors of the form $\pi(\|\theta\|^2)$ under the following assumptions (see Theorem 5.7 and also Fourdrinier and Strawderman 2008a, 2010).

Assumption 1

(1) $\pi'(\|\theta\|^2) \;\; \leq 0$ i.e. $\pi(\|\theta\|^2)$ is unimodal;
(2) $\Delta\pi(\|\theta\|^2) \;\; \leq 0$ i.e. $\pi(\|\theta\|^2)$ is superharmonic;
(3) $\Delta\pi(\|\theta\|^2)$ is nondecreasing in $\|\theta\|^2$.

Note that Condition (2) in fact implies Condition (1) by the mean value property of superharmonic functions. Several examples of priors which satisfy Assumption 1 have been given in Sect. 5.4: Examples 5.8, 5.9 and 5.10.

Our main result below is that a generalized Bayes estimator of θ for a density (6.33), a prior (6.34) and the loss (6.16) is minimax under weak moment conditions and conditions on b, provided the prior satisfies the Assumptions above. We remind the reader that, according to Proposition 6.3, the generalized Bayes estimator is independent of the sampling density, f, provided the assumption (6.35) holds. Hence, each such estimator is simultaneously generalized Bayes and minimax for the entire class of spherically symmetric distributions.

Before developing our minimaxity result, we give a theorem which guarantees the risk finiteness of the generalized Bayes estimators.

Theorem 6.3 *Assume that* π *satisfies Assumption 1 and that* $b > -(k/2+1)$. *Then the generalized Bayes estimator associated to* π *has finite risk.*

Proof According to (6.36), the risk finiteness condition (6.17) is satisfied as soon as

$$\begin{aligned} E_{\theta,\eta}&\left[\left\| \frac{\int_{\mathbb{R}^p} (\theta - X)\, \frac{\pi(\|\theta\|^2)}{(\|X-\theta\|^2+\|U\|^2)^{(p+k)/2+b+2}}\, d\theta}{\int_{R^p} \frac{\pi(\|\theta\|^2)}{(\|X-\theta\|^2+\|U\|^2)^{(p+k)/2+b+2}}\, d\theta} \right\|^2\right] \\ &\leq E_{\theta,\eta}\left[\frac{\int_{R^p} \|\theta - X\|^2\, \frac{\pi(\|\theta\|^2)}{(\|X-\theta\|^2+\|U\|^2)^{(p+k)/2+b+2}}\, d\theta}{\int_{R^p} \frac{\pi(\|\theta\|^2)}{(\|X-\theta\|^2+\|U\|^2)^{(p+k)/2+b+2}}\, d\theta} \right] \\ &< \infty\,. \end{aligned} \tag{6.42}$$

Note that, for any $(x, u) \in \mathbb{R}^p \times \mathbb{R}^k$ and for any nonnegative function h on $\mathbb{R}_+ \times \mathbb{R}_+$ (see Lemma 1.4),

$$\int_{R^p} \pi(\|\theta\|^2)\, h(\|x-\theta\|^2, \|u\|^2)\, d\theta$$
$$= \int_0^\infty \int_{S_{R,x}} \pi(\|\theta\|^2)\, d\mathscr{U}_{R,x}(\theta)\, \sigma(S)\, R^{p-1}\, h(R^2, \|u\|^2)\, dR\,, \tag{6.43}$$

where $\mathscr{U}_{R,x}$ is the uniform distribution on the sphere $S_{R,x}$ of radius R and centered at x and $\sigma(S)$ is the area of the unit sphere. Through the change of variable $R = \sqrt{v}$, the right hand side of (6.43) can be written as

$$\int_0^\infty \mathscr{S}_\pi(\sqrt{v}, x)\, v^{p/2-1}\, h(v, \|U\|^2)\, dv,$$

where

$$\mathscr{S}_\pi(\sqrt{v}, x) = \frac{\sigma(S)}{2} \int_{S_{\sqrt{v},x}} \pi(\|\theta\|^2)\, d\mathscr{U}_{\sqrt{v},x}(\theta)$$

is nonincreasing in v by the superharmonicity of $\pi(\|\theta\|^2)$.

Now we can express the last quantity in brackets in (6.42) as

$$\frac{\int_o^\infty \mathscr{S}_\pi(\sqrt{v}, x)\, \frac{v^{p/2}}{(v+\|u\|^2)^{(p+k)/2+b+2}} dv}{\int_o^\infty \mathscr{S}_\pi(\sqrt{v}, x)\, \frac{v^{p/2-1}}{(v+\|u\|^2)^{(p+k)/2+b+2}} dv} = E_1[v] \le E_2[v], \tag{6.44}$$

where E_1 is the expectation with respect to the density $f_1(v)$ proportional to

$$\mathscr{S}_\pi(\sqrt{v}, x)\, \frac{v^{p/2-1}}{(v+\|u\|^2)^{(p+k)/2+b+2}},$$

and E_2 is the expectation with respect to the density $f_2(v)$ proportional to

$$\frac{v^{p/2-1}}{(v+\|u\|^2)^{(p+k)/2+b+2}}.$$

Indeed the ratio $f_2(v)/f_1(v)$ is nondecreasing by the monotonicity of $\mathscr{S}_\pi(\sqrt{v}, x)$. In (6.44), $E_2[v]$ is

$$E_2[v] = \frac{\int_0^\infty \frac{v^{p/2}}{(v+\|u\|^2)^{(p+k)/2+b+2}}\, dv}{\int_0^\infty \frac{v^{p/2-1}}{(v+\|u\|^2)^{(p+k)/2+b+2}}\, dv}$$

$$= \|u\|^2 \frac{\int_0^\infty \frac{v^{p/2}}{(v+1)^{(p+k)/2+b+2}}\, dv}{\int_0^\infty \frac{v^{p/2-1}}{(v+1)^{(p+k)/2+b+2}}\, dv}$$

$$= \|u\|^2 \frac{B(p/2+1, k/2+b+1)}{B(p/2, k/2+b+2)},$$

which is finite for $k/2 + b + 1 > 0$.

Finally the expectations in (6.42) are bounded above by $K\, E_{\theta,\eta}[\|U\|^2]$ where K is a constant, and hence are finite. □

We will need the following result which is essentially a reexpression of Lemma 5.6.

Lemma 6.5 *Let $m(x, \|u\|^2)$ and $M(x, \|u\|^2)$ be as defined in* (6.40) *and* (6.41) *and let $\cdot$ be the inner product in $\mathbb{R}^p$. Then we have*

(1)

$$x \cdot \nabla_x m(x, \|u\|^2) = -2 \int_0^\infty H(v, \|x\|^2)\, v^{p/2}\, \varphi'(v + \|u\|^2)\, dv,$$

and

$$x \cdot \nabla_x M(x, \|u\|^2) = \int_0^\infty H(v, \|x\|^2)\, v^{p/2}\, \varphi(v + \|u\|^2)\, dv,$$

where, for $v > 0$,

$$H(v, \|x\|^2) = \lambda(B) \int_{B_{\sqrt{v},x}} x \cdot \theta\, \pi'(\|\theta\|^2)\, d\mathcal{V}_{\sqrt{v},x}(\theta) \tag{6.45}$$

and $\mathcal{V}_{\sqrt{v},x}$ is the uniform distribution on the ball $B_{\sqrt{v},x}$ of radius $\sqrt{v}$ centered at x and $\lambda(B)$ is the volume of the unit ball;

(2) *For any $x \in \mathbb{R}^p$, the function $H(v, \|x\|^2)$ in* (6.45) *is nondecreasing in v provided that $\Delta\pi(\|\theta\|^2)$ is nondecreasing in $\|\theta\|^2$. (Assumption* 1 (3)*);*

(3) *For any $v > 0$ and any $x \in \mathbb{R}^p$, the function $H(v, \|x\|^2)$ in* (6.45) *is nonpositive provided $\pi'(\|\theta\|^2) \le 0$. (Assumption* 1 (1)*).*

Given these preliminaries, we present our main result.

Theorem 6.4 *Suppose that π satisfies Assumption* 1. *Then the generalized Bayes estimator associated to $\pi(\|\theta\|^2)\, \eta^b$ is minimax provided that $b \ge \frac{2p-k-2}{4}$ and the assumptions of Theorem* 6.3 *are satisfied.*

Proof It suffices to show that $\mathcal{O}g(X, \|U\|^2)$ in (6.38), with $m(X, \|U\|^2)$ and $M(X, \|U\|^2)$ given respectively by (6.39) and (6.41), is non positive since the assumptions guarantee that the generalized Bayes estimator δ is of the form $\delta(X, \|U\|^2) = X + \nabla_X M(X, \|U\|^2)/m(X, \|U\|^2)$ and has finite risk.

Due to the superharmonicity of $\pi(\|\theta\|^2)$, for any $(x, u) \in \mathbb{R}^p \times \mathbb{R}^k$, we have $\Delta_x M(x\|u\|^2) \le 0$ so that

$$\mathscr{O}g(x, \|u\|^2) \le -2\, \frac{\nabla_x m(x, \|u\|^2)^{\mathrm{T}} \nabla_x M(x, \|u\|^2)}{m^2(x, \|u\|^2)} + \frac{k-2}{\|u\|^2} \left\| \frac{\nabla_x M(x, \|u\|^2)}{m(x, \|u\|^2)} \right\|^2 + 2 \frac{\partial}{\partial s} \left\| \frac{\nabla_x M(x, s)}{m(x, s)} \right\|^2 \Bigg|_{s=\|u\|^2}.$$

Note that

$$\begin{aligned} m^2(x, s)\, \frac{\partial}{\partial s} \left\| \frac{\nabla_x M(x, s)}{m(x, s)} \right\|^2 &= \frac{\partial}{\partial s} \|\nabla_x M(x, s)\|^2 + \|\nabla_x M(x, s)\|^2\, m^2(x, s)\, \frac{\partial}{\partial s} \frac{1}{m^2(x, s)} \\ &\le \frac{\partial}{\partial s} \|\nabla_x M(x, s)\|^2 + (p + k + 2b + 4)\, \frac{1}{s}\, \|\nabla_x M(x, s)\|^2, \end{aligned}$$

since

$$\begin{aligned} \frac{\partial}{\partial s} \frac{1}{m^2(x, s)} &= \frac{-2}{m^3(x, s)} \int_{\mathbb{R}^p} \frac{-[(p+k)/2 + b + 2]}{(\|x-\theta\|^2 + s)^{(p+k)/2+b+3}}\, \pi(\|\theta\|^2)\, d\theta \\ &= \frac{p+k+2b+4}{m^3(x, s)}\, \frac{1}{s} \int_{\mathbb{R}^p} \frac{s}{\|x-\theta\|^2 + s}\, \frac{1}{(\|x-\theta\|^2 + s)^{(p+k)/2+b+2}}\, \pi(\|\theta\|^2)\, d\theta \\ &\le \frac{p+k+2b+4}{m^2(x, s)}\, \frac{1}{s}. \end{aligned}$$

Therefore

$$\begin{aligned} m^2(x, s)\, \mathscr{O}g(x, s) \le & -2\, \nabla_x m(x, s)^{\mathrm{T}} \nabla_x M(x, s) \\ & + \frac{k - 2 + 2(p + k + 2b + 4)}{s}\, \|\nabla_x M(x, s)\|^2 \\ & + 2 \frac{\partial}{\partial s}\, \|\nabla_x M(x, s)\|^2 . \end{aligned} \tag{6.46}$$

As $m(x, s)$ and $M(x, s)$ depend on x only through $\|x\|^2$, it is easy to check that (as in Fourdrinier and Strawderman 2008a)

$$\nabla_x m(x, s)^{\mathrm{T}} \nabla_x M(x, s) = \frac{x^{\mathrm{T}} \nabla_x m(x, s)\, x^{\mathrm{T}} \nabla_x M(x, s)}{\|x\|^2}$$

and

$$\|\nabla_x M(x,s)\|^2 = \frac{(x^{\mathrm{T}}\nabla_x M(x,s))^2}{\|x\|^2}\,.$$

Thus the right hand side of (6.46) will be nonpositive as soon as

$$-\,2x^{\mathrm{T}}\nabla_x m(x,s) + \frac{2p+3k+4b+6}{s}\,x^{\mathrm{T}}\nabla_x M(x,s) + 4\frac{\partial}{\partial s}x^{\mathrm{T}}\nabla_x M(x,s) \geq 0, \tag{6.47}$$

since, according to Lemma 6.5, the common factor $x^{\mathrm{T}}\nabla_x M(x,s)$ is nonpositive. Using again Lemma 6.5, the left hand side of (6.47) equals

$$\begin{aligned}
&4\int_0^\infty H(v,\|x\|^2)\,v^{p/2}\,\varphi'(v+s)\,dv \\
&\qquad + \frac{2p+3k+4b+6}{s}\int_0^\infty H(v,\|x\|^2)\,v^{p/2}\,\varphi(v+s)\,dv \\
&\qquad\qquad + 4\int_o^\infty H(v,\|x\|^2)v^{p/2}\varphi'(v+s)dv \\
&= \int_0^\infty v^{p/2}\varphi(v+s)\left\{8\,E\left[H(v,\|x\|^2)\,\frac{\varphi'(v+s)}{\varphi(v+s)}\right]\right. \\
&\qquad\qquad \left. + \frac{2\,p+3\,k+4\,b+6}{s}\,E\left[H(v,\|x\|^2)\right]\right\} dv\,,
\end{aligned} \tag{6.48}$$

where E denotes the expectation with respect to the density proportional to $v \longmapsto v^{p/2}\,\varphi(v+s)$.

As

$$\frac{\varphi'(v+s)}{\varphi(v+s)} = \frac{-((p+k)/2+b+2)}{v+s} \tag{6.49}$$

is nondecreasing in v and, according to Lemma 6.5, $H(v,\|x\|^2)$ is also nondecreasing in v, the first expectation in (6.48) satisfies

$$E\left[H(v,\|x\|^2)\,\frac{\varphi'(v+s)}{\varphi(v+s)}\right] \geq E\left[H(v,\|x\|^2)\right]E\left[\frac{\varphi'(v+s)}{\varphi(v+s)}\right]$$

by the covariance inequality. Therefore Inequality (6.47) will be satisfied as soon as

$$8\,E\left[\frac{\varphi'(v+s)}{\varphi(v+s)}\right] + \frac{2p+3k+4b+6}{s} \leq 0, \tag{6.50}$$

since $H(v, \|x\|^2) \leq 0$ by Lemma 6.5.

From (6.49) we have

$$E\left[\frac{\varphi'(v+s)}{\varphi(v+s)}\right] = -\big((p+k)/2+b+2\big)\, E\left[\frac{1}{v+s}\right] \tag{6.51}$$

$$= -\big((p+k)/2+b+2\big)\, \frac{\int_o^\infty \frac{1}{v+s}\, v^{p/2} \frac{1}{(v+s)^{(p+k)/2+b+2}}\, dv}{\int_0^\infty v^{p/2} \frac{1}{(v+s)^{(p+k)/2+b+2}}\, dv}$$

$$= -\big((p+k)/2+b+2\big)\, \frac{1}{s}\, \frac{\int_0^\infty \frac{z^{p/2}}{(z+1)^{(p+k)/2+b+3}}\, dz}{\int_0^\infty \frac{z^{p/2}}{(z+1)^{(p+k)/2+b+2}}\, dz}$$

$$= -\big((p+k)/2+b+2\big)\, \frac{1}{s}\, \frac{B(p/2+1, k/2+b+2)}{B(p/2+1, k/2+b+1)},$$

where $B(\alpha, \beta)$ is the beta function with parameters $\alpha > 0$ and $\beta > 0$. Then (6.51) becomes

$$E\left[\frac{\varphi'(v+s)}{\varphi(v+s)}\right] = -\frac{((p+k)/2+b+2)}{s}\, \frac{\Gamma((k/2+b+2))}{\Gamma((p+k)/2+b+3)}$$

$$= \frac{\Gamma((p+k)/2+b+2)}{\Gamma(k/2+b+1)}\, \frac{-(k/2+b+1)}{s}. \tag{6.52}$$

It follows from (6.52) that (6.50) reduces to

$$b \geq \frac{2p-k-2}{4},$$

which is the condition given in the theorem. □

The condition on b in Theorem 6.4 can be alternatively expressed as $k \geq 2p - 4b - 2$ which dictates that the dimension, k, of the residual vector, U, increases with the dimension, p, of θ. This dependence can be (essentially) eliminated provided the generalized Bayes estimator in Proposition 6.3 satisfies the following assumption.

Assumption 2 The function $g(x, \|u\|^2)$ in (6.36) can be expressed as

$$g(x, \|u\|^2) = \frac{\nabla_x M(x, \|u\|^2)}{m(x, \|u\|^2)} = -\frac{r(\|x\|^2, \|u\|^2)\, \|u\|^2}{\|x\|^2}\, x,$$

where $r(\|x\|^2, \|u\|^2)$ is nonnegative and nonincreasing in $\|u\|^2$.

Assumption 2 is satisfied, for example, by the generalized Bayes estimator corresponding to the prior on (θ, η) proportional to $\pi(\|\theta\|^2) = \big(1/\|\theta\|^2\big)^{-b/2}\, \eta^a$ for

$0 < b \le p-2$ and $a > -\frac{k}{2} - \frac{b}{2} - 2$, in which case the function $r(\|x\|^2, \|u\|^2) = \phi\left(\|x\|^2/\|u\|^2\right)$, where $\phi(t)$ is increasing in t, and hence $r(\|x\|^2, \|u\|^2)$ is decreasing in $\|u\|^2$ (see, Maruyama 2003b).

We have the following corollary.

Corollary 6.2 *Suppose π satisfies Assumptions* 1 *and the assumptions of Theorem* 6.4 *and suppose also that the generalized Bayes estimator (which does not depend on the underlying density f) satisfies Assumption* 2. *Then the generalized Bayes estimator is minimax provided $b \ge -(k+2)/4$.*

Proof Assumption 2 guarantees that

$$\frac{\partial}{\partial s}\left(\frac{1}{s^2}\left\|\frac{\nabla_x M(x,s)}{m(x,s)}\right\|^2\right) = \frac{\partial}{\partial s}\left(\frac{r^2(\|x\|^2, s)}{\|x\|^2}\right) \le 0.$$

Since

$$\begin{aligned}\frac{\partial}{\partial s}\left\|\frac{\nabla_x M(x,s)}{m(x,s)}\right\|^2 &= \frac{\partial}{\partial s}\left(\frac{s^2}{s^2}\left\|\frac{\nabla_x M(x,s)}{m(x,s)}\right\|^2\right)\\ &= \frac{2}{s}\left\|\frac{\nabla_x M(x,s)}{m(x,s)}\right\|^2 + s^2\frac{\partial}{\partial s}\left(\frac{1}{s^2}\left\|\frac{\nabla_x M(x,s)}{m(x,s)}\right\|^2\right),\end{aligned}$$

the inequality for $\mathscr{O}g(X, \|U\|^2)$ in the proof of Theorem 6.4 can be replaced by

$$\mathscr{O}g(x, \|u\|^2) \le -2\,\frac{\nabla_x m(x, \|u\|^2)^{\mathrm{T}}\nabla_x M(x, \|u\|^2)}{m^2(x, \|u\|^2)} + \frac{k+2}{\|u\|^2}\left\|\frac{\nabla_x M(x, \|u\|^2)}{m(x, \|u\|^2)}\right\|^2.$$

It follows that inequality condition (6.47) becomes

$$-2\,x^{\mathrm{T}}\nabla_x m(x,s) + \frac{k+2}{s}\,x^{\mathrm{T}}\nabla_x M(x,s) \ge 0,$$

and that inequality condition (6.50) becomes

$$4\,E\left[\frac{\varphi'(v+s)}{\varphi(v+s)}\right] + \frac{k+2}{s} \le 0,$$

which, by (6.52), becomes

$$4\left[-\left(\frac{k/2+b+1}{s}\right)\right] + \frac{k+2}{s} \le 0,$$

which is equivalent to $b \ge -(k+2)/4$. □

6.4 The Unknown Covariance Matrix Case

In this section, we consider estimation of the mean vector in the case of elliptically symmetric distribution with an unknown nonsingular scale matrix. Most of the material of this section is taken from Fourdrinier et al. (2003). We assume there is sufficient data in the form of residual vectors to estimate the unknown covariance matrix. In the canonical form of this model, X, $V_1, \ldots, V_{n-1}$ are n random vectors in $\mathbb{R}^p$ with joint density of the form

$$|\Sigma|^{-n/2} f\left((x-\theta)^{\mathrm{T}}\Sigma^{-1}(x-\theta) + \sum_{j=1}^{n-1} V_j^{\mathrm{T}}\Sigma^{-1}V_j\right) \tag{6.53}$$

where the $p \times 1$ location vector θ and the $p \times p$ scale matrix Σ are unknown. Note occasionally we will absorb the normalizing factor $|\Sigma^{-1}|^{n/2}$ in the function f. If both θ and Σ are unknown, X and $S = \sum_{j=1}^{n-1} V_j\, V_j^{\mathrm{T}} = V\, V^{\mathrm{T}}$ are minimal sufficient statistics. Throughout this section, we assume that $p \le n-1$ so that S is invertible.

The canonical form (6.53) arises through an $n \times n$ orthogonal transformation of

$$(Y_1, \ldots, Y_n) \sim |\Sigma|^{-n/2} f\left(\sum_{j=1}^{n}(Y_i-\theta)^{\mathrm{T}}\Sigma^{-1}(Y_i-\theta)\right)$$

as in the case of an i.i.d. sample of size n from a $\mathcal{N}_p(\theta, \Sigma)$ distribution.

To show this reduction to the canonical form define the $p \times n$ matrices $\mathbf{Y} = (Y_1 : \ldots : Y_n)$ for $Y_i \in \mathbb{R}^p$ and $\Theta = (\theta : \ldots : \theta)$. Let P be an $n \times n$ orthogonal matrix such that the first row of P is $\mathbf{1}_n^{\mathrm{T}}/\sqrt{n}$, where $\mathbf{1}_n^{\mathrm{T}} = (1, \ldots, 1)$ is the $1 \times n$ row vector of ones. Let the $p \times n$ matrices $\mathbf{X} = (X_1 : \ldots : X_n)$ and $\boldsymbol{\nu}\mathrm{T}{=}(\nu_1{:}\ldots{:}\nu_n)$ be defined through $\mathbf{X}^{\mathrm{T}} = P\,\mathbf{Y}^{\mathrm{T}}$ and $\boldsymbol{\nu}^{\mathrm{T}} = P\,\Theta^{\mathrm{T}}$. Then

$$\begin{aligned}
\sum_{i=1}^{n}(Y_i-\theta)^{\mathrm{T}}\Sigma^{-1}(Y_i-\theta) &= \mathrm{tr}\left\{(\mathbf{Y}-\Theta)^{\mathrm{T}}\,(\mathbf{Y}-\Theta)\,\Sigma^{-1}\right\} \\
&= \mathrm{tr}\left\{(\mathbf{Y}-\Theta)^{\mathrm{T}}\,P\,P^{\mathrm{T}}\,(\mathbf{Y}-\Theta)\,\Sigma^{-1}\right\} \\
&= \mathrm{tr}\left\{(\mathbf{X}-\boldsymbol{\nu})^{\mathrm{T}}\,(\mathbf{X}-\boldsymbol{\nu})\,\Sigma^{-1}\right\} \\
&= \sum_{i=1}^{n}(X_i-\nu_i)^{\mathrm{T}}\Sigma^{-1}(X_i-\nu_i) \\
&= (X_1-\theta)^{\mathrm{T}}\Sigma^{-1}(X_1-\theta) + \sum_{i=2}^{n} X_i^{\mathrm{T}}\Sigma^{-1}X_i\,,
\end{aligned}$$

since $\boldsymbol{\nu}^{\mathrm{T}} = P\,\Theta^{\mathrm{T}} = (\theta : 0, \ldots : 0)^{\mathrm{T}}$ because the ith column of Θ is $\theta_i\,\mathbf{1}_n$ and since $P_1^{\mathrm{T}}\,\mathbf{1}_n = 1$ and $P_i^{\mathrm{T}}\,\mathbf{1}_n = 0$ for $i = 2, \ldots, n$, where P_i^{T} is the ith row of P.

Letting $X = X_1$ and $V_{i-1} = X_i$ for $i = 2, \ldots, n$ and noting that the Jacobian of the transformation $\mathbf{Y} \mapsto \mathbf{X} \mapsto (X_1, V_1, \ldots, V_{n-1})$ is 1, the density of $(X_1, V_1, \ldots, V_{n-1})$ is given by (6.53) (see also e.g. Rao 1973; Muirhead 1982 or Anderson 1984).

There is an obvious connection with the canonical form of the general linear model given in Sect. 4.5. Indeed, if $\Sigma = \sigma^2 I_p$, the density (6.53) becomes

$$\sigma^{-pn/2} f\left(\frac{(x-\theta)^{\mathrm{T}}(x-\theta) + \sum_{j=1}^{n-1} V_j^{\mathrm{T}} V_j}{\sigma^2}\right).$$

So, if $U_{ij} = V_{ij}$ and $U = (U_{12}, \ldots, U_{1p}, U_{21}, \ldots, U_{2p}, U_{n-11}, \ldots, U_{n-1p})$ then $(X, U) \sim SS_{p,(n-1)p}(\theta, 0)$. This model is also related to the general (normal) multivariate linear model $Y_{n\times m} = X_{n\times p}\,\beta_{p\times m} + \epsilon_{n\times m}$ where $\epsilon_{i\times m} \sim \mathcal{N}_m(0, \Sigma)$, $i = 1, \ldots, n$ are independent, X is a known design matrix and β is a matrix of unknown regression parameters.

We consider the problem of estimating θ with the invariant loss

$$L(\theta, \delta) = (\delta - \theta)^{\mathrm{T}} \Sigma^{-1} (\delta - \theta). \tag{6.54}$$

Recall that the usual estimator $\delta_0(X) = X$ is minimax provided $E_{0,I}[\|X\|^2] < \infty$ (where $E_{\theta,\Sigma}$ denotes the expectation with respect to the density in (6.53)). Note that, when Σ is a covariance matrix, this expectation is necessarily finite and equal to p. Moreover X is typically admissible when $p \leq 2$ and inadmissible when $p \geq 3$.

We concentrate on the case $p \geq 3$ and construct a class of estimators, depending on the sufficient statistics (X, S), of the form

$$\delta(X, S) = X + g(X, S), \tag{6.55}$$

where $S = \sum_{i=1}^{n-1} V_i V_i^{\mathrm{T}}$, which dominate $\delta_0(X) = X$ simultaneously under loss (6.54), for the entire class of distributions defined in (6.53) such that $E_{0,I}[\|X\|^2] < \infty$. Note that, although the loss in (6.54) is invariant, the estimate in (6.55) may not be equivariant (except for $\delta_0(X)$).

The risk difference $\Delta_{\theta,\Sigma}$ between $\delta(X, S)$ given in (6.55) and $\delta_0(X) = X$ equals

$$\begin{aligned} \Delta_{\theta,\Sigma} &= R\big(\theta, \delta(X, S)\big) - R\big(\theta, \delta_0(X)\big) \qquad (6.56) \\ &= E_{\theta,\Sigma}\big[2g^{\mathrm{T}}(X, S)\Sigma^{-1}(X - \theta)\big] + E_{\theta,\Sigma}\big[g^{\mathrm{T}}(X, S)\Sigma^{-1}g(X, S)\big], \end{aligned}$$

provided $E_{\theta,\Sigma}\big[g^{\mathrm{T}}(X, S)\Sigma^{-1}g(X, S)\big] < \infty$.

We first give a lemma which expresses the two terms in the last expression of (6.56) as expectations $E^*_{\theta,\Sigma}$ with respect to the distribution

$$C^{-1} F\left((x-\theta)^{\mathrm{T}}\Sigma^{-1}(x-\theta) + \sum_{j=1}^{n-1} V_j^{\mathrm{T}}\Sigma^{-1}V_j\right)$$

where F and C are defined as

$$F(t) = \frac{1}{2} \int_t^{\infty} f(s) ds$$

and

$$C = \int_{\mathbb{R}^p \times \cdots \times \mathbb{R}^p} F\left((x-\theta)^{\mathrm{T}} \Sigma^{-1} (x-\theta) + \sum_{j=1}^{n-1} V_j^{\mathrm{T}} \Sigma^{-1} V_j \right) dx\, dv_1 \cdots dv_{n-1}.$$

To this end, we will use the following notations. For any matrix M, ∇_M is interpreted as the matrix with components $(\nabla_M)_{ij} = \partial / \partial M_{ij}$. The differential operator for a symmetric matrix S is $\mathscr{D}_S = \left(\frac{1}{2} (1 + \delta_{ij}) (\nabla_S)_{ij} \right)$ and Haff differential operator is defined, for any $p \times p$ matrix function of a symmetric matrix S, say $H(S)$, to be

$$D_{1/2}^*(H(S)) = \mathrm{tr}\big(\mathscr{D}_S\, H(S)\big) = \sum_{i=1}^{p} \frac{\partial H_{ii}(S)}{\partial S_{ii}} + \frac{1}{2} \sum_{i \neq j} \frac{\partial H_{ij}(S)}{\partial S_{ij}}. \tag{6.57}$$

Lemma 6.6 *Let* $(X, V) = (X, V_1, \ldots, V_{n-1})$ *be a* $p \times n$ *random matrix with density* (6.53) *where* $p \leq n - 1$ *and let* $S = V\, V^{\mathrm{T}}$.

(1) *Suppose* $g(x, s)$ *is a weakly differentiable function in* x *for each* s *such that the expectation* $E_{\theta,\Sigma}\big[g^{\mathrm{T}}(X, S)(X, S)\Sigma^{-1}(X - \theta)\big]$ *exists. Then*

$$E_{\theta,\Sigma}\big[g^{\mathrm{T}}(X, S)\Sigma^{-1}(X - \theta)\big] = C\, E_{\theta,\Sigma}^*\big[\mathrm{div}_X g(X, S)\big] \tag{6.58}$$

where $\mathrm{div}_x\, g(x, s)$ *is the divergence of* $g(x, s)$ *with respect to* x.

(2) *Suppose* $T(x, s)$ *is a* $p \times p$ *matrix function weakly differentiable in* v_i $(i = 1, \ldots, n-1)$ *for any* x *and such that the expectation* $E_{\theta,\Sigma}\big[\mathrm{tr}\left(T(X, S)\right)\Sigma^{-1}\big]$ *exists. Then*

$$\begin{aligned} &E_{\theta,\Sigma}\big[\mathrm{tr}\left(T(X, S)\Sigma^{-1}\right)\big] \\ &= C\, E_{\theta,\Sigma}^*\big[2\, D_{1/2}^* T(X, S) + (n - p - 2)\, \mathrm{tr}\,(S^{-1} T(X, S))\big] \\ &= C\, E_{\theta,\Sigma}^*\big[\mathrm{tr}(V\, \nabla_{V^{\mathrm{T}}}\, \{S^{-1} T(X, S)\}^{\mathrm{T}}) + (n - 1)\, \mathrm{tr}\,(S^{-1} T(X, S))\big]. \end{aligned} \tag{6.59}$$

The proof of Lemma 6.6 is given at the end of this section. The two expressions in (6.58) follow from equality between the two integrand terms thanks to the link between the differential operators $D_{1/2}^*$ and $\mathrm{tr}(V\, \nabla_{V^{\mathrm{T}}})$ established in Proposition 6.5 (also given at the end of this section).

Note that, when $X, V_1, \ldots, V_{n-1}$ are independent normal vectors with covariance Σ, then $f = F$ and therefore $E_{\theta,\Sigma}[\ \ ] = E_{\theta,\Sigma}^*[\ \ ]$. Hence for Lemma 6.6, the identity in (6.58) essentially reduces to Stein's lemma (Stein 1981), and the identity in (6.59) corresponds to a result of Stein (1977a) and Haff (1979), known as the Stein-Haff identity.

Applying (6.58) to the first term in (6.56) and (6.59) to the second term in (6.56) with $T(x, s) = g(x, s)g'(x, s)$, noting that

$$g^{\mathrm{T}}(x, s)\Sigma^{-1}g(x, s) = \mathrm{tr}\left(g(x, s)g^{\mathrm{T}}(x, s)\Sigma^{-1}\right)$$

gives immediately the following theorem.

Theorem 6.5 *Assume that* $g(x, s)$ *and* $T(x, s) = g(x, s)g^{\mathrm{T}}(x, s)$ *satisfy the assumptions of Lemma* 6.6. *Assume also that* $E_{0,\Sigma}[\|X\|^2] < \infty$ *and* $E_{\theta,\Sigma}\left[g^{\mathrm{T}}(X, S)\Sigma^{-1}g(X, S)\right] < \infty$. *Then the risk difference* $\Delta_{\theta,\Sigma}$ *in* (6.56) *between* $\delta(X, S) = X + g(X, S)$ *and* $\delta_0(X) = X$ *equals*

$$C\, E^*_{\theta,\Sigma}\Big[2\,\mathrm{div}_X g(X, S) + 2\, D^*_{1/2}\left(g(X, S)g^{\mathrm{T}}(X, S)\right) + (n-p-2)\, g^{\mathrm{T}}(X, S)\, S^{-1}g(X, S)\Big]. \tag{6.60}$$

A sufficient condition for $\delta(X, S)$ *to be minimax is that, for all* x *and* s,

$$2\,\mathrm{div}_x g(x, s) + 2\, D^*_{1/2}\big(g(x, s)g^{\mathrm{T}}(x, s)\big) + (n-p-2)\, g^{\mathrm{T}}(x, s)s^{-1}g(x, s) \le 0 \tag{6.61}$$

or, equivalently,

$$2\,\mathrm{div}_x g(x, s) + \mathrm{tr}(v\, \nabla_{v^{\mathrm{T}}}\, \{s^{-1}g(x, s)g^{\mathrm{T}}(x, s)\}^{\mathrm{T}} + (n-1)\, g^{\mathrm{T}}(x, s)s^{-1}\, g(x, s) \le 0\,, \tag{6.62}$$

where $V = (V_1, \ldots, V_{n-1})$ *is a* $p \times (n-1)$ *matrix and* $S = V V^{\mathrm{T}}$. *Furthermore* $\delta(X, S)$ *dominates* $\delta_0(X)$ *as soon as* (6.61) *or* (6.62) *is satisfied with strict inequality on a set of positive measure.*

Note that in the normal case $E^*_{\theta,\Sigma}[\ \ ] = E_{\theta,\Sigma}[\ \ ]$ so that the left-hand side of (6.61) is an unbiased estimator of the risk difference between $\delta(X, S)$ and $\delta_0(X)$. Perhaps, most importantly, observe that the theorem leads to an extremely strong robustness property for estimators satisfying (6.61). Namely, any such estimator is minimax and, as soon as strict inequality occurs on a set of positive measure in (6.61), dominates $\delta_0(X)$ for the entire class of elliptically symmetric distributions (6.53). This property is analogous to the robustness property mentioned in Sect. 6.1 in the case of spherically symmetric distributions. The following corollary gives a general class of examples of minimax estimates which dominate $\delta_0(X)$ uniformly for densities of the form (6.53).

Corollary 6.3 *Assume that* $E_{0,\Sigma}[\|X\|^2] < \infty$ *and* $E_{\theta,\Sigma}\left[\frac{\|X\|^2}{(X^{\mathrm{T}}S^{-1}X)^2}\right] < \infty$. *Let* $\delta(X, S) = (1 - r(X^{\mathrm{T}}S^{-1}X)/X^{\mathrm{T}}S^{-1}X)\, X$ *where* $r(\cdot)$ *is a nondecreasing function bounded between 0 and* $2(p-2)/(n-p+2)$. *Then* $\delta(X, S)$ *is minimax for any density of the form* (6.53). *Furthermore* $\delta(X, S)$ *dominates* $\delta_0(X)$ *as soon as either* r *is strictly increasing or bounded away for 0 and* $\frac{2(p-2)}{n-p+2}$ *on a set of positive measure.*

Proof Setting

$$g(x, s) = -\frac{r(x^{\mathrm{T}}s^{-1}x)}{x^{\mathrm{T}}s^{-1}x}\, x,$$

we have

$$\mathrm{div}_x g(x, s) = -\left[(p-2)\frac{r(x^{\mathrm{T}}s^{-1}x)}{xs^{-1}x} + 2r'(x^{\mathrm{T}}s^{-1}x)\right]$$

by routine calculations. Now we have

$$\begin{aligned}
&D^*_{1/2}\big(g(x, s)g^{\mathrm{T}}(x, s)\big)\\
&= \sum_{i=1}^{p} \frac{\partial}{\partial s_{ii}}\left[\frac{r^2(x^{\mathrm{T}}s^{-1}x)}{(x^{\mathrm{T}}s^{-1}x)^2}\right]x_i^2 + \frac{1}{2}\sum_{i\neq j}\frac{\partial}{\partial s_{ij}}\left[\frac{r^2(x^{\mathrm{T}}s^{-1}x)}{(x^{\mathrm{T}}s^{-1}x)^2}\right]x_i x_j\\
&= \frac{2(x^{\mathrm{T}}s^{-1}x)^2 r(x^{\mathrm{T}}s^{-1}x)r'(x; s^{-1}x) - 2(x^{\mathrm{T}}s^{-1}x)r^2(x^{\mathrm{T}}s^{-1}x)}{(x^{\mathrm{T}}s^{-1}x)^4}\\
&\quad\times\left\{\sum_{i=1}^{p}\frac{\partial}{\partial s_{ii}}(x^{\mathrm{T}}s^{-1}x)X_i^2 + \frac{1}{2}\sum_{i\neq j}\frac{\partial}{\partial s_{ij}}(x^{\mathrm{T}}s^{-1}x)x_i x_j\right\}.
\end{aligned} \tag{6.63}$$

Using the fact that

$$\frac{\partial}{\partial s_{ij}}(x^{\mathrm{T}}s^{-1}x) = -(2-\delta_{ij})(x^{\mathrm{T}}s^{-1})_i(x^{\mathrm{T}}s^{-1})_j$$

it follows that the bracketed term in (6.63) equals

$$\begin{aligned}
&-\left\{\sum_{i=1}^{p}(x^{\mathrm{T}}s^{-1})_i^2 x_i^2 + \frac{1}{2}\sum_{i\neq j}2(x^{\mathrm{T}}s^{-1})_i(x^{\mathrm{T}}s^{-1})_j x_i x_j\right\}\\
&= -\sum_{1\le i,j\le p}(x^{\mathrm{T}}s^{-1})_i(x^{\mathrm{T}}s^{-1})_j x_j\\
&= -\left(\sum_{i=1}^{p}(x^{\mathrm{T}}s^{-1})_i X_i\right)^2\\
&= -(x^{\mathrm{T}}s^{-1}x)^2
\end{aligned}$$

and hence

$$D^*_{1/2}\big(g(x, s)g^{\mathrm{T}}(x, s)\big) = -2\left\{r(x^{\mathrm{T}}s^{-1}x)r'(x^{\mathrm{T}}sx) - \frac{r^2(x^{\mathrm{T}}s^{-1}x)}{x^{\mathrm{T}}s^{-1}x}\right\}.$$

Finally it is clear that

$$g^{\mathrm{T}}(x,s)s^{-1}g(x,s) = \frac{r^2(x^{\mathrm{T}}s^{-1}x)}{x^{\mathrm{T}}s^{-1}x}$$

so that the left-hand side of (6.61) equals

$$\begin{aligned}
&-2\left\{(p-2)\frac{r(x^{\mathrm{T}}s^{-1}x)}{x^{\mathrm{T}}s^{-1}x} + 2r'(x^{\mathrm{T}}s^{-1}x)\right\} + (n-p-2)\frac{r^2(x^{\mathrm{T}}s^{-1}x)}{x^{\mathrm{T}}s^{-1}x} \\
&-4\left\{r(x^{\mathrm{T}}s^{-1}x)r'(x^{\mathrm{T}}s^{-1}x) - \frac{r^2(x^{\mathrm{T}}s^{-1}x)}{x^{\mathrm{T}}s^{-1}x}\right\} \\
&= \frac{r(x^{\mathrm{T}}s^{-1}x)}{x^{\mathrm{T}}s^{-1}x}\left\{-2(p-2) + (n-p+2)r(x^{\mathrm{T}}s^{-1}x)\right\} \\
&\quad -4r'(x^{\mathrm{T}}s^{-1}x)\left\{1 + r(x^{\mathrm{T}}s^{-1}x)\right\} \\
&\leq 0,
\end{aligned} \tag{6.64}$$

according to the assumptions on $r(\cdot)$.

Hence the minimaxity of $\delta(X,S)$ follows. The domination result follows as well since strict inequality in (6.64) holds on a set of positive measure under the additional assumptions. □

***Proof of Lemma* 6.6** (Part 1) By definition, we have

$$E_\theta\left[g(X,S)^{\mathrm{T}}\Sigma^{-1}(X-\theta)\right] = \int_{\mathbb{R}^p\times\cdots\times\mathbb{R}^p}\int_{\mathbb{R}^p} g(x,s)^{\mathrm{T}}\Sigma^{-1}(x-\theta)$$
$$f\left((x-\theta)^{\mathrm{T}}\Sigma^{-1}(x-\theta) + \sum_{j=1}^{n-1} v_j^{\mathrm{T}}\Sigma^{-1}v_j\right) dx\, dv_1 \ldots dv_{n-1}\,.$$

Now applying the integration-by-slice in Lemma A.2 in Appendix A.5 with $\varphi(x) = \sqrt{(x-\theta)^{\mathrm{T}}\Sigma^{-1}(x-\theta)}$ to the inner most integral

$$\begin{aligned}
&I(v_1,\ldots,v_{n-1}) \\
&= \int_{\mathbb{R}^p} g(x,s)^{\mathrm{T}}\Sigma^{-1}(x-\theta)\, f\left((x-\theta)^{\mathrm{T}}\Sigma^{-1}(x-\theta) + \sum_{j=1}^{n-1} v_j^{\mathrm{T}}\Sigma^{-1}v_j\right) dx
\end{aligned}$$

gives

$$\nabla\varphi(x) = \frac{\Sigma^{-1}(x-\theta)}{\sqrt{(x-\theta)^{\mathrm{T}}\Sigma^{-1}(x-\theta)}}$$

and

$$
\begin{aligned}
&I(v_1, \ldots, v_{n-1}) \\
&= \int_0^\infty f\left(R^2 + \sum_{j=1}^{n-1} v_j^{\mathrm{T}} \Sigma^{-1} v_j\right) \int_{[\varphi=R]} \frac{g(x,s)^{\mathrm{T}} \Sigma^{-1}(x-\theta)}{\|\nabla\varphi(x)\|} d\sigma_R(x)\, dR \\
&= \int_0^\infty f\left(R^2 + \sum_{j=1}^{n-1} v_j^{\mathrm{T}} \Sigma^{-1} v_j\right) \int_{[\varphi=R]} g(x,s)^{\mathrm{T}} \sqrt{(x-\theta)^{\mathrm{T}} \Sigma^{-1}(x-\theta)} \\
&\quad \times \frac{\nabla\varphi(x)}{\|\nabla\varphi(x)\|}\, d\sigma_R(x)\, dR\,,
\end{aligned}
$$

according to the expression of $\nabla\varphi(x)$. Then, as $\sqrt{(x-\theta)^{\mathrm{T}} \Sigma^{-1}(x-\theta)} = R$ on $[\varphi = R]$, it follows using Stokes' theorem that

$$
\begin{aligned}
&I(v_1, \ldots, v_{n-1}) = \\
&\int_0^\infty R\, f\left(R^2 + \sum_{j=1}^{n-1} v_j^{\mathrm{T}} \Sigma^{-1} v_j\right) \int_{[\varphi=R]} g(x,s)\, \frac{\nabla\varphi(x)}{\|\nabla\varphi(x)\|}\, d\sigma_R(x)\, dR = \\
&\int_0^\infty R\, f\left(R^2 + \sum_{j=1}^{n-1} v_j^{\mathrm{T}} \Sigma^{-1} v_j\right) \int_{[\varphi\le R]} \mathrm{div}_x\, g(x,s)\, dx\, dR\,.
\end{aligned}
$$

Now, using Fubini's theorem gives

$$
\begin{aligned}
&I(v_1, \ldots, v_{n-1}) = \\
&\int_{\mathbb{R}^p} \mathrm{div}_x g(x,s) \int_{\sqrt{(x-\theta)^{\mathrm{T}} \Sigma^{-1}(x-\theta)}}^\infty R\, f\left(R^2 + \sum_{j=1}^{n-1} v_j^{\mathrm{T}} \Sigma^{-1} v_j\right) dR\, dx = \\
&\int_{\mathbb{R}^p} \mathrm{div}_x\, g(x,s)\, \frac{1}{2} \int_{(x-\theta)^{\mathrm{T}} \Sigma^{-1}(x-\theta)}^\infty f\left(r + \sum_{j=1}^{n-1} v_j^{\mathrm{T}} \Sigma^{-1} v_j\right) dr\, dx = \\
&\int_{\mathbb{R}^p} \mathrm{div}_x\, g(x,s) F\left((x-\theta)^{\mathrm{T}} \Sigma^{-1}(x-\theta) + \sum_{j=1}^{n-1} v_j^{\mathrm{T}} \Sigma^{-1} v_j\right) dx\,, \qquad (6.65)
\end{aligned}
$$

through the change of variable $r = R^2$ and by definition of the function F.

Finally integrating (6.65) with respect to the v_j gives an expression for the expectation $E_\theta[g(X, S)^{\mathrm{T}} \Sigma^{-1}(X - \theta)]$ and yields (6.58).

(Part 2) First note that, setting $G = S^{-1} T(X, S)$, we have

$$\operatorname{tr}\left(T(X, S)\Sigma^{-1}\right) = \operatorname{tr}\left(\Sigma^{-1} S\, G(X, S)\right).$$

Then, as $V = (V_1, \ldots, V_{n-1})$ and $S = V V^{\mathrm{T}}$, we have

$$\begin{aligned}
\operatorname{tr}\Big(\Sigma^{-1} S\, G(X, S)\Big) &= \operatorname{tr}\Big(G(X, S)\, \Sigma^{-1} S\Big) \\
&= \operatorname{tr}\left(G(X, S)\, \Sigma^{-1} \sum_{i=1}^{n-1} V_i\, V_i^{\mathrm{T}}\right) \\
&= \sum_{i=1}^{n-1} \operatorname{tr}\Big(V_i^{\mathrm{T}}\, G(X, S)\, \Sigma^{-1}\, V_i\Big) \\
&= \sum_{i=1}^{n-1} V_i^{\mathrm{T}}\, G(X, S)\, \Sigma^{-1}\, V_i\,. \qquad (6.66)
\end{aligned}$$

Now, according to Part 1 of Lemma 6.6 where the roles of X and θ are played by V_i and 0 respectively, it follows from (6.66) that

$$\begin{aligned}
E_{\theta,\Sigma}\left[\operatorname{tr}\Big(\Sigma^{-1} S\, G(X, S)\Big)\right] &= C \sum_{i=1}^{n-1} E^*_{\theta,\Sigma}\left[\operatorname{div}_{V_i}(G^{\mathrm{T}}(X, S)\, V_i)\right] \\
&= C\, E^*_{\theta,\Sigma}[A_1 + A_2]\,, \qquad (6.67)
\end{aligned}$$

where

$$\begin{aligned}
A_1 &= \sum_{i=1}^{n-1} \sum_{j=1}^{p} \sum_{m=1}^{p} \frac{\partial V_{mi}}{\partial V_{ji}}\, G^{\mathrm{T}}_{jm} \\
&= \sum_{i=1}^{n-1} \sum_{j=1}^{p} \sum_{m=1}^{p} \delta_{jm}\, G^{\mathrm{T}}_{jm}(X, S) \\
&= (n-1) \sum_{j=1}^{p} G^{\mathrm{T}}_{jj}(X, S) \\
&= (n-1)\operatorname{tr}(G(X, S)) \qquad (6.68)
\end{aligned}$$

and

$$
\begin{aligned}
A_2 &= \sum_{i=1}^{n-1} \sum_{j=1}^{p} \sum_{m=1}^{p} V_{mi} \, \frac{\partial G^{\mathrm{T}}_{jm}(X,S)}{\partial V_{ji}} \\
&= \sum_{i=1}^{n-1} \sum_{m=1}^{p} V_{mi} \sum_{j=1}^{p} \frac{\partial G^{\mathrm{T}}_{jm}(X,S)}{\partial V_{ji}} \\
&= \sum_{i=1}^{n-1} \sum_{m=1}^{p} V_{mi} (\nabla_{V^{\mathrm{T}}} G^{\mathrm{T}}(X,S))_{im} \\
&= \sum_{m=1}^{p} (V \nabla_{V^{\mathrm{T}}} G^{\mathrm{T}}(X,S))_{mm} \\
&= \mathrm{tr}(V \, \nabla_{V^{\mathrm{T}}} \, G^{\mathrm{T}}(X,S)) \, .
\end{aligned}
\tag{6.69}
$$

Finally, combining (6.67), (6.68) and (6.69), we obtain the second formula in (6.59).

As for the first formula in (6.59), it follows directly from the link between the differential expressions $D^*_{1/2} S \, G(X,S)$ and $\mathrm{tr}\big(V \, \nabla_{V^{\mathrm{T}}} \, G^{\mathrm{T}}(X,S)\big)$ given in Proposition 6.5 below, whose proof is given in Appendix A.7. □

Proposition 6.5 (Fourdrinier et al. 2016) *For any $p \times p$ matrix function $G(x,s)$ weakly differentiable with respect to s for any x,*

$$
2 \, D^*_{1/2}\big(S \, G(X,S)\big) = (p+1) \, \mathrm{tr}(G(X,S)) + \mathrm{tr}\big(V \, \nabla_{V^{\mathrm{T}}} \, G^{\mathrm{T}}(X,S)\big) \, . \tag{6.70}
$$

6.5 Shrinkage Estimators for Concave Loss in the Presence of a Residual Vector

In this section, we consider the case of concave loss and illustrate that certain classes of shrinkage estimators which properly use the residual vector have the strong robustness property of dominating the usual unbiased estimator uniformly over the class of spherically symmetric distributions, simultaneously for a broad class of concave loss functions. It extends and broadens the results of Sect. 5.5 to the residual vector case. We follow closely the development in Brandwein and Strawderman (1991a).

Specifically, let (X, U) be a $p + k$ dimensional vector with mean vector $(\theta, 0)$, where the dimensions of X and θ are equal to p and the dimensions of the residual vector U and its mean vector, 0, are equal to k, that is, $(X, U) \sim SS_{p+k}(\theta, 0)$. The loss function we consider is

$$
L(\theta, \delta) = \ell(||\theta - \delta||^2), \tag{6.71}
$$

for $\ell(t)$ a nonnegative concave monotone nondecreasing function.

The estimators we consider will be of the now familiar form

$$\delta(X, \|U\|^2) = X + a(S/(k+2))g(X), \tag{6.72}$$

where $S = ||U||^2$, and $g(\cdot)$ maps $\mathbb{R}^p$ into $\mathbb{R}^p$.

The following result, extracted from the development in Theorem 5.5 due to Brandwein and Strawderman (1991a) is basic to the development of this section.

Lemma 6.7 (Brandwein and Strawderman 1991a) *Let $X \sim SS_p(\theta)$, for $p \geq 4$ and let $g(X)$ map R^p into $\mathbb{R}^p$ be weakly differentiable, and such that*

(1) $||g(X)||^2/2 \leq -h(X) \leq -\nabla^{\mathrm{T}} g(X)$,
(2) *$-h(X)$ is superharmonic and $E_\theta[R^2 h(W)|R]$ is a nondecreasing function of R, where W has a uniform distribution on the sphere of radius R centered at θ.*

Then $E_\theta[||X+ag(X)-\theta||^2 - ||X-\theta||] \leq E[(-2a^2/r^2+2a/p)E_\theta[r^2 h(W)|r^2]]$, where $r^2 = ||X-\theta||^2$.

We will also need the following well known result (see e.g. the discussion at the end of Sect. 1.2).

Lemma 6.8 *Suppose $(X, U) \sim SS_{p+k}(\theta, 0)$. Then the random variable $\beta = ||X - \theta||^2/(||X-\theta||^2+S)$ has a $Beta(p/2, k/2)$ distribution, independent of $R^2 = ||X-\theta||^2 + S$, where $S = ||U||^2$.*

The main result is the following.

Theorem 6.6 *Suppose $(X, U) \sim SS_{p+k}(\theta, 0)$, that loss is given by loss (6.71) and that the estimator $\delta(X, S)$ is given by (6.72). Then $\delta(X, S)$ dominates the unbiased estimator X, provided that*

(1) *$g(X)$ satisfies assumptions (1) and (2) of Lemma 6.7,*
(2) *the concave nondecreasing function $\ell(t)$ also satisfies $t^\alpha \ell'(t)$ is nondecreasing,*
(3) $0 < a \leq (p-2-2\alpha)/p$.

Note first, by concavity of $\ell(\cdot)$, that $\ell(t) \leq \ell(y) + (t-y)\ell'(y)$. Hence the risk satisfies

$$\begin{aligned}
R(\theta, \delta) &= E[\ell(||X + \frac{aSg(X)}{k+2} - \theta||^2)] \\
&\leq E[\ell(||X-\theta||^2) + \ell'(||X-\theta||^2)(||X + \frac{aSg(X)}{k+2} - \theta||^2 - ||X-\theta||^2)] \\
&= R(\theta, X) + E[\ell'(||X-\theta||^2)(||X + \frac{aSg(X)}{k+2} - \theta||^2 - ||X-\theta||^2)].
\end{aligned}$$

It suffices to prove the second term in the above expression is negative. Now, let $r^2 = ||X-\theta||^2$, $R^2 = ||X-\theta||^2 + S$ (where $S = ||U||^2 = R^2 - r^2$), and note that the conditional distribution of X given r and R is $SS_p(\theta)$. Then it follows, using

Lemma 6.7 that

$$
\begin{aligned}
&E[\ell'(||X-\theta||^2)(||X+\frac{aSg(X)}{k+2}-\theta||^2-||X-\theta||^2)] \\
&= E[\ell'(r^2)E[||X+\frac{aSg(X)}{k+2}-\theta||^2-||X-\theta||^2|R,r]] \\
&\leq E[\ell'(r^2)E[(2(\frac{aS}{(k+2)r})^2-2\frac{aS}{(k+2)p})E_\theta[-r^2h(W)|r^2]|R,r]].
\end{aligned}
$$

Now using Lemma 6.8, this last expression may be written as

$$
\begin{aligned}
&2\,E\left[\ell'(\beta R^2)\left(\left\{\frac{a(1-\beta)R^2}{k+2}\right\}^2\frac{1}{\beta R^2}-\frac{a(1-\beta)R^2}{(k+2)p}\right)\right. \\
&\qquad\qquad \left.\times\, E_\theta[-\beta R^2 h(W)|\beta R^2]|R]\right] \\
&=\frac{2a}{k+2}E\left[(R^2(\beta R^2)^\alpha\ell'(\beta R^2)(\beta R^2)^{-\alpha}(1-\beta)\left(\frac{(1-\beta)a}{\beta(k+2)}-\frac{1}{p}\right)\right. \\
&\qquad\qquad \left.\times\, E_\theta[-\beta R^2 h(W)|\,\beta R^2]|\,R]\right].
\end{aligned}
$$

Next, for fixed R, by assumption (2) of Lemma 6.7 $E_\theta[-\beta R^2 h(W) \mid \beta R^2]$ is nonnegative and nondecreasing in β and by assumption (6.71) so is $\beta^\alpha\ell'(\beta R^2)$. Also $(1-\beta)/\beta$ is decreasing in β. Hence it follows from the covariance inequality (and independence of β and R) that the previous expression is less than or equal to

$$
\begin{aligned}
&\frac{2a}{k+2}E\left[[E_\theta[-\beta R^2 h(W)|\beta R^2]R^2(R^2\beta)^\alpha\ell'(\beta R^2)|R]E[\beta^\alpha(1-\beta)]\right. \\
&\qquad\qquad \left.\times\left(\frac{a(1-\beta)}{\beta(k+2)}-\frac{1}{p}\right)\right].
\end{aligned}
$$

Since the first expectation in this term is nonnegative, it suffices that the second expectation is negative. But this is equivalent to

$$
0\leq a\leq\frac{k+2}{p}E[\beta^\alpha(1-\beta)]/E[(\beta^\alpha(1-\beta)^2)/\beta]=(p-2-2\alpha)/p,
$$

which completes the proof. □

For the loss $L(\theta,\delta) = ||\theta-\delta||^q$, $\ell(t) = t^{q/2}$, it follows that $t^\alpha\ell'(t) = (q/2)t^{\alpha+q/2-1}$ is nondecreasing for $\alpha\geq 1-q/2$. Thus, the following corollary is immediate.

Corollary 6.4 *Under the loss* $L(\theta, \delta) = ||\theta - \delta||^q$, *for* $p > 4$ *and* $0 < q \leq 2$, *the estimator in Theorem* 6.6 *dominates* X *for* $0 < a \leq (p-4+2q)/p$ *simultaneously for all spherically symmetric distributions with finite second moment. It does so simultaneously for all such losses for* $0 < a \leq (p-4)/p$.

Note that the range of shrinkage constants for which domination holds includes $a = 1/2$ as soon as $p \geq 8$. For the usual James-Stein estimator,

$$\delta(X) = (1 - a(2(p-2)S)/((k+2)||X||^2))X, \tag{6.73}$$

the uniformly optimal constant for quadratic loss ($\ell(\cdot) = 1$) is $a = 1/2$ and hence this optimal estimator improves for all such l_q losses simultaneously for $p \geq 8$.

Chapter 7
Restricted Parameter Spaces

7.1 Introduction

In this chapter, we will consider the problem of estimating a location vector which is constrained to lie in a convex subset of $\mathbb{R}^P$. Estimators that are constrained to a set should be constrasted to the shrinkage estimators discussed in Sect.2.4.4 where one has "vague knowledge" that a location vector is in or near the specified set and consequently wishes to shrink toward the set but does not wish to restrict the estimator to lie in the set. Much of the chapter is devoted to one of two types of constraint sets, balls, and polyhedral cones. However, Sect.7.2 is devoted to general convex constraint sets and more particularly to a striking result of Hartigan (2004) which shows that in the normal case, the Bayes estimator of the mean with respect to the uniform prior over any convex set, $\mathscr{C}$, dominates X for all $\theta \in \mathscr{C}$ under the usual quadratic loss $\|\delta - \theta\|^2$.

Section 7.3 considers the situation where X is normal with a known scale but the constraint set is a ball, B, of known radius centered at a known point in $\mathbb{R}^p$. Here again, a natural estimator to dominate is the projection onto the ball $P_B X$. Hartigan's result of course applies and shows that the Bayes estimate corresponding to the uniform prior dominates X, but a finer analysis lead to domination over $P_B X$ (provided the radius of the ball is not too large relative to the dimension) by the Bayes estimator corresponding to the uniform prior on the sphere of the same radius.

Section 7.4 will consider estimation of a normal mean vector restricted to a polyhedral cone, $\mathscr{C}$, in the normal case under quadratic loss. Both the cases of known and unknown scale are treated. Special methods need to be developed to handle this restriction because the shrinkage functions considered are not necessarily weakly differentiable. Hence the methods of Chap.4 are not directly applicable. A version of Stein's lemma is developed for positively homogeneous sets which allows the analysis to proceed.

In general, if the constraint set, $\mathscr{C}$, is convex, a natural alternative to the UMVUE X, is $P_c X$ the projection of X onto $\mathscr{C}$. Our methods lead to Stein type shrinkage

D. Fourdrinier et al., *Shrinkage Estimation*, Springer Series in Statistics,
https://doi.org/10.1007/978-3-030-02185-6_7

estimators that shrink $P_c X$ which dominate $P_c X$, and hence X itself, when $\mathscr{C}$ is a polyhedral cone.

Section 7.5 is devoted to the case of a general spherically symmetric distribution with a residual vector when the mean vector is restricted to a polyhedral cone. As in Sect.7.4, the potential nondifferentiability of the shrinkage factors is a complication. We develop a general method that allows the results of Sect.7.4 for the normal case to be extended to the general spherically symmetric case as long as a residual vector is available. This method also allows for an alternative development of some of the results of Chap.6 that rely on an extension of Stein's lemma to the general spherical case.

7.2 Normal Mean Vector Restricted to a Convex Set

In this section, we treat the case $X \sim \mathscr{N}_p(\theta, \sigma^2 I_p)$ where σ^2 is known and where the unknown mean θ is restricted to lie in a convex set $\mathscr{C} \subseteq \mathbb{R}^p$ (with nonempty interior and sufficiently regular boundary), and where the loss is $L(\theta, \delta) = \|\delta - \theta\|^2$. We show that the (generalized) Bayes estimator with respect to the uniform prior distribution on $\mathscr{C}$, say $\pi(\theta) = \mathbb{1}_{\mathscr{C}}(\theta)$, dominates the usual (unrestricted) estimator $\delta_0(X) = X$. At this level of generality the result is due to Hartigan (2004) although versions of the result (in $\mathbb{R}^1$) date back to Katz (1961). We follow the discussion in Marchand and Strawderman (2004).

Theorem 7.1 (Hartigan 2004) *Let $X \sim \mathscr{N}_p(\theta, \sigma^2 I_p)$ with σ^2 known and $\theta \in \mathscr{C}$, a convex set with nonempty interior and sufficiently regular boundary $\partial\mathscr{C}$ ($\partial\mathscr{C}$ is Lipshitz of order 1 suffices). Then the Bayes estimator, $\delta_U(X)$ with respect to the uniform prior on $\mathscr{C}$, $\pi(\theta) = \mathbb{1}_{\mathscr{C}}(\theta)$, dominates $\delta_0(X) = X$ with respect to quadratic loss.*

Proof Without loss of generality, assume $\sigma^2 = 1$. Recall from (1.20) that the form of the Bayes estimator is $\delta_U(X) = X + \nabla m(X)/m(X)$ where, for any $x \in \mathbb{R}^p$,

$$m(x) \propto \int_{\mathscr{C}} \exp\left(-\frac{1}{2}\|x - \nu\|^2\right) d\nu.$$

The difference in risk between δ_U and δ_0 is $R(\theta, \delta_U) - R(\theta, \delta_0)$

$$\begin{aligned} R(\theta, \delta_U) - R(\theta, \delta_0) &= E_\theta\left[\left\|X + \frac{\nabla m(X)}{m(X)} - \theta\right\|^2 - \|X - \theta\|^2\right] \\ &= E_\theta\left[\frac{\|\nabla m(X)\|^2}{m^2(X)} + 2\frac{\nabla m(X)^{\mathrm{T}}(X - \theta)}{m(X)}\right]. \end{aligned} \tag{7.1}$$

Hartigan's clever development proceeds by applying Stein's Lemma 2.3 to only half of the cross product term in order to cancel the squared norm term in the above.

Indeed, since

$$E_\theta\left[(X-\theta)^{\mathrm{T}}\left(\frac{\nabla m(X)}{m(X)}\right)\right] = E_\theta\left[\operatorname{div}\left(\frac{\nabla m(X)}{m(X)}\right)\right]$$
$$= E_\theta\left[\frac{\Delta m(X)}{m(X)} - \frac{\|\nabla m(X)\|^2}{m^2(X)}\right],$$

(7.1) then becomes

$$R(\theta,\delta_U) - R(\theta,\delta_0) = E_\theta\left[\frac{\Delta m(X) + (X-\theta)^{\mathrm{T}}\nabla m(X)}{m(X)}\right]$$
$$= E_\theta\left[\frac{H(X,\theta)}{m(X)}\right] \tag{7.2}$$

with $H(x,\theta) = \Delta m(x) + (x-\theta)^{\mathrm{T}}\nabla m(x)$. Hence it suffices to show $H(x,\theta) \le 0$ for all $\theta \in \mathscr{C}$ and $x \in \mathbb{R}^p$. Using the facts that

$$\nabla_x \exp\left(-\frac{1}{2}\|x-\nu\|^2\right) = -\nabla_\nu \exp\left(-\frac{1}{2}\|x-\nu\|^2\right)$$

and

$$\Delta_x \exp\left(-\frac{1}{2}\|x-\nu\|^2\right) = \Delta_\nu \exp\left(-\frac{1}{2}\|x-\nu\|^2\right),$$

it follows that

$$H(x,\theta) \propto \Delta_x \int_{\mathscr{C}} \exp\left(-\frac{1}{2}\|x-\nu\|^2\right) d\nu + (x-\theta)'\nabla_x \int_{\mathscr{C}} \exp\left(\frac{1}{2}\|x-\nu\|^2\right) d\nu$$
$$= \int_{\mathscr{C}} \left[\Delta_\nu \exp\left(-\frac{1}{2}\|x-\nu\|^2\right) - (x-\theta)'\nabla_\nu \exp\left(-\frac{1}{2}\|x-\nu\|^2\right)\right] d\nu$$
$$= \int_{\mathscr{C}} \operatorname{div}_\nu \left[\nabla_\nu \exp\left(-\frac{1}{2}\|x-\nu\|^2\right) - (x-\theta)\exp\left(-\frac{1}{2}\|x-\nu\|^2\right)\right] d\nu$$
$$= \int_{\mathscr{C}} \operatorname{div}_\nu \left[(\theta-\nu)\exp\left(-\frac{1}{2}\|x-\nu\|^2\right)\right] d\nu.$$

By Stokes' theorem (see Sect.A.5 of the Appendix) this last expression can be expressed as

$$\int_{\partial\mathscr{C}} \eta^{\mathrm{T}}(\nu)(\theta-\nu)\exp\left(-\frac{1}{2}\|x-\nu\|^2\right) d\sigma(\nu)$$

where $\eta(\nu)$ is the unit outward normal to $\partial\mathscr{C}$ at ν and σ is the surface area Lebesgue measure on $\partial\mathscr{C}$. Finally, since $\mathscr{C}$ is convex and $\theta \in \mathscr{C}$, the angle between $\eta(\nu)$ and $\theta - \nu$ is obtuse for $\nu \in \partial\mathscr{C}$ and so $\eta^{\mathrm{T}}(\nu)(\theta - \nu) \le 0$, for all $\theta \in \mathscr{C}$ and $\nu \in \partial\mathscr{C}$, which implies the risk difference in (7.2) is nonpositive. □

Note that, if θ is in the interior of $\mathscr{C}$, $\mathscr{C}^0$, then $\eta'(\nu)(\theta - \nu)$ is strictly negative for all $\nu \in \partial\mathscr{C}$, and hence, $R(\theta, \delta_U) - R(\theta\delta_0) < 0$ for all $\theta \in \mathscr{C}^0$. However, if $\mathscr{C}$ is a pointed cone at θ_0, then $\eta^{\mathrm{T}}(\nu)(\theta_0 - \nu) \equiv 0$ for all $\nu \in \partial\mathscr{C}$ and $R(\theta_0, \delta_U) = R(\theta_0, \delta_0)$.

Note also that, if $\mathscr{C}$ is compact, the uniform prior on $\mathscr{C}$ is proper, and hence, $\delta_U(X)$ not only dominates $\delta_0(X)$ (on $\mathscr{C}$) but is also admissible for all p. On the other hand, if $\mathscr{C}$ is not compact, it is often (typically for $p \ge 3$) the case that δ_U is not admissible and alternative shrinkage estimators may be desirable.

Furthermore, it may be argued in general, that a more natural basic estimator which one should seek to dominate is $P_c X$, the projection of X onto $\mathscr{C}$ which is the MLE. We consider this problem for the case where $\mathscr{C}$ is a ball in Sect.7.3 and where $\mathscr{C}$ is a polyhedral cone in Sect.7.4. □

7.3 Normal Mean Vector Restricted to a Ball

When the location parameter $\theta \in \mathbb{R}^p$ is restricted, the most common constraint is a ball, that is, to a set for which $\|\theta\|$ is bounded above by some constant R. In this setting Bickel (1981) noted that, by an invariance argument and analyticity considerations, the minimax estimate is Bayes with respect to a unique spherically symmetric least favorable prior distribution concentrating on a finite number of spherical shells. This result extends what Casella and Strawderman (1981) obtained in the univariate case. Berry (1990) specified that when R is small enough, the corresponding prior is supported by a single spherical shell. In this section we address the issues of minimax estimation under a ball constraint.

Let $X \sim \mathcal{N}_p(\theta, \sigma^2 I_p)$, with unknown mean $\theta = (\theta_1, \ldots, \theta_p)$ and known σ^2, and with the additional information that $\sum_{i=1}^{p} (\theta_i - \tau_i)^2/\sigma^2 \le R^2$ where $\tau_1, \ldots, \tau_p, \sigma^2, R$ are known. From a practical point of view, a constraint as the one above signifies that the squared standardized deviations $|(\theta_i - \tau_i)/\sigma|^2$ are on average bounded by R^2/p. We are concerned here with estimating θ under quadratic loss $L(\theta, \delta) = \|\delta - \theta\|^2$. Without loss of generality, we proceed by setting $\sigma^2 = 1$ and $\tau_i = 0, i = 1, \ldots, p$, so that the constraint is the ball $B_R = \{\theta \in \mathbb{R}^p \mid \|\theta\| \le R\}$.

Since the usual minimax estimators take on values outside of B_R with positive probability, they are neither admissible nor minimax when θ is restricted to B_R. The argument given by Berry (1990) is the following. As for inadmissibility, it can be seen that these estimators are dominated by their truncated versions. Thus, the unbiased estimator X is dominated by the MLE $\delta_{MLE}(X) = (R/\|X\| \wedge 1)X$ (which is the truncation of X on B_R). Now, if an estimator which takes on values outside of B_R with positive probability were minimax, its truncated version would be minimax as well, with a strictly smaller risk. This is a contradiction since the risk function is continuous and attains its maximum in B_R. Further $\delta_{MLE}(X)$ is not admissible since, due to its non differentiability, it is not a generalized Bayes estimator. See Sect.3.4. For further discussions on this issue related to inadmissibility of estimators taking values on the boundary of a convex parameter space, see the review paper of Marchand and Strawderman (2004) and the monograph of van Eeden (2006).

As alternative estimators to $\delta_{MLE}(X)$, the Bayes estimators are attractive since they may have good frequentist performances in addition to their Bayesian property. Two natural estimators are the Bayes estimators with respect to the uniform distribution on the ball B_R and the uniform distribution on its boundary, the sphere $S_R = \{\theta \in \mathbb{R}^p \mid \|\theta\| = R\}$. We will see that the latter is particularly interesting.

The model is dominated by the Lebesgue measure on $\mathbb{R}^p$ and has likelihood L given by

$$\forall x \in \mathbb{R}^p \quad \forall \theta \in \mathbb{R}^p \quad L(x, \theta) = \frac{1}{(2\pi)^{p/2}} \exp\left(-\frac{1}{2}\|x - \theta\|^2\right). \tag{7.3}$$

Hence, if the prior distribution is the uniform distribution $\mathscr{U}_R$ on the sphere S_R, the marginal distribution has density m with respect to the Lebesgue measure on $\mathbb{R}^p$ given by

$$\begin{aligned} \forall x \in \mathbb{R}^p \quad m(x) &= \int_{S_R} L(x, \theta) d\mathscr{U}_R(\theta) \\ &= \frac{1}{(2\pi)^{p/2}} \exp\left(-\frac{1}{2}\|x\|^2\right) \exp\left(-\frac{1}{2}R^2\right) \int_{S_R} \exp(x^{\mathrm{T}}\theta) d\mathscr{U}_R(\theta), \end{aligned} \tag{7.4}$$

after expanding the likelihood in (7.3). Also, the posterior distribution given $x \in \mathbb{R}^p$ has density $\pi(\theta|x)$ with respect to the prior distribution $\mathscr{U}_R$ given by

$$\forall \theta \in S_R \quad \pi(\theta|x) = \frac{L(x, \theta)}{m(x)} = \frac{\exp(x^{\mathrm{T}}\theta)}{\int_{S_R} \exp(x^{\mathrm{T}}\theta) d\mathscr{U}_R(\theta)}, \tag{7.5}$$

thanks to the second expression of $m(x)$ in (7.4). As the loss is quadratic, the Bayes estimator δ_R is the posterior mean, that is,

$$\forall x \in S_R \quad \delta_R(x) = \int_{S_R} \theta \pi(\theta|x) d\mathscr{U}_R(\theta) = \frac{\int_{S_R} \theta \exp(x^{\mathrm{T}}\theta) d\mathscr{U}_R(\theta)}{\int_{S_R} \exp(x^{\mathrm{T}}\theta) d\mathscr{U}_R(\theta)}. \tag{7.6}$$

The Bayes estimator in (7.6) can be expressed through the modified Bessel function I_ν, solutions of the differential equation $z^2\varphi''(z) + z\varphi'(z) - (z^2 + \nu^2)\varphi(z) = 0$ with $\nu \geq 0$. More precisely, we will use the integral representation of the modified Bessel function

$$I_\nu(z) = \frac{(z/2)^\nu}{\pi^{1/2}\Gamma(\nu+1/2)} \int_0^\pi \exp(z\cos t)\sin^{2\nu} t dt \tag{7.7}$$

from which we may derive the formula

$$I'_\nu(z) = \frac{\nu}{z} I_\nu(z) + I_{\nu+1}(z). \tag{7.8}$$

Using the parametrization in terms of polar coordinates, we can see from the proof of Lemma 1.4 that, for any function h integrable with respect to $\mathscr{U}_R$,

$$\int_{S_R} h(\theta) d\mathscr{U}_R(\theta) = \frac{1}{\sigma(S)} \int_V h(\varphi_R(t_1,\dots,t_{p-1})) \sin^{p-2} t_1 \dots \sin t_{p-2} dt_1,\dots,dt_{p-1}$$

where $\sigma(S)$ is the area measure of the unit sphere and, as in (1.9), where $V = (0,\pi)^{p-2} \times (0, 2\pi)$ and for $(t_1,\dots,t_{n-1}) \in V$, $\varphi_R(t_1,\dots,t_{p-1}) = (\theta_1,\dots,\theta_p)$ with

$$\begin{aligned}
\theta_1 &= R\sin t_1 \sin t_2 \dots \sin t_{p-2} \sin t_{p-1}\\
\theta_2 &= R\sin t_1 \sin t_2 \dots \sin t_{p-2} \cos t_{p-1}\\
\theta_3 &= R\sin t_1 \sin t_2 \dots \cos t_{p-2}\\
&\vdots\\
\theta_{p-1} &= R\sin t_1 \cos t_2\\
\theta_p &= R\cos t_1.
\end{aligned}$$

Setting $h(\theta) = \exp(x^{\mathrm{T}}\theta)$ and choosing the angle between x and $\theta \in S_R$ as the first angle t_1 gives

$$\int_{S_R} \exp(x^{\mathrm{T}}\theta) d\mathscr{U}_R(\theta) = \frac{K}{\sigma(S)} \int_0^\pi \exp(\|x\| R \cos t_1) \sin^{p-2} t_1 dt_1$$

where

$$K = \int_{V^{\mathrm{T}}} \sin^{p-3} t_2 \dots \sin t_{p-2} dt_2,\dots,dt_{p-1}$$

with $V^{\mathrm{T}} = (0,\pi)^{p-3} \times (0,2\pi)$.

Therefore, according to (7.7), the marginal in (7.4) is proportional to

$$m_R(\|x\|) = \exp\!\left(-\frac{1}{2}\|x\|^2\right) \exp\left(-\frac{1}{2}R^2\right) \frac{I_{(p-2)/2}(\|x\|R)}{(\|x\|R)^{(p-2)/2}}, \tag{7.9}$$

the proportionality constant being independent of R. Then the Bayes estimator $\delta_R(X)$ can be obtained thanks to (1.20), that is, for any $x \in \mathbb{R}^p$,

$$\delta_R(x) = x + \nabla \log m(x).$$

As only the quantities depending on x matter, we have using (7.9), for any $x \in \mathbb{R}^p$,

$$\begin{aligned}
\delta_R(x) &= x + \nabla \log m_R(\|x\|) \\
&= -\frac{p-2}{2}\frac{\nabla(\|x\|R)}{\|x\|R} + \frac{\nabla I_{(p-2)/2}(\|x\|R)}{I_{(p-2)/2}(\|x\|R)} \\
&= -\frac{p-2}{2}\frac{x}{\|x\|^2} + \frac{\frac{x}{\|x\|}\left[\frac{p-2}{2\|x\|} I_{(p-2)/2}(\|x\|R) + R I_{p/2}(\|x\|R)\right]}{I_{(p-2)/2}(\|x\|R)} \\
&= \frac{R I_{p/2}(\|x\|R)}{I_{(p-2)/2}(\|x\|R)} \frac{x}{\|x\|},
\end{aligned} \tag{7.10}$$

where (7.8) has been used for the second to last equality.

Thus, according to (7.10), the Bayes estimator is expressed through a ratio of modified Bessel functions, that is, denoting by $\rho_\nu(t) = I_{\nu+1}/I_\nu$ with $t > 0$ and $\nu > -1/2$,

$$\delta_R(x) = \frac{R}{\|x\|} \rho_{p/2-1}(R\|x\|)x. \tag{7.11}$$

Before proceeding, we give two results from Marchand and Perron (2001).

(i) For sufficiently small R, say $R \le c_1(p)$, all Bayes estimators δ_π with respect to an orthogonally invariant prior π (supported on B_R) dominate $\delta_{MLE}(X)$;
(ii) The Bayes estimator $\delta_R(X)$ with respect to the uniform prior on the sphere S_R dominates $\delta_{MLE}(X)$ whenever $R \le \sqrt{p}$.

Note that Marchand and Perron (2002) extend the result in (ii) showing that domination of $\delta_R(X)$ over $\delta_{MLE}(X)$ subsists for some $m_0(p)$ such that $m_0(p) \ge \sqrt{p}$ and for $R \le m_0(p)$.

Various other dominance results, such as those pertaining to a fully uniform prior on B_R and other absolutely continuous priors are also available from Marchand and Perron (2001), but we will focus here on results (i) and (ii) above, following Fourdrinier and Marchand (2010).

With respect to important properties of $\delta_R(X)$, we point out that it is the optimal equivariant estimator for $\theta \in S_R$, and thus necessarily improves upon $\delta_{MLE}(X)$ on S_R. Furthermore, $\delta_R(X)$ also represents the Bayes estimator which expands greatest, or shrinks the least towards the origin (i.e., $\|\delta_\pi\| \le \|\delta_R(X)\|$ for all π supported on B_R; Marchand and Perron 2001). Despite this, as expanded upon below, $\delta_R(X)$

still shrinks more than $\delta_{MLE}(X)$ whenever $R \leq \sqrt{p}$, but not otherwise with the consequence of increased risk at $\theta = 0$ and failure to dominate $\delta_{MLE}(X)$ for large R. With the view of seeking dominance for a wider range of values of R, for potentially modulating these above effects by introducing more (but not too much) shrinkage, we consider the class of uniform priors supported on spheres S_α of radius α; $0 \leq \alpha \leq R$; about the origin, and their corresponding Bayes estimators δ_α. The choice is particularly interesting since the amount of shrinkage is calibrated by the choice of α (as formalized below), with the two extremes $\delta_R(X) \equiv \delta_R(X)$, and $\delta_0 \equiv 0$ (e.g., prior degenerate at 0). Moreover, knowledge about dominance conditions for the estimators δ_α may well lead, through further analytical risk and unbiased estimates of risk comparisons (e.g., Marchand and Perron (2001), Lemma 5 and the Remarks that follow), to implications relative to other Bayes estimators such as the fully uniform on B_R prior Bayes estimator.

Using Stein's unbiased estimator of risk technique, Karlin's sign change arguments, and a conditional risk analysis, Fourdrinier and Marchand (2010) obtain, for a fixed (R, p), necessary and sufficient conditions on α for δ_α to dominate $\delta_{MLE}(X)$.

Theorem 7.2

(a) *An unbiased estimator of the difference in risks*

$$\Delta_\alpha(\|\theta\|) = R(\theta, \delta_\alpha) - R(\theta, \delta_{MLE}(X))$$

is given by $D_\alpha(\|X\|) = D_{\alpha,1}(\|X\|)\ \mathbb{1}[0 \leq \|X\| \leq R] + D_{\alpha,2}(\|X\|)\ \mathbb{1}[\|X\| > R]$, *with*

$$\begin{aligned} D_{\alpha,1}(r) &= 2\alpha^2 + r^2 - 2p - 2\alpha r \rho_{p/2-1}(\alpha r) - \alpha^2 \rho^2_{p/2-1}(\alpha r), \text{ and} \\ D_{\alpha,2}(r) &= 2\alpha^2 - m^2 - \alpha^2 \rho^2_{p/2-1}(\alpha r) + 2Rr\{1 - \frac{\alpha}{R}\rho_{p/2-1}(\alpha r)\} - 2(p-1)\frac{R}{r}. \end{aligned}$$

(b) *For* $p \geq 3$, *and* $0 \leq \alpha \leq R$, $D_\alpha(r)$ *changes signs as a function of* r *according to the order:* (i) $(-, +)$ *whenever* $\alpha \leq \sqrt{p}$, *and* (ii) $(+, -, +)$ *whenever* $\alpha > \sqrt{p}$.

(c) *For* $p \geq 3$ *and* $0 \leq \alpha \leq R$, *the estimator* δ_α *dominates* $\delta_{MLE}(X)$ *if and only if*

 (i) $\Delta_\alpha(R) \leq 0$ *whenever* $\alpha \leq \sqrt{p}$ *or*
 (ii) $\Delta_\alpha(0) \leq 0$ *and* $\Delta_\alpha(R) \leq 0$, *whenever* $\alpha > \sqrt{p}$.

Proof

(a) Writing $\delta_{MLE}(x) = x + g_{MLE}(x)$ with $g_{MLE}(x) = (R/r - 1)x\ \mathbb{1}_{[r>R]}$ (with $r = \|x\|$) note that g_{MLE} is weakly differentiable. Then we have

$$2\text{div}g_{MLE}(x) + \|g_{MLE}(x)\|^2 = \left\{2(p-1)\frac{R}{r} - 2p + (R-r)^2\right\} \mathbb{1}_{(R,\infty)}(r)$$

and, by virtue of Stein's identity, $R(\theta, \delta_{MLE}) = E_\theta[\eta_{MLE}(X)]$ with

$$\eta_{MLE}(x) = p\,\mathbb{1}_{[0,R]}(r) + \left\{2(p-1)\frac{R}{r} - p + (R-r)^2\right\} \mathbb{1}_{(R,\infty)}(r). \qquad (7.12)$$

Analogously, as derived by Berry (1990), the representations of δ_α and $\frac{d}{dt}\rho_\nu(t)$ given in (7.11) and Lemma A.8, along with (2.3), permit us to write $R(\theta, \delta_\alpha) = E_\theta[\lambda_\alpha(X)]$ with

$$\lambda_\alpha(x) = 2\alpha^2 + r^2 - p - 2\alpha r \rho_{p/2-1}(\alpha r) - \alpha^2 \rho^2_{p/2-1}(\alpha r). \qquad (7.13)$$

Finally, the given expression for the unbiased estimator $D_\alpha(\|X\|)$ follows directly from (7.12) and (7.13).

(b) We begin with three intermediate observations which are proven below.

- (I) The sign changes of $D_{\alpha,1}(r)$; $r \in [0, R]$; are ordered according to one of the five following combinations: $(+)$, $(-)$, $(-,+)$, $(+,-)$, $(+,-,+)$;
- (II) $\lim_{r\to R^+}\{D_\alpha(r)\} = \lim_{r\to R^-}\{D_\alpha(r)\} + 2$;
- (III) the function $D_{\alpha,2}(r)$; $r \in (R, \infty)$ is either positive, or changes signs once from $-$ to $+$.

From properties (I), (II) and (III), we deduce that the sign changes of $D_\alpha(r)$ $r \in (0,\infty)$; an everywhere continuous function except for the jump discontinuity at R; are ordered according to one of the three following combinations: $(+)$, $(-,+)$, $(+,-,+)$. Now, recall that δ_α is a Bayes and admissible estimator of θ under squared error loss. Therefore, among the combinations above, $(+)$ is not possible since this would imply that δ_α is dominated by δ_{MLE} in contradiction with its admissibility. Finally, the two remaining cases are distinguished by observing that, $D_\alpha(0) = 2\alpha^2 - 2p \le 0$ if and only if $\alpha \le \sqrt{p}$.

Proof of (I): Begin by making use of Lemma A.8 to differentiate $D_{\alpha,1}$ and obtain:

$$r^{-1}\, D'_{\alpha,1}(r) = 2 - 2\frac{\alpha}{r}\rho_{p/2-1}(\alpha r) - 2\alpha^2 \rho'_{p/2-1}(\alpha r) - 2\alpha^3 \rho'_{p/2-1}(\alpha r)\frac{\rho_{p/2-1}(\alpha r)}{r}.$$

Since, the quantities $r^{-1}\rho_{p/2-1}(\alpha r)$ and $\rho'_{p/2-1}(\alpha r)$ are positive and decreasing in r by virtue of Lemma A.8, $r^{-1}\, D'_{\alpha,1}(r)$ is necessarily increasing in r, $r \in [0, R]$. Hence, $D'_{\alpha,1}(\cdot)$ has, on $[0, R]$, sign changes ordered as either: $(+)$, $(-)$, or $(-,+)$. Finally, observe as a consequence that $D_{\alpha,1}(\cdot)$ has at most two sign changes on $[0, R]$, and furthermore that, among the six possible combinations, the combination $(-,+,-)$ is not consistent with the sign changes of $D'_{\alpha,1}$.

Proof of (II): Follows by a direct evaluation of $D_{\alpha,1}(R)$ and $D_{\alpha,2}(R)$ which are given in part (a) of this lemma.

Proof of (III): First, one verifies from (7.13), part (a) of Lemma A.8, and part (c) of Lemma A.9 that $\lim_{r\to\infty} D_{\alpha,2}(r)$ is $+\infty$, for $\alpha < R$; and equal to $p-1$ if $\alpha = R$. Moreover, part (a) also permits us to express $D_{\alpha,2}(r)$; $r > R$; as $(1 - \frac{\alpha}{R}\,\rho_{p/2-1}(\alpha r))\sum_{i=1}^{3} H_i(\alpha, r)$ with

$$H_1(\alpha, r) = 2rR\left\{1 - \frac{(1 - \frac{\alpha^2}{R^2})R}{r\{1 - \frac{\alpha}{R}\rho_{p/2-1}(\alpha r)\}}\right\},$$

$$H_2(\alpha, r) = \frac{-2(p-1)R}{r\,\{\,1 - \frac{\alpha}{R}\rho_{p/2-1}(\alpha r)\,\}}$$

and

$$H_3(\alpha, r) = R^2 + \alpha R \rho_{p/2-1}(\alpha r).$$

Hence, to establish property (III), it will suffice to show that each one of the functions $H_i(\alpha, \cdot)$, $i = 1, 2, 3$, is increasing on (R, ∞) under the given conditions on (p, α, R). The properties of Lemma A.8 clearly demonstrate that $H_3(\alpha, \cdot)$ is increasing, and it is the same for $H_2(\alpha, \cdot)$ given also Lemmas A.8 and A.9 since

$$r(1 - \frac{\alpha}{R}\rho_{p/2-1}(\alpha r)) = r(1 - \rho_{p/2-1}(\alpha r)) + r\,(1 - \frac{\alpha}{R})\,\rho_{p/2-1}(\alpha r).$$

Finally, for the analysis of $H_1(\alpha, r), r > R$, begin by differentiation and a rearrangement of terms to obtain

$$\frac{\partial}{\partial r} H_1(\alpha, r) \geq 0 \Leftrightarrow T(R) \geq 0$$

where, for $r > R \geq \alpha$,

$$T(R) = (R - \alpha\rho_{p/2-1}(\alpha r))^2 - \alpha^2(R^2 - \alpha^2)\rho'_{p/2-1}(\alpha r).$$

But notice that $T(\alpha) = \alpha^2(1 - \rho_{p/2-1}(\alpha r))^2 \geq 0$, and

$$\begin{aligned}
\frac{1}{2}\frac{\partial T(R)}{\partial R} &= (R - \alpha\rho_{p/2-1}(\alpha r)) - R\alpha^2\rho'_{p/2-1}(\alpha r) \\
&\geq (\alpha - \alpha\rho_{p/2-1}(\alpha r)) - R\alpha^2\frac{1 - \rho_{p/2-1}(\alpha r)}{\alpha r} \\
&= \alpha(1 - \rho_{p/2-1}(\alpha r))\left(1 - \frac{R}{r}\right) \\
&\geq 0,
\end{aligned}$$

by Lemma A.9, part (b), since $r \geq R \geq \alpha$. The above establishes that $T(R) \geq T(\alpha) \geq 0$ for all $R \geq \alpha$, that $H_1(\alpha, r)$ increases in r, and completes the proof of the Theorem.

(c) The probability distribution of $\|X\|^2$ is $\chi^2_p(\lambda^2)$, so that the potential sign changes of $\Delta_\alpha(\lambda) = E_\lambda[D_\alpha(\|X\|)]$ are controlled by the variational properties of $D_\alpha(\cdot)$ in terms of sign changes (e.g., Brown et al. 1981). Therefore, in situation (i) with $\alpha \le \sqrt{p}$, it follows from part (b) of Lemma 7.2 that, as $\Delta_\alpha(\cdot)$ varies on $[0, \infty]$ (or $[0, R]$), the number of sign changes is at most one, and that such a change must be from $-$ to $+$. Therefore, since δ_α is admissible; and that the case $\Delta_\alpha(\lambda) \ge 0$ for all $\lambda \in [0, R]$ is not possible[1]; we must have indeed that $\Delta_\alpha(\cdot) \le 0$ on $[0, R]$ if and only if $\Delta_\alpha(R) \le 0$ establishing (i). A similar line of reasoning implies the result in (ii) as well. □

We refer to Fourdrinier and Marchand (2010) for other results. In particular, large sample determinations of these conditions are provided. Both cases where all such δ_α's, or no such δ_α's dominate δ_{MLE} are elicited. As a particular case, they establish that the boundary uniform Bayes estimator δ_R dominates δ_{MLE} if and only if $R \le k(p)$ with $\lim_{p\to\infty} k(p)/\sqrt{p} = \sqrt{2}$, improving on the previously known sufficient condition of Marchand and Perron (2001) for which $k(p) \ge \sqrt{p}$. Finally, they improve upon a universal dominance condition due to Marchand and Perron, by establishing that all Bayes estimators δ_π with π spherically symmetric and supported on the parameter space dominate δ_{MLE} whenever $R \le c_1(p)$ with $\lim_{p\to\infty} c_1(p)/\sqrt{p} = \sqrt{1/3}$.

See Marchand and Perron (2005) for analogous results for other spherically symmetric distributions including multivariate t.

Other significant contributions to the study of minimax estimation of a normal mean restricted to an interval or a ball of radius R, were given by Bickel (1981) and Levit (1981). These contributions consisted of approximations to the minimax risk and least favourable prior for large R under squared error loss. In particular, Bickel showed that for $p = 1$, as $R \to \infty$, the least favourable distributions rescaled to $[-1, 1]$ converge weakly to a distribution with density $\cos^2(\pi x/2)$, and that the minimax risks behave like $1 - \pi^2/(8R^2) + o(R^{-2})$. There is also a substantial literature on efficiency comparisons of minimax procedures and affine linear minimax estimators for various models, and restricted parameter spaces; see Donoho et al. (1990) and Johnstone and MacGibbon (1992) and the references therein.

Finally, we observe that the loss function plays a critical role. In the case where $p = 1$ and loss is absolute error $|d - \theta|$, $\delta_{MLE}(X)$ is admissible. See Isawa and Moritani (1997) and Kucerovsky et al. (2009).

7.4 Normal Mean Vector Restricted to a Polyhedral Cone

In this section, we consider first the case when $X \sim \mathcal{N}_p(\theta, \sigma^2 I_p)$ where σ^2 is known and θ is restricted to a polyhedral cone $\mathscr{C}$ and where the loss is $\|\delta - \theta\|^2$. Later in this Section, we will consider the case where σ^2 is unknown and, in

[1]The risks of δ_α and δ_{MLE} cannot match either, since a linear combination of these two distinct estimators would improve on δ_α.

Sect.7.5, the general spherically symmetric case with a residual vector. The reader is referred to Fourdrinier et al. (2006) for more details.

A natural estimator in this problem is $\delta_{\mathscr{C}}(X) = P_{\mathscr{C}}X$, the projection of X onto the cone $\mathscr{C}$. The estimator $\delta_{\mathscr{C}}$ is the MLE and dominates X which is itself minimax provided $\mathscr{C}$ has a nonempty interior. Our goal will be to dominate $\delta_{\mathscr{C}}$ and therefore also $\delta_0(X) = X$.

We refer the reader to Stoer and Witzgall (1970) and Robertson et al. (1988) for an extended discussion of polyhedral cones. Here is a brief summary. A polyhedral cone $\mathscr{C}$ is defined as the intersection of a finite number of half spaces, that is,

$$\mathscr{C} = \{x \mid a_i^{\mathrm{T}} x \leq 0, i = 1, \ldots, m\} \tag{7.14}$$

for n fixed vectors $a_1, \ldots, a_m$ in $\mathbb{R}^p$.

It is positively homogeneous, closed and convex, and, for each $x \in \mathbb{R}^p$, there exists a unique point $P_{\mathscr{C}}x$ in $\mathscr{C}$ such that $\|P_{\mathscr{C}}x - x\| = \inf_{y \in \mathscr{C}} \|y - x\|$.

We assume throughout that $\mathscr{C}$ has a nonempty interior, $\mathscr{C}^o$ so that $\mathscr{C}$ may be partitioned into $\mathscr{C}_i$, $i = 0, \ldots, m$, where $\mathscr{C}_0 = \mathscr{C}^o$ and $\mathscr{C}_i$, $i = 1, \ldots, m$, are the relative interiors of the proper faces of $\mathscr{C}$. For each set $\mathscr{C}_i$, let $D_i = P_{\mathscr{C}}^{-1}\mathscr{C}_i$ (the pre-image of $\mathscr{C}_i$ under the projection operator $P_{\mathscr{C}}$ and $s_i = \dim \mathscr{C}_i$). Then $D_i, i = 0, \ldots, m$ form a partition of $\mathbb{R}^p$, where $D_0 = \mathscr{C}_0$.

For each $x \in C_i$, we have $P_{\mathscr{C}}x = P_i x$ where P_i is the orthogonal linear projection onto the s_i−dimensional subspace L_i spanned by $\mathscr{C}_i$. Also for each such x, the orthogonal projection onto $L_i^{\perp}$, is equal to $P_{\mathscr{C}^*}x$ where $\mathscr{C}^* = \{y \mid x^{\mathrm{T}} y \leq 0\}$ is the polar cone corresponding to $\mathscr{C}$. Additionally, if $x \in D_i$, then $a P_i x + P_i^{\perp} x \in D_i$ for all $a > 0$, so D_i is positively homogeneous in $P_i x$ for each fixed $P_i^{\perp} x$ (see Robertson et al. 1988, Theorem 8.2.7). Hence we may express

$$\delta_{\mathscr{C}}(X) = \sum_{i=0}^{m} \mathbb{1}_{D_i}(X) P_i X. \tag{7.15}$$

The problem of dominating $\delta_{\mathscr{C}}$ is relatively simple in the case where $\mathscr{C}$ has the form $\mathscr{C} = \mathbb{R}_+^k \oplus \mathbb{R}^{p-k}$ where $\mathbb{R}_+^k = \{(x_1, \ldots, x_k) \mid x_i \geq 0,\ i = 1, \ldots, k\}$. In this case,

$$\delta_{\mathscr{C}}(X)_i = \begin{cases} X_i & \text{if } X_i \geq 0 \\ 0 & \text{if } X_i < 0 \end{cases} \text{for } i = 1, \ldots k \text{ and } \delta_{\mathscr{C}}(X)_i = X_i \text{ for } i = k+1, \ldots, p.$$

Furthermore, $\delta_{\mathscr{C}}(X)$ is weakly differentiable and the techniques of Chap.3 (i.e. Stein's lemma) are available.

As a simple example, suppose $\mathscr{C} = \mathbb{R}_+^p$, i.e. all coordinates of θ are nonnegative. Then $\delta_{\mathscr{C}_i}(X) = X_i + \partial_i(X) \quad i = 1, \ldots, p$ where

$$\partial_i(X) = \begin{cases} -X_i & \text{if } X_i < 0 \\ \ \ 0 & \text{if } X_i \geq 0. \end{cases}$$

Also, we may rewrite (7.15) as $X_+ = \sum_{i=1}^{2^p} \mathbb{1}_{O_i}(X) P_i X$ where $O_1 = \mathbb{R}_+^p$, and O_j, for $j > 1$, represent the other $2^p - 1$ orthants and P_i is the projection of X onto the space generated by the face of O_1 adjacent to O_i.

Then a James-Stein type shrinkage estimator that dominates X_+ is given by

$$\delta(X) = \sum_{i=1}^{2^p} \left(1 - \frac{c_i}{\|X_+\|^2}\right) X_+ \mathbb{1}_{O_i}(X)$$

where $c_i = (s_i - 2)_+$ and s_i is the number of positive coordinates in O_i. Note that shrinkage occurs only in those orthants such that $s_i \geq 3$.

The proof of domination follows essentially by the usual argument of Chap.3, Sect.2.4, applied separately to each orthant since X_+ and $X_+/\|X_+\|^2$ are weakly differentiable in O_i and

$$\nabla_X \frac{X_+}{\|X_+\|^2} \mathbb{1}_{O_i}(X) = \frac{s_i - 2}{\|X_+\|^2} \mathbb{1}_{O_i}(X),$$

provided $s_i > 2$. Note also that c_i may be replaced by any value between 0 and $2(s_i - 2)_+$.

Difficulties arise when the cone $\mathscr{C}$ is not of the form $\mathscr{C} = \mathbb{R}_+^k \oplus \mathbb{R}^{p-k}$ because the estimator $P_{\mathscr{C}} X$ may not be weakly differentiable (see Appendix A.1). In this case, a result of Sengupta and Sen (1991) can be used to give an unbiased estimator of the risk. Here is a version of their result.

Lemma 7.1 (Sengupta and Sen 1991) *Let $X \sim \mathscr{N}_p(\theta, \sigma^2 I_p)$ and $\mathscr{C}$ a positively homogeneous set. Then for every absolutely continuous function $h(\cdot)$ from $\mathbb{R}_+$ to $\mathbb{R}$ such that* $\lim_{y \to 0, \infty} h(y) y^{k+p/2} e^{-y/2} = 0$ *for all $k \geq 0$ and $E_\theta[h^2(\|X\|^2)\|X\|^2] < \infty$ we have*

$$\begin{aligned} E_\theta[h(\|X\|^2) X^{\mathrm{T}}(X - \theta) \mathbb{1}_{\mathscr{C}}(X)] &= \sigma^2 E_\theta[2\|X\|^2 h'(\|X\|^2) + p h(\|X\|^2) \mathbb{1}_{\mathscr{C}}(X)] \\ &= \sigma^2 E_\theta[\operatorname{div}(h(\|X\|^2) X) \mathbb{1}_{\mathscr{C}}(X)]. \end{aligned}$$

Note that for $\mathscr{C} = \mathbb{R}^p$, Lemma 7.1 follows from Stein's lemma with $g(X) = h(\|X\|^2) X$ provided $E[h(\|X\|^2)\|X\|^2] < \infty$. The possible non-weak differentiability of the function $h(\|X\|^2) X \mathbb{1}_{\mathscr{C}}(X)$ prevents a direct use of Stein's lemma for general $\mathscr{C}$.

Proof of Lemma 7.1 Note first that, if, for any θ, $E_\theta[\|g(X)\|] < \infty$, then

$$E_0\left[g(X) e^{X^{\mathrm{T}}\theta/\sigma^2}\right] = \sum_{k=0}^{\infty} E_0\left[\frac{g(X)(X^{\mathrm{T}}\theta/\sigma^2)^k}{k!}\right]$$

by the dominated convergence theorem. Without loss of generality, assume $\sigma^2 = 1$ and let

$$\begin{aligned}
A_\theta &= E_\theta[h(\|X\|^2)X^{\mathrm{T}}(X-\theta)\mathbb{1}_{\mathscr{C}}(X)] \qquad (7.16)\\
&= (2\pi)^{-p/2}e^{-\|\theta\|^2/2}\int_{\mathbb{R}^p} e^{-\|X\|^2/2}e^{X^{\mathrm{T}}\theta}h(\|X\|^2)(\|X\|^2 - X^{\mathrm{T}}\theta)\mathbb{1}_{\mathscr{C}}(X)dx\\
&= (2\pi)^{-p/2}e^{-\|\theta\|^2/2}\sum_{k=0}^{\infty} E_0\left[h(\|X\|^2)(\|X\|^2 - X^{\mathrm{T}}\theta)\frac{(X^{\mathrm{T}}\theta)^k}{k!}\mathbb{1}_{\mathscr{C}}(X)\right]\\
&= (2\pi)^{-p/2}e^{-\|\theta\|^2/2}\sum_{k=0}^{\infty}\frac{1}{k!}E_0\left[h(\|X\|^2)\mathbb{1}_{\mathscr{C}}(X)(X^{\mathrm{T}}\theta)^k(\|X\|^2 - k)\right]\\
&= (2\pi)^{-p/2}e^{-\|\theta\|^2/2}\sum_{k=0}^{\infty}\frac{1}{k!}E_0\left[h(\|X\|^2)\mathbb{1}_{\mathscr{C}}(X)\left(\frac{X^{\mathrm{T}}\theta}{\|X\|}\right)^k(\|X\|^{k+2} - k\|X\|^k)\right].
\end{aligned}$$

By the positive homogeneity of $\mathscr{C}$, we have $\mathbb{1}_{\mathscr{C}}(X) = \mathbb{1}_{\mathscr{C}}(X/\|X\|)$ and, by the independence of $\|X\|$ and $X/\|X\|$ for $\theta = 0$, we have

$$\begin{aligned}
A_\theta = (2\pi)^{-p/2}e^{-\|\theta\|^2/2}\sum_{k=0}^{\infty}\frac{1}{k!}E_0\left[\left(\frac{X^{\mathrm{T}}\theta}{\|X\|}\right)^k\mathbb{1}_{\mathscr{C}}\left(\frac{X}{\|X\|}\right)\right]&\\
\times E_0\left[h(\|X\|^2)\left(\|X\|^{k+2} - k\|X\|^k\right)\right]& \qquad (7.17)
\end{aligned}$$

When $\theta = 0$, $\|X\|^2$ has a central Chi-square distribution with p degrees of freedom. Thus, with $d = 1/2^{p/2}\Gamma(p/2)$, we have

$$\begin{aligned}
E_0[h(\|X\|^2)(\|X\|^{k+2} - k\|X\|^k)] &= d\int_0^\infty y^{p/2-1}h(y)(y^{(k+2)/2} - ky^{k/2})e^{-y/2}dy\\
&= d\int_0^\infty y^{(p+k)/2}h(y)e^{-y/2}dy - dk\int_0^\infty y^{(p+k)/2-1}h(y)e^{-y/2}dy
\end{aligned}$$

Integrating by parts, the first integral gives

$$\begin{aligned}
&\int_0^\infty y^{(p+k)/2}h(y)e^{-y/2}dy\\
&= 2\left[\int_0^\infty \frac{p+k}{2}y^{(p+k)/2-1}h(y)e^{-y/2}dy + \int_0^\infty y^{(p+k)/2}h'(y)e^{-y/2}dy\right]
\end{aligned}$$

and thus combining gives

$$E_0[h(\|X\|^2)(\|X\|^{k+2} - k\|X\|^k)] = d\int_0^\infty y^{k/2}[2yh'(y) + ph(y)]y^{(p-2)/2}e^{-y/2}dy$$
$$= E_0[(2\|X\|^2h'(\|X\|^2) + ph(\|X\|^2))\|X\|^k].$$

Thus (7.17) becomes

$$A_\theta = (2\pi)^{-p/2}e^{-\|\theta\|^2/2}\sum_{k=0}^\infty \frac{1}{k!}E_0\left(\frac{X^{\mathrm{T}}\theta}{\|X\|}\right)^k \mathbb{1}_{\mathscr{C}}(X)\left(\frac{X^{\mathrm{T}}\theta}{\|X\|}\right)$$
$$\times E_0[(2\|X\|^2h'(\|X\|^2) + ph(\|X\|^2))\|X\|^k]$$
$$= (2\pi)^{-p/2}e^{-\|\theta\|^2/2}\sum_{k=0}^\infty \frac{1}{k!}E_0[(X^{\mathrm{T}}\theta)^k\{2\|X\|^2h'(\|X\|^2) + ph(\|X\|^2)\}\mathbb{1}_{\mathscr{C}}(X)]$$
$$= E_\theta[\{2\|X\|^2h'(\|X\|^2) + ph(\|X\|^2)\}\mathbb{1}_{\mathscr{C}}(X)]$$

where the final identity follows by the dominated convergence theorem. □

General dominating estimators will be obtained by shrinking each P_iX in (7.15) on the set D_i. Recall that each D_i has the property that, if $x \in D_i$, then $aP_ix + P_i^\perp x \in D_i$ for all $a > 0$. The next result extends Lemma 7.1 to apply to projections P_i onto sets which have this conditional homogeneity property.

Lemma 7.2 *Let* $X \sim \mathscr{N}_p(\theta, \sigma^2 I_p)$ *and* P *be a linear orthogonal projection of rank* s. *Further, let* D *be a set such that, if* $x = Px + P^\perp x \in D$, *then* $aPx + P^\perp x \in D$ *for all* $a > 0$. *Then, for any absolutely continuous function* $h(\cdot)$ *on* $\mathbb{R}_+$ *into* $\mathbb{R}$ *such that* $\lim_{y\to 0,\infty} h(y)y^{(j+s)/2}e^{-y/2} = 0$ *for all* $j \geq 0$, *we have*

$$E_\theta[(X-\theta)^{\mathrm{T}}PXh(\|PX\|^2)\mathbb{1}_D(X)]$$
$$= \sigma^2E_\theta[\{2\|PX\|^2h'(\|PX\|^2) + sh(\|PX\|^2)\}\mathbb{1}_D(X)].$$

Proof By assumption $(Y_1, Y_2) = (PX, P^\perp X) \sim (\mathscr{N}_p(\eta_1, \sigma^2 P), \mathscr{N}_p(\eta_2, \sigma^2 P^\perp))$ where $(P\theta, P^\perp\theta) = (\eta_1, \eta_2)$. Also

$$A(\theta) = E_\theta[(X-\theta)^{\mathrm{T}}PXh(\|PX\|^2)\mathbb{1}_D(X)]$$
$$= E_\theta[(PX - P\theta)^{\mathrm{T}}PXh(\|PX\|^2)\mathbb{1}_D(X)]$$
$$= E_{\eta_1\eta_2}[(Y_1 - \eta_1)^{\mathrm{T}}Y_1h(\|Y_1\|^2)\mathbb{1}_{D'}(Y_1, Y_2)]$$

where

$$D' = \{(Y_1, Y_2)|(Y_1, Y_2) = (PX, P^\perp X) \in D\}.$$

On conditioning on Y_2 (which is independent of Y_1), and applying Lemma 7.1 to Y_1, we have

$$
\begin{aligned}
A(\theta) &= E_{\eta_2}[E_{\eta_1}[(Y_1 - \eta_1)^{\mathrm{T}} Y_1 h(\|Y_1\|^2) \mathbb{1}_{D'}(Y_1, Y_2)|Y_2]] \\
&= \sigma^2 E_{\eta_2}[E_{\eta_1}[\{2\|Y_1\|^2 h'(\|Y_1\|^2) + sh(\|Y_1\|^2)\} \mathbb{1}_{D'(Y_1,Y_2)}|Y_2]] \\
&= \sigma^2 E[\{2\|PX\|^2 h'(\|PX\|^2) + sh(\|PX\|^2)\} \mathbb{1}_D(X)].
\end{aligned}
$$

□

Now we use Lemma 7.2 to obtain the main domination result of this section.

Theorem 7.3 *Let $X \sim \mathcal{N}_p(\theta, \sigma^2 I_p)$ where σ^2 is known and θ is restricted to lie in the polyhedral cone $\mathscr{C}$, (7.14), with nonempty interior. Then, under loss $L(\theta, d) = \|d - \theta\|^2/\sigma^2$, the estimator*

$$
\delta(X) = \sum_{i=0}^{m} \left(1 - \sigma^2 \frac{r_i(\|P_i X\|^2)(s_i - 2)_+}{\|P_i X\|^2}\right) P_i X \mathbb{1}_{D_i}(X) \tag{7.18}
$$

dominates the rule $\delta_{\mathscr{C}}(X)$ given by (7.15) provided $0 < r_i(t) < 2$, $r_i(\cdot)$ is absolutely continuous, and $r_i'(t) \geq 0$, for each $i = 0, 1, \ldots, m$.

Proof The difference in risk between δ and $\delta_{\mathscr{C}}$ can be expressed as

$$
\begin{aligned}
\Delta(\theta) &= R(\theta, \delta) - R(\theta, \delta_{\mathscr{C}}) \\
&= \sum_{i=0}^{m} E_\theta \left[\sigma^2 \frac{r_i^2(\|P_i X\|^2)((s_i - 2)_+)^2}{\|P_i X\|^2}\right. \\
&\qquad \left. -2 \frac{r_i(\|P_i X\|^2)(s_i - 2)_+}{\|P_i X\|^2} (P_i X)^{\mathrm{T}} (P_i X - \theta)\right] \mathbb{1}_{D_i}(X).
\end{aligned} \tag{7.19}
$$

Now apply Lemma 7.2 (noting that $(P_i X)^{\mathrm{T}}(P_i X - \theta) = (P_i X)^{\mathrm{T}}(X - \theta)$) to each summand in the second term to get

$$
\begin{aligned}
\Delta(\theta) &= \sigma^2 \sum_{i=0}^{m} E_\theta \left[\frac{r_i^2(\|P_i X\|^2)((s_i - 2)_+)^2}{\|P_i X\|^2}\right. \\
&\qquad \left. - 2 \frac{r_i(\|P_i X\|^2)(s_i - 2)_+}{\|P_i X\|^2} - 4 r_i'(\|P_i X\|^2)(s_i - 2)_+\right] \mathbb{1}_{D_i}(X) \\
&\leq 0|
\end{aligned} \tag{7.20}
$$

since each $r_i'(\cdot) \geq 0$ and $0 < r_i(\cdot) < 2$. □

As noted in Chap.3, the case of an unknown σ^2 is easily handled provided an independent statistic $S \sim \sigma^2 \chi_k^2$ is available. For completeness we give the result for this case in the following theorem.

Theorem 7.4 *Suppose $X \sim \mathcal{N}_p(\theta, \sigma^2 I_p)$ and $S \sim \sigma^2 \chi^2_k$ with X independent of S. Let the loss be $\|d - \theta\|^2/\sigma^2$. Suppose that θ is restricted to the polyhedral cone $\mathscr{C}$,* (7.14), *with non-empty interior. Then the estimator*

$$\delta(X, S) = \sum_{i=0}^{m} \left(1 - \left(\frac{S}{k+2} \right) \frac{r_i(\|P_i X\|^2)(s_i - 2)_+}{\|P_i X\|^2} \right) P_i X \mathbb{1}_{D_i}(X) \qquad (7.21)$$

dominates $\delta_{\mathscr{C}}(X)$ given in (7.15) *provided $0 < r_i(\cdot) < 2$ and $r_i(\cdot)$ is absolutely continuous with $r_i'(\cdot) \geq 0$, for $i = 0, \ldots, m$.*

Many of the classical problems in ordered inference are examples of restrictions to polyhedral cones. Here are a few examples.

Example 7.1 (Orthant Restrictions) Estimation problems where k of the coordinate means are restricted to be greater than (or less than) a given set constants, can be transformed easily into the case where these same components are restricted to be positive. This is essentially the case for $\mathscr{C} = \mathbb{R}^k_+ \oplus \mathbb{R}^{p-k}$ mentioned earlier.

Example 7.2 (Ordered Means) The restrictions that $\theta_1 \leq \theta_2 \leq \ldots \leq \theta_p$ (or that a subset are so ordered) is a common example in the literature and corresponds to the finite set of half space restrictions $\theta_2 \geq \theta_1, \theta_3 \geq \theta_2, \ldots, \theta_p \geq \theta_{p-1}$.

Example 7.3 (Umbrella Ordering) The ordering $\theta_1 \leq \theta_2 \leq \ldots \leq \theta_k \geq \theta_{k+1} \geq \theta_{k+2}, \ldots, \theta_{p-1} \geq \theta_p$ corresponds to the polyhedral cone generated by the half space restrictions

$$\theta_2 - \theta_1 \geq 0, \theta_3 - \theta_2 \geq 0, \ldots, \theta_k - \theta_{k-1} \geq 0, \theta_{k+1} - \theta_k \leq 0, \ldots, \theta_p - \theta_p \leq 0.$$

In some examples, such as Example 7.1, it is relatively easy to specify P_i and D_i. In others, such as Example 7.2 and 7.3 it is more complicated. The reader is referred to Robertson et al. (1988) and references therein for further discussion of this issue.

7.5 Spherically Symmetric Distribution with a Mean Vector Restricted to a Polyhedral Cone

This Section is devoted to proving the extension of Theorem 7.4 to the case of a spherically symmetric distribution when a residual vector is present. Specifically we assume that $(X, U) \sim SS(\theta, 0)$ where $\dim X = \dim \theta = p$, $\dim U = \dim 0 = k$ and where θ is restricted to lie in a polyhedral cone, $\mathscr{C}$, with non-empty interior. Recall that the shrinkage functions in the estimator (7.21) are not necessarily weakly differentiable because of the presence of the indicator functions $I_{D_i}(X)$. Hence the methods of Chap.4 are not immediately applicable.

The following theorem develops the required tools for the desired extension of Theorem 7.4. It also allows for an alternative approach to the results in Sect.6.1 as well.

Theorem 7.5 (Fourdrinier et al. 2006) *Let* $(X, U) \sim \mathscr{N}_{p+k}((\theta, 0), \sigma^2 I_{p+k})$ *and assume* $f : \mathbb{R}^p \to \mathbb{R}$ *and* $g : \mathbb{R}^p \to \mathbb{R}^p$ *are such that*

$$E_{\theta,0}[(X-\theta)^{\mathrm{T}} g(X)] = \sigma^2 E_{\theta,0}[f(X)]$$

where both expected values exist for all $\sigma^2 > 0$*. Then, if* $(X, U) \sim SS_{p+k}(\theta, 0)$*, we have*

$$E_{\theta,0}[\|U\|^2 (X-\theta)^{\mathrm{T}} g(X)] = \frac{1}{k+2} E_{\theta,0}[\|U\|^4 f(X)]$$

provided either expected value exists.

Proof As (X, U) is normal, $X \sim \mathscr{N}_p(\theta, \sigma^2 I)$ and $\|U\|^2 \sim \sigma^2 \chi^2_k$ are independent, using $E[\|U\|^2] = k\sigma^2$ and $E[\|U\|^4] = \sigma^4 k(k+2)$, we have, for each fixed σ^2,

$$\begin{aligned} E_{\theta,0}[\|U\|^2 (X-\theta)^{\mathrm{T}} g(X)] &= k\sigma^2 E_{\theta,0}[(X-\theta)^{\mathrm{T}} g(X)] \\ &= k\sigma^4 E_{\theta,0}[f(X)] \\ &= \frac{1}{k+2} E_{\theta,0}[\|U\|^4 f(X)] \end{aligned} \tag{7.22}$$

For each θ (considered fixed), $\|X-\theta\|^2 + \|U\|^2$ is a complete sufficient statistic for $(X, U) \sim \mathscr{N}_{p+k}((\theta, 0), \sigma^2 I)$. Now noting

$$E_{\sigma^2}[E[\|U\|^2 (X-\theta)^{\mathrm{T}} g(X) \mid \|X-\theta\|^2 + \|U\|^2]] = E_{\theta,0}[\|U\|^2 (X-\theta)^{\mathrm{T}} g(X)]$$

and

$$\frac{1}{k+2} E_{\sigma^2}[\|U\|^4 f(X)] = \frac{1}{k+2} E_{\sigma^2}[E[\|U\|^4 f(X) \mid \|X-\theta\|^2 + \|U\|^2]]$$

it follows from (7.22) and the completeness of $\|X-\theta\|^2 + \|U\|^2$ that

$$\begin{aligned} &E[\|U\|^2 (X-\theta)^{\mathrm{T}} g(X) \mid \|X-\theta\|^2 + \|U\|^2] \\ &= \frac{1}{k+2} E[\|U\|^2 (X-\theta)^{\mathrm{T}} g(X) \mid \|X-\theta\|^2 + \|U\|^2] \end{aligned} \tag{7.23}$$

almost everywhere. We show at the end of this section that the functions in (7.23) are both continuous in $\|X-\theta\|^2 + \|U\|^2$, and hence, they are in fact equal everywhere.

Since the conditional distribution of (X, U) conditional on $\|X-\theta\|^2 + \|U\|^2 = R^2$ is uniform on the sphere centered at $(\theta, 0)$ for all spherically symmetric distributions (including the normal), the result follows on integration of (7.23) with respect to the radial distribution of (X, U). □

The main result of this section results from an application of Theorem 7.3 to the development of the proof of Theorem 7.4.

Theorem 7.6 *Let* $(X, U) \sim SS_{p+k}(\theta, 0)$ *and let* θ *be restricted to the polyhedral cone* $\mathscr{C}$, (7.14), *with nonempty interior. Then, under loss* $L(\theta, d) = \|d - \theta\|^2$, *the estimator*

$$\delta(X, U) = \sum_{i=0}^{m} \left(1 - \frac{\|U\|^2}{k+2} \frac{(s_i - 2)_+ r_i(\|P_i X\|^2)}{\|P_i X\|^2}\right) P_i X \mathbb{1}_{D_i}(X) \tag{7.24}$$

dominates $P_{\mathscr{C}} X = \delta_{\mathscr{C}}(X)$, *given in* (7.15) *provided,* $0 < r_i(\cdot) < 2$, $r_i(\cdot)$ *is absolutely continuous and* $r_i'(\cdot) \geq 0$ *for* $i = 0, \ldots, m$.

Proof The key observation is that, in passing from (7.19) to (7.20) in the proof of Theorem 7.3, we used Lemma 7.2 and the fact that $P_i X^{\mathrm{T}}(P X_i - \theta) = P_i X^{\mathrm{T}}(X - \theta)$ to establish that

$$E\left[\frac{r_i(\|P_i X\|^2)(s_i - 2)_+}{\|P_i X\|^2}(P_i X)^{\mathrm{T}}(P_i X - \theta)\mathbb{1}_{D_i}(X)\right]$$
$$= \sigma^2 E\left[\frac{r_i(\|P_i X\|^2)((s_i - 2)_+)^2}{\|P_i X\|^2} + 2r_i'(\|P_i X\|^2)(s_i - 2)_+ \mathbb{1}_{D_i}(X)\right].$$

Hence, by Theorem 7.5,

$$E\left[\frac{\|U\|^2}{k+2}\frac{r_i(\|P_i X\|^2)(s_i - 2)_+}{\|P_i X\|^2}(P_i X)^{\mathrm{T}}(P_i X - \theta)\mathbb{1}_{D_i}(X)\right]$$
$$= \sigma^2 E\left[\frac{\|U\|^4}{(k+2)^2}\left\{\frac{r_i(\|P_i X\|^2)((s_i - 2)_+)^2}{\|P_i X\|^2} + 2r_i'(\|P_i X\|^2)(s_i - 2)_+\right\}\mathbb{1}_{D_i}(X)\right].$$

It follows then, as in the proof of Theorem 7.3,

$$\begin{aligned} R(\theta, \delta(X, U)) - R(\theta, \delta_{\mathscr{C}}) = \sum_{i=0}^{m} E_\theta &\left[\frac{\|U\|^4}{(k+2)^2}\left\{\frac{r_i^2(\|P_i X\|^2)((s_i - 2)_+)^2}{\|P_i X\|^2}\right.\right. \\ &\left.\left. - \left\{\left(2\frac{r_i(\|P_i X\|^2)(s_i - 2)_+}{\|P_i X\|^2} + 4r_i'(\|P_i X\|^2)\right)(s_i - 2)_+\right\}\mathbb{1}_{D_i}(X)\right] \\ &\leq 0. \end{aligned} \tag{7.25}$$

□

Theorem 7.5 is an example of a meta result which follows from Theorem 7.6, and states roughly that, if one can find an estimator $X + \sigma^2 g(X)$ that dominates X for each σ^2 using a Stein-type differential equality in the normal case, then $X + \|U\|^2/(k+2)g(X)$ will dominate X in the general spherically symmetric case, $(X, U) \sim SS_{1+k}(\theta, U)$, under $L(\theta, \delta) = \|\delta - \theta\|^2$. The "proof" goes as follows.

Suppose one can show also that $E[(X-\theta)^{\mathrm{T}} g(X)] = \sigma^2 E[f(X)]$ in the normal case, and also that $\|g(x)\|^2 + 2f(x) \leq 0$, for any $x \in \mathbb{R}^p$. Then, in the normal case,

$$R(\theta, X - \sigma^2 g(X)) - R(\theta, X) = \sigma^4 E[\|g(X)\|^2 + 2f(X)] \leq 0.$$

Using Theorem 7.5 (and assuming finiteness of expectations), it follows in the general case that

$$R\left(\theta, X + \frac{\|U\|^2}{k+2} g(X)\right) - R(\theta, X) = E\left[\frac{\|U\|^4}{(k+2)^2}\{\|g(X)\|^2 + 2f(X)\}\right] \leq 0.$$

In this Section, application of the above meta-result had the additional complication of a separate application (to $P_i X$ instead of X) on each D_i but the basic idea is the same. The results of Chap.6 which rely on extending a version of Stein's lemma to the general spherically symmetric case can be proved in the same way.

We close this Section with a result that implies the claimed continuity of the conditional expectations in (7.23).

Lemma 7.3 *Let $(X, U) \sim SS_{p+k}(\theta, 0)$ and let $\alpha \in N$. Assume $\varphi(\cdot)$ is such that for any $R > 0$, the conditional expectation*

$$f(R) = E_{(\theta,0)}[\|U\|^\alpha \varphi(X) \mid \|X - \theta\|^2 + \|U\|^2 = R^2]$$

exists. Then the function f is continuous on $\mathbb{R}_+$.

Proof Assume without loss of generality that $\theta = 0$ and $\varphi(\cdot) \geq 0$. Since the conditional distribution or (X, U) conditional on $\|X\|^2 + \|U\|^2 = R^2$ is the uniform distribution U_R on the sphere $S_R = \{y \in R^{p+k} / \|y\| = R\}$ centered at 0 with radius R, we have

$$f(R) = \int_{S_R} \|u\|^\alpha \varphi(x) d\mathscr{U}_R(x, u).$$

Since $\|u\|^2 = R^2 - \|x\|^2$ for any $(x, u) \in S_R$ and X has distribution concentrated on the ball $B_r = \{x \in \mathbb{R}^p | \|x\| \leq R\}$ in $\mathbb{R}^p$ with density proportional to $R^{2-(p+k)}((R^2 - \|x\|^2)^{k/2-1})$ we have that $R^{p+k-2} f(R)$ is proportional to

$$g(R) = \int_{B_R} (R^2 - \|x\|^2)^{(k+\alpha)/2-1} \varphi(x) dx.$$

$$= \int_0^R \int_{S_r} (R^2 - \|x\|^2)^{(k+\alpha)/2-1} \varphi(x) d\sigma_r(x) dr$$

$$= \int_0^R (R^2 - r^2)^{(k+\alpha)/2-1} H(r) dr$$

where

$$H(r) = \int_{S_r} \varphi(x) d\sigma_r(x)$$

and where σ_r is the area measure on the sphere S_r. Since $H(\cdot)$ and $(k+\alpha)/2 - 1$ are non-negative, the family of integrable functions $r \to K(R, r) = (R^2 - r^2)^{(k+\alpha)/2-1} H(r) I_{[0,R]}(r)$, indexed by R, is nondecreasing in R and bounded above for $R < R_0$ by the integrable function $K(R_0, r)$. Then the continuity of $g(R)$, and hence of $f(R)$, is guaranteed by the dominated convergence theorem. □

Note that the continuity of (7.23) is not necessary for the application to $(X, U) \sim SS_{p+k}(\theta, 0)$ if (X, U) has a density, since then equality a.e. suffices.

Chapter 8
Loss and Confidence Level Estimation

8.1 Introduction

Suppose X is an observation from a distribution P_θ parameterized by an unknown parameter θ. In classical decision theory, after selecting an estimation procedure $\varphi(X)$ of θ, it is typical to evaluate it through a criterion, i.e. a loss, $L(\theta, \varphi(X))$, which represents the cost incurred by the estimator $\varphi(X)$ when the unknown parameter equals θ. In the long run, as it depends on the particular value of X, this loss cannot be appropriate to assess the performance of the estimator φ. Indeed, to be valid (in the frequentist sense), a global evaluation of such a statistical procedure should be based on all the possible observations. Consequently, it is common to report the risk $R(\theta, \varphi) = E_\theta[L(\theta, \varphi(X))]$ as a gauge of the efficiency of φ (E_θ denotes expectation with respect to P_θ). Thus, we have at our disposal a measure of the long run performance of $\varphi(X)$ for each value of θ. However, although this notion of risk can effectively be used in comparing $\varphi(X)$ with other estimators, it is inaccessible since θ is unknown. A common and, in principle, accessible, frequentist risk assessment is the maximum risk $\bar{R}_\varphi = \sup_\theta R(\theta, \varphi)$.

By construction, this last report on the estimation procedure is non-data-dependent (as we were guided by a global notion of the accuracy of $\varphi(X)$). However there exist situations where the fact that the observation X has a particular value x may influence the judgment on a statistical procedure. A particularly clarifying example is given by the following simple confidence interval estimation (which can also be seen as a loss estimation problem). Assume that the observation is a pair (X_1, X_2) of independent copies of a random variable X satisfying, for $\theta \in \mathbb{R}$,

$$P[X = \theta - 1] = P[X = \theta + 1] = \frac{1}{2} .$$

Then it is clear that the confidence interval for θ defined by

D. Fourdrinier et al., *Shrinkage Estimation*, Springer Series in Statistics,
https://doi.org/10.1007/978-3-030-02185-6_8

$$I(X_1, X_2) = \left\{ \theta \in \mathbb{R} \;\middle|\; \left| \frac{X_1 + X_2}{2} - \theta \right| < \frac{1}{2} \right\}$$

satisfies

$$\mathbb{1}_{[I(X_1,X_2)\ni\theta]} = \begin{cases} 1 \text{ if } X_1 \neq X_2 \\ \\ 0 \text{ if } X_1 = X_2 \end{cases}$$

so that it suffices to observe $(X_1, X_2) = (x_1, x_2)$ in order to know exactly whether $I(x_1, x_2)$ contains θ or not.

The previous (somewhat ad hoc) example indicates that data-dependent reports are relevant. In our estimation context when $X = x$, note that, if it were available (but θ is unknown), it would be the loss $L(\theta, \varphi(x))$ itself that should serve as a perfect measure of the accuracy of φ. It is then natural to estimate $L(\theta, \varphi(x))$ by a data-dependent estimator $\delta(X)$, a new estimator called a loss estimator which will serve as a data-dependent report (instead of $\bar{R}_\varphi$). This is a conditional approach in the sense that accuracy assessment is made on a data-dependent quantity, the loss, instead of the risk.

Remark 8.1 Throughout this chapter, we will typically use $\varphi(X)$ to denote the estimator of the unknown parameter, θ, and $\delta(X)$ to denote the corresponding estimator of loss, $L(\theta, \varphi(X))$.

To evaluate the extent to which $\delta(X)$ successfully estimates $L(\theta, \varphi(X))$, another loss is required and it has become standard to use the squared error

$$L^*(\theta, \varphi(X), \delta(X)) = (\delta(X) - L(\theta, \varphi(X)))^2 \,, \tag{8.1}$$

for simplicity. In so far as we are thinking in terms of long-run frequencies, we adopt a frequentist approach to evaluating the performance of L^* by averaging over the sampling distribution of X given θ, that is, by using a new notion of risk

$$\mathscr{R}(\theta, \varphi, \delta) = E_\theta[L^*(\theta, \varphi(X), \delta(X))] = E_\theta[(\delta(X) - L(\theta, \varphi(X)))^2] \,. \tag{8.2}$$

As $\bar{R}_\varphi$ reports on the worst situation (the maximum risk), we may hope that a competitive data-dependent report $\delta(X)$ improves on $\bar{R}_\varphi$ under new risk (8.1), that is, for all θ, satisfies

$$\mathscr{R}(\theta, \varphi, \delta) \leq \mathscr{R}(\theta, \varphi, \bar{R}_\varphi) \,. \tag{8.3}$$

More generally, a reference loss estimator δ_0 will be dominated by a competitive estimator δ if, for all θ,

$$\mathscr{R}(\theta, \varphi, \delta) \leq \mathscr{R}(\theta, \varphi, \delta_0) \,, \tag{8.4}$$

with strict inequality for some θ.

Note that, unlike the usual estimation setting where the quantity of interest is a function of the parameter θ, loss estimation involves a function of both θ and X (the data). This feature may make the statistical analysis more difficult but it is clear that the usual notions of minimaxity, admissibility, etc, and their methods of proof can be directly adapted to that situation. Also, although frequentist interpretability was evoked above, it is easily seen that a Bayesian approach could be naturally based on the usual Bayes estimator φ_B of θ and the posterior loss $\delta_B(X) = E[L(\theta, \varphi_B)|X]$.

The problem of estimating a loss function has been considered by Sandved (1968) who developed a notion of an unbiased estimator of $L(\theta, \varphi(X))$ in various settings. However the underlying conditional approach traces back to Lehmann and Sheffé (1950) who estimated the power of a statistical test. Kiefer, in a series of papers (1975, 1976, 1977), developed conditional and estimated confidence theories through frequentist interpretability. A subjective Bayesian approach was compared by Berger (1985b,c,d) with the frequentist one.

We propose the following definition of an unbiased estimate of loss.

Definition 8.1 $\delta(X)$ is an unbiased estimator of the loss $L(\theta, \varphi(X))$, if $E_\theta[\delta(x)] = E_\theta[L(\theta, \varphi(X))]$ for all $\theta \in \Omega$. Hence an unbiased estimator of loss is also an unbiased estimator of risk.

Johnstone (1988) considered the (in)admissibility of unbiased estimators of loss for the maximum likelihood estimator $\varphi_0(X) = X$ and for the James-Stein estimator $\varphi^{JS}(X) = (1 - (p-2)/||X||^2)X$ of a p-variate normal mean θ (with cov$X = I_p$) based on Stein's lemma (Theorem 2.1). For $\varphi_0(X) = X$, the unbiased estimator of the quadratic loss $L(\theta, \varphi_0(X)) = ||\varphi_0(X) - \theta||^2$ which satisfies, for all θ,

$$E_\theta[\delta_0] = E_\theta[L(\theta, \varphi_0(X))] = R(\theta, \varphi_0)\,, \tag{8.5}$$

is $\delta_0 = \bar{R}_\varphi = p$ (where we assume $\sigma^2 = 1$). Johnstone proved that (8.3) is satisfied with the competitive estimator $\delta(X) = p - 2(p-4)/||X||^2$ when $p \geq 5$, with the risk difference between δ_0 and δ being expressed as $-4(p-4)^2 E_\theta[1/||X||^4]$.

For the James-Stein estimator φ^{JS}, the unbiased estimator of loss, from Corollary 2.1 (3), is itself data-dependent and equal to $\delta_0^{JS}(X) = p - (p-2)^2/||X||^2$. Jonhstone showed that improvement on δ_0^{JS} can be obtained with $\delta^{JS}(X) = p - (p-2)^2/||X||^2 + 2\,p/||X||^2$ when $p \geq 5$, with strict inequality in (8.4) for all θ since the difference in risk between δ^{JS} and δ_0^{JS} equals $-4\,p^2\,E_\theta[1/||X||^4]$.

In Sect. 8.2, we develop the quadratic loss estimation problem for a multivariate normal mean. After a review of the basic ideas, a new class of loss estimators is constructed in Sect. 8.2.1. In Sect. 8.2.2, we turn our focus on some interesting and surprising behavior of Bayesian assessments; this paradoxical result is illustrated in a general inadmissibility theorem. Section 8.3 is devoted to the case where the variance is unknown. Extensions to the spherical case are given in Sect. 8.4. In Sect. 8.4.1, we consider the general case of a spherically symmetric distribution around a fixed vector $\theta \in \mathbb{R}^p$. In Sect. 8.4.2, these ideas are then generalized to the case where a residual vector is available. Section 8.5 discusses some connections of loss estimation with model selection. Section 8.6 covers topics in confidence set assessment, while Sect. 8.7 presents material on a dimension cut-off phenomenon for improved loss estimation associated with differential operators. We conclude by mentioning a number of applied and theoretical developments of loss estimation not covered in this overview.

8.2 Quadratic Loss Estimation: Multivariate Normal with Known Variance

8.2.1 Dominating Unbiased Estimators of Loss

Let X be a random vector having a multivariate normal distribution $\mathcal{N}_p(\theta, I_p)$ with unknown mean θ and identity covariance matrix I_p. To estimate θ, the observable X is itself a reference estimator, being both MLE and unbiased. It is convenient to write any estimator of θ as $\varphi(X) = X + g(X)$ for a certain function g from $\mathbb{R}^p$ into $\mathbb{R}^p$. Under squared error loss $||\varphi(X) - \theta||^2$, the (quadratic) risk of φ is defined by

$$R(\theta, \varphi) = E_\theta[||\varphi(X) - \theta||^2] \tag{8.6}$$

where E_θ denotes the expectation with respect to $\mathcal{N}_p(\theta, I_p)$.

Clearly, the risk of the MLE X equals p and $\varphi(X)$ will be a reasonable estimator only if its risk is finite. As seen in Chap. 2,

$$E_\theta[||g(X)||^2] < \infty \tag{8.7}$$

is a necessary and sufficient condition for this finiteness. We assume this in what follows. Note again that in this chapter, the estimators of θ with typically be denoted by $\varphi(X)$ while the estimators of loss will typically be denoted by $\delta(X)$. We will largely focus on improving some reference estimator of loss $\delta_0(X)$ for a fixed estimator $\varphi(X)$ of θ.

Recall from Corollary 2.1 (3) that if $X \sim \mathcal{N}(\theta, I_p)$ that an, in fact the unique, unbiased estimator of loss for an estimator of θ of the form $\varphi(X) = X + g(X)$ is given by

$$\delta_0(X) = p + 2 \operatorname{div} g(X) + ||g(X)||^2. \tag{8.8}$$

Hence for the UMVUE, MLE, MRE $\varphi(X) = X$ it follows that $\delta_0(X) = p$ and for the James-Stein estimator, $\delta_a^{JS}(X) = \left(1 - a/\|X\|^2\right) X$ in (2.13), the proof of Theorem 2.2 shows that $\delta_0(X) = p + (a^2 - 2a(p-2))/||X||^2$.

Any competitive loss estimator $\delta(X)$ can be written as $\delta(X) = \delta_0(X) - \gamma(X)$ for a certain function $\gamma(X)$, which can be interpreted as a correction to $\delta_0(X)$. If the MLE is concerned (that is, if $g(X) = 0$), we may expect that an improvement on $\delta_0(X) = p$ would be obtained with a function $\gamma(X)$ satisfying the requirement expressed by condition (8.3). Note also that, similarly to the finiteness risk condition (8.7), we will require that

$$E_\theta[\gamma^2(X)] < \infty \tag{8.9}$$

to assure that the risk of $\delta(X)$ is finite.

Using straightforward algebra, the risk difference (under loss (8.1)) $\mathscr{D}(\theta, \varphi, \delta) = \mathscr{R}(\theta, \varphi, \delta) - \mathscr{R}(\theta, \varphi, \delta_0)$ simplifies in

$$\mathscr{D}(\theta, \varphi, \delta) = E_\theta[\gamma^2(X) - 2\,\gamma(X)\,\delta_0(X)] + 2\,E_\theta[\gamma(X)\,||\varphi(X) - \theta||^2]\,. \tag{8.10}$$

Conditions for which $\mathscr{D}(\theta, \varphi, \delta) \leq 0$ will be formulated after finding, along the lines of Stein's techniques used above, an unbiased estimate of the term $\gamma(X)\,||\varphi(X) - \theta||^2$ in the last expectation. This is given in the next lemma. Recall, for a function g from $\mathbb{R}^p$ into $\mathbb{R}^p$, that Stein's lemma (Theorem 2.1) states that

$$E_\theta[(X-\theta)^{\mathrm{T}} g(X)] = E_\theta[\operatorname{div} g(X)]\,, \tag{8.11}$$

provided that these expectations exist.

Lemma 8.1 *Let $X \sim \mathscr{N}_p(\theta, I_p)$ and γ be a twice weakly differentiable function such that $E_\theta[||X-\theta||^2\,|\gamma(X)|] < \infty$. Then*

$$E_\theta[||X-\theta||^2\,\gamma(X)] = E_\theta[p\,\gamma(X) + \Delta\gamma(X)]\,.$$

Proof Writing

$$||X-\theta||^2\,\gamma(X) = (X-\theta)^{\mathrm{T}}(X-\theta)\,\gamma(X) \tag{8.12}$$

naturally leads to an iteration of Stein's identity and involves the twice weak differentiability of γ, we have

$$\begin{aligned} E_\theta[||X-\theta||^2\,\gamma(X)] &= E_\theta[\operatorname{div}((X-\theta)^{\mathrm{T}}\,\gamma(X))] \\ &= E_\theta[p\,\gamma(X) + (X-\theta)^{\mathrm{T}}\,\nabla\gamma(X)]\,, \end{aligned} \tag{8.13}$$

by the property of the divergence. Then again, applying Stein's identity to the last term in (8.13) gives

$$E_\theta[(X-\theta)^{\mathrm{T}}\,\nabla\gamma(X)] = E_\theta[\operatorname{div}(\nabla\gamma(X))] = E_\theta[\Delta\gamma(X)] \tag{8.14}$$

by definition of the Laplacian . Finally, gathering (8.13) and (8.14), we obtain that

$$E_\theta[||X-\theta||^2\,\gamma(X)] = E_\theta[p\,\gamma(X) + \Delta\gamma(X)]\,, \tag{8.15}$$

which completes the proof. □

We are now in a position to provide an unbiased estimator of $\mathscr{D}(\theta, \varphi, \delta)$. Its nonpositivity will be a sufficient condition for $\mathscr{D}(\theta, \varphi, \delta) \leq 0$ and hence for δ to improve on δ_0.

Lemma 8.2 *Suppose γ is a twice weakly differentiable function and all expectations are finite. Then*

$$\mathscr{D}(\theta, \varphi, \delta) = E_\theta[\gamma^2(X) + 4\,\nabla\gamma(X)^{\mathrm{T}}\,g(X) + 2\,\Delta\gamma(X)]. \tag{8.16}$$

Note that, when $X \sim \mathscr{N}_p(\theta, \sigma^2 I_p)$, then if the correction term is replaced by $\sigma^2\,\gamma(X)$ the risk difference $\mathscr{D}(\theta, \varphi, \delta)$ becomes replaced by $\sigma^4\,\mathscr{D}(\theta, \varphi, \delta)$. For notation simplicity we will restrict attention to $X \sim \mathscr{N}_p(\theta, I_p)$ but the theorems in this section remain valid with the change $\gamma(X) \mapsto \sigma^2\,\gamma(X)$.

Proof Note that

$$\begin{aligned} ||\varphi(X)-\theta||^2 &= ||X+g(X)-\theta||^2 \\ &= ||g(X)||^2 + 2\,(X-\theta)^{\mathrm{T}}\,g(X) + ||X-\theta||^2 \end{aligned}$$

so that, according to (8.11) and (8.15),

$$\begin{aligned} E_\theta[||\varphi(X)-\theta||^2\,\gamma(X)] = E_\theta[&\gamma(X)\,||g(X)||^2 + 2\,\mathrm{div}(\gamma(X)\,g(X)) \\ &+ p\,\gamma(X) + \Delta\gamma(X)]\,. \end{aligned}$$

Therefore, as

$$\mathrm{div}(\gamma(X)\,g(X)) = \gamma(X)\,\mathrm{div} g(X) + \nabla\gamma(X)^{\mathrm{T}}\,g(X)$$

and $\delta_0(X) = p + 2\,\mathrm{div}\,g(X) + ||g(X)||^2$, the risk difference $\mathscr{D}(\theta, \varphi, \delta)$ in (8.10) reduces to (8.16). □

It follows from Lemma 8.2 that a sufficient condition for $\mathscr{D}(\theta, \varphi, \delta)$ to be nonpositive is

$$\gamma^2(x) + 4\,\nabla\gamma(x)^{\mathrm{T}}\,g(x) + 2\,\Delta\gamma(x) \leq 0 \tag{8.17}$$

for any $x \in \mathbb{R}^p$. Note that applying Lemma 8.2 to $\varphi(X) = X$ and $\delta_0(X) = p$, $\gamma(X) = -2(p-4)/\|X\|^2$ gives Johnstone's (1988) result mentioned above. This choice of γ also gives Jonhstone's result for an improved loss estimator based on the

James-Stein estimator. We will also give some further comments on these examples after Theorem 8.1.

How can one determine a "best" correction γ satisfying (8.17)? The following theorem provides a way to associate to the function g a suitable correction γ that satisfies (8.17) in the case where $g(x)$ is of the form $g(x) = \nabla m(x)/m(x)$ for a certain nonnegative function m. This is the case, as we saw in Chaps. 1 and 3 when φ is a Bayes estimator of θ related to a prior π, the function m being the corresponding marginal (see also Brown 1971). Through the choice of m, Bock (1988) showed that such estimators constitute a wide class of estimators of θ (which are called pseudo-Bayes estimators when the function m does not correspond to a true prior π).

Theorem 8.1 *Let m be a nonnegative function that is also superharmonic (respectively subharmonic) on $\mathbb{R}^p$ such that $\nabla m/m \in W^{1,1}_{loc}(\mathbb{R}^p)$ (see Appendix* A.1*). Let ξ be a real valued function, strictly positive and strictly subharmonic (respectively superharmonic) on $\mathbb{R}^p$, and such that*

$$E_\theta\left[\left(\frac{\Delta\xi(X)}{\xi(X)}\right)^2\right] < \infty . \tag{8.18}$$

Assume also that there exists a constant $K > 0$ such that, for any $x \in \mathbb{R}^p$,

$$m(x) > K\frac{\xi^2(x)}{|\Delta\xi(x)|} \tag{8.19}$$

and let

$$K_0 = \inf_{x\in\mathbb{R}^p} m(x)\,\frac{|\Delta\xi(x)|}{\xi^2(x)} .$$

Then the unbiased loss estimator δ_0 of the estimator φ of θ defined by $\varphi(X) = X + \nabla m(X)/m(X)$ is dominated by the estimator $\delta = \delta_0 - \gamma$, where the correction term γ is given, for any $x \in \mathbb{R}^p$ such that $m(x) \neq 0$, by

$$\gamma(x) = -\alpha\,\mathrm{sgn}(\Delta\xi(x))\,\frac{\xi(x)}{m(x)} , \tag{8.20}$$

as soon as $0 < \alpha < 2\,K_0$.

Proof The domination condition will be obtained by proving that the risk difference is less than zero. We only consider the case where m is superharmonic and ξ is strictly subharmonic, the case where m is subharmonic and ξ is strictly superharmonic being similar.

First, note that the finiteness risk condition (8.9) is guaranteed by Condition (8.18) and the fact that (8.19) implies that, for any $x \in \mathbb{R}^p$,

$$\gamma^2(x) = \alpha^2 \frac{\xi^2(x)}{m^2(x)} \leq \frac{\alpha^2}{K_0^2} \left(\frac{\Delta \xi(x)}{\xi(x)} \right)^2 .$$

Also note that, for a shrinkage function g of the form $g(x) = \nabla m(x)/m(x)$, the left hand side of (8.17) can be expressed as (since $\Delta(m\,\gamma) = m\,\Delta\gamma + 2\,(\nabla m)^{\mathrm{T}}\,\nabla\gamma + \gamma\,\Delta m$)

$$\mathscr{D}\gamma(x) \equiv \gamma^2(x) + 2 \left\{ \frac{\Delta(m(x)\,\gamma(x))}{m(x)} - \gamma(x) \frac{\Delta m(x)}{m(x)} \right\} \tag{8.21}$$

and hence, for γ in (8.20), as

$$\mathscr{D}\gamma(x) = \alpha^2 \frac{\xi^2(x)}{m^2(x)} + 2\,\alpha \left\{ -\frac{\Delta \xi(x)}{m(x)} + \frac{\xi(x)\,\Delta m(x)}{m^2(x)} \right\} . \tag{8.22}$$

Now, since m is superharmonic and ξ is positive, it follows from (8.22) that

$$\mathscr{D}\gamma(x) \leq \frac{\alpha}{m(x)} \left\{ \frac{\alpha\,\xi^2(x)}{m(x)} - 2\,\Delta\xi(x) \right\}$$

and hence, by the subharmonicity of ξ, (8.19) and the definition of K_0, that

$$\mathscr{D}\gamma(x) < \frac{\alpha}{m(x)} \{\alpha - 2\,K_0\} \frac{\xi^2(x)}{m(x)} . \tag{8.23}$$

Finally, since $0 < \alpha < 2\,K_0$, Inequality (8.23) gives $\mathscr{D}\gamma(x) < 0$, which is the desired result. □

As an example, consider $m(x) = 1/||x||^{p-2}$, which is the fundamental harmonic function that is superharmonic on the entire space $\mathbb{R}^p$ (see du Plessis 1970). Then $g(x) = -(p-2)/||x||^2$ and $\varphi(X)$ is the James-Stein estimator whose unbiased estimator of loss is $\delta_0(X) = p - (p-2)^2/||X||^2$. For any $x \neq 0$, choosing the function $\xi(x) = 1/||x||^p$ gives rise to $\Delta\xi(x) = 2\,p/||x||^{p+2} > 0$, and hence, to

$$\frac{\xi^2(x)}{|\Delta\xi(x)|} = \frac{1}{2\,p} \frac{1}{||x||^{p-2}} ,$$

which means that Condition (8.19) is satisfied with $K = 1/2\,p$. Also, we have

$$\left(\frac{\Delta\xi(x)}{\xi(x)} \right)^2 = \frac{4\,p^2}{||x||^4} ,$$

which implies that condition (8.18) is satisfied for $p \geq 5$. Now it is clear that the constant K_0 is equal to $2\,p$ and that the correction term γ in (8.20) equals $\gamma(x) = -\alpha/||x||^2$ for any $x \neq 0$. Finally, Theorem 8.1 guarantees that an improved loss estimator over the unbiased estimator of loss $\delta_0(X)$ is $\delta(X) = \delta_0(X) + \alpha/||X||^2$ for $0 < \alpha < 4\,p$, which is Johnstone's (1988) result for the James-Stein estimator.

Similarly Johnstone's result for $\varphi(X) = X$ can be constructed with $m(x) = 1$ (which is both subharmonic and superharmonic) and with the choice of the superharmonic function $\xi(x) = 1/||x||^2$, for which $K_0 = 2\,(p - 4)$, so that $\delta(x) = p - \alpha/||x||^2$ dominates p for $0 < \alpha < 4\,(p - 4)$ and $p \geq 5$.

A possible shortcoming with the improved estimator in (8.20) is that it may be negative, which is undesirable since we are estimating a nonnegative quantity. A simple remedy to this problem is to use a positive-part estimator. If we define the positive-part $\delta^+ = \max\{\delta, 0\}$, the loss difference between δ^+ and δ is $(\delta - L(\theta, \varphi))^2 - (\delta^+ - L(\theta, \varphi))^2 = (\delta^2 - 2\delta L(\theta, \varphi))\mathbb{1}_{\delta \leq 0}$. Hence it is always nonnegative. Therefore, the risk difference is positive, which implies that δ^+ dominates δ. It would be of interest to find an estimator that dominates δ^+.

In the context of variance estimation, despite warnings on its inappropriate behavior (Stein 1964 and Brown 1968), the decision theoretic approach to the normal variance estimation is typically based on the standardized quadratic loss function where overestimation of the variance is much more severely penalized than underestimation, thus leading to presumably too small estimates. Similarly, in loss estimation under quadratic loss, overestimation of the loss is also much more severely penalized than underestimation. A possible alternative to quadratic loss would be a Stein-type loss. Suppose $\varphi(X)$ is an estimator of θ under $\| \theta - \varphi(X) \|^2$ and let $\delta(X)$ be an estimator of $\| \theta - \varphi(X) \|^2$ for $\delta(X) > 0$. Then we can define the Stein-type loss for evaluating $\delta(X)$ as

$$L(\theta, \varphi(X), \delta(X)) = \frac{\| \theta - \varphi(X) \|^2}{\delta(X)} - \log \frac{\| \theta - \varphi(X) \|^2}{\delta(X)} - 1. \tag{8.24}$$

The analysis of loss estimates under a Stein-type loss is more challenging, but can be carried out using the integration by parts tools developed in this section.

We have shown that the unbiased estimator of loss can be dominated under certain conditions. Often one may wish to add a frequentist-validity constraint to a loss estimation problem. Specifically in our problem, the frequentist-validity constraint for some estimator δ would be $E_\theta[\delta(X)] \geq E_\theta[\delta_0(X)]$ for all θ. Kiefer (1977) suggested that conditional and estimated confidence assessments should be conservatively biased; the average reported loss should be greater than the average actual loss. Under such a frequentist-validity condition, Lu and Berger (1989) give improved loss estimators for several of the most important Stein-type estimators. One of their estimators is a generalized Bayes estimator, suggesting that Bayesians and frequentists can potentially agree on a conditional assessment of loss.

8.2.2 Dominating the Posterior Risk

In the previous sections, we have seen that the unbiased estimator of loss can often be dominated. When a (generalized) Bayes estimator of θ is available, incorporating the same prior information for estimating the loss of this Bayesian estimator is coherent, and we may expect that the corresponding Bayes estimator is a good candidate to improve on the unbiased estimator of loss. However, somewhat surprisingly, Fourdrinier and Strawderman (2003) found in the normal setting that the unbiased estimator often dominates the corresponding generalized Bayes estimator of loss for priors that give minimax estimators in the original point estimation problem. In particular, they give a class of priors for which the generalized Bayes estimator of θ is admissible and minimax, but for which the unbiased estimator of loss dominates the generalized Bayes estimator of loss. They also give a general inadmissibility result for a generalized Bayes estimator of loss. While much of their focus is on pseudo-Bayes estimators, in this section, we concentrate on their results on generalized Bayes estimators.

Suppose X is distributed as $\mathcal{N}_p(\theta, I_p)$ and the loss function is $L(\theta, \varphi(X)) = \|\varphi(X) - \theta\|^2$ where we are estimating θ with the estimator $\varphi(X)$. For a given generalized prior π, we denote the generalized marginal by m and the generalized Bayes estimator of θ by

$$\varphi_m(X) = X + \frac{\nabla m(X)}{m(X)}. \tag{8.25}$$

Then (see Theorem 8.1 or Stein 1981) the unbiased estimator of the risk of $\varphi_m(X)$ is

$$\delta_0(X) = p + 2\frac{\Delta m(x)}{m(X)} - \frac{\|\nabla m(X)\|^2}{m^2(X)}, \tag{8.26}$$

while the posterior risk of $\varphi_m(X)$ is (see (1.20) for $\sigma^2 = 1$)

$$\delta_m(X) = p + \frac{\Delta m(X)}{m(X)} - \frac{\|\nabla m(X)\|^2}{m^2(X)}. \tag{8.27}$$

It is interesting to note that (8.26) and (8.27) differ only by the factor 2 in the middle term.

Domination of $\delta_0(X)$ over $\delta_m(X)$ may be obtained thanks to the fact $(\Delta m(X)/m(X))^2 - 2\,\Delta^{(2)}m(X)/m(X)$ is an unbiased estimator of their risk difference, where $\Delta^{(2)}m = \Delta(\Delta m)$ is the bi-Laplacian of m (see Fourdrinier and Strawderman 2003). That is,

$$\mathcal{R}(0, \varphi_m, \delta_0) - \mathcal{R}(\theta, \varphi_m, \delta_m) = E_\theta\left[\left(\frac{\Delta m(X)}{m(X)}\right)^2 - 2\,\frac{\Delta^{(2)}m(X)}{m(X)}\right] \tag{8.28}$$

Thus, the above domination of the posterior risk as an estimator of loss by the unbiased estimator of loss will occur as soon as

$$\left(\frac{\Delta m(X)}{m(X)}\right)^2 - 2\,\frac{\Delta^{(2)} m(X)}{m(X)} \leq 0\,. \tag{8.29}$$

Applicability of that last condition is underlined by the remarkable fact that if the prior π satisfies (8.29), that is, if

$$\left(\frac{\Delta \pi(\theta)}{\pi(\theta)}\right)^2 - 2\,\frac{\Delta^{(2)} \pi(\theta)}{\pi(\theta)} \leq 0\,, \tag{8.30}$$

then (8.29) is satisfied for the marginal m.

As an example, Fourdrinier and Strawderman (2003) considered the prior $\pi(\theta) = (\|\theta\|^2/2 + a)^{-b}$ (where $a \geq 0$ and $b \geq 0$) and showed that, if $p \geq 2(b+3)$, then (8.30) holds and hence δ_u dominates δ_m. Since π is integrable if and only if $b > \frac{p}{2}$ (for $a > 0$), the prior π is improper whenever this condition for domination of δ_u over δ_m holds. Of course, whenever π is proper, the Bayes estimator δ_m is admissible provided its Bayes risk is finite.

Inadmissibility of the generalized Bayes loss estimator is not exceptional. Thus, in Fourdrinier and Strawderman (2003), the following general inadmissibility result is given; its proof is parallel to the proof of Theorem 8.1.

Theorem 8.2 *Under the conditions of Theorem* 8.1, δ_m *is inadmissible and a class of dominating estimators is given by*

$$\delta_m(X) + \alpha \ \text{sgn}\,(\Delta\xi(X))\frac{\xi(X)}{m(X)} \quad \textit{for} \ \ 0 < \alpha < 2\,K_0\,.$$

Note that Theorem 8.2 gives conditions for improvement on δ_m while Theorem 8.1 looks for improvements on δ_0. As we saw, δ_0 often dominates δ_m.

In Fourdrinier and Strawderman (2003), it is suggested that the inadmissibility of the generalized Bayes (or pseudo-Bayes) estimator is due to the fact that the loss function $(\delta(x) - \|\varphi(x) - \theta\|^2)^2$ may be inappropriate. The possible deficiency of this loss is illustrated by the following simple result concerning estimation of the square of a location parameter in $\mathbb{R}$.

Suppose $X \in \mathbb{R} \sim f\left((X-\theta)^2\right)$ such that $E_\theta[X^4] < \infty$. Consider estimation of θ^2 under the loss $(\delta - \theta^2)^2$. The generalized Bayes estimator δ_π of θ^2 with respect to the uniform prior $\pi(\theta) \equiv 1$ is given by

$$\delta_\pi(X) = \frac{\int \theta^2 f\left((X-\theta)^2\right) d\theta}{\int f\left((X-\theta)^2\right) d\theta} = X^2 + E_0\left[X^2\right].$$

Since this estimator has constant bias $2\,E_0[X^2]$, it is dominated by the unbiased estimator $X^2 - E_0[X^2]$ (the risk difference is $4\,(E_0[X^2])^2$). Hence δ_π is inadmissible for any $f(\cdot)$ such that $E_\theta[X^4] < \infty$.

8.2.3 Examples of Improved Estimators

In this subsection, we give some examples of Theorems 8.1 and 8.2. Although the shrinkage factor in Theorems 8.1 and 8.2 are the same, in the examples below we will only focus on improvements of posterior risk.

As an application of Theorem 8.2, let $\xi_b(x) = \left(\|x\|^2 + a\right)^{-b}$ (with $a \geq 0$ and $b \geq 0$). It can be shown that $\Delta\xi_b(x) < 0$ for $a \geq 0$ and $0 < 2\,(b+1) < p$. Also $\Delta\xi_b(x) > 0$ if $a = 0$ and $2(b+1) > p$. Furthermore,

$$\frac{\xi_b^2(x)}{|\Delta\xi_b(x)|} = \frac{1}{2\,b\,\left|p - 2(b+1)\frac{\|x\|^2}{\|x\|^2+a}\right|}\;\frac{1}{(\|x\|^2+a)^{b-1}}.$$

(I) Suppose that $0 < 2\,(b+1) < p$ and $a \geq 0$. Then

$$\frac{\xi_b^2(x)}{|\Delta\xi_b(x)|} \leq \frac{1}{2\,b(p-2(b+1))}\;\frac{1}{(\|x\|^2+a)^{b-1}}$$

and $E_\theta\left[\left(\frac{\Delta\xi_b(X)}{\xi_b(X)}\right)^2\right] < \infty$ since it is proportional to $E_\theta\left[\frac{1}{(\|X\|^2+a)^2}\right]$, which is finite for $a > 0$ or for $a = 0$ and $p > 4$.

Suppose that $m(x)$ is greater than or equal to some multiple of $\left(\frac{1}{(\|x\|^2+a)}\right)^{b-1}$ or equivalently,

$$m(x) \geq \frac{k}{2\,b(p-2(b+1))}\left(\frac{1}{\|x\|^2+a}\right)^{b-1} \tag{8.31}$$

for some $k > 0$. Theorem 8.2 implies that $\delta_m(X)$ is inadmissible and is dominated by

$$\delta_m(X) - \frac{\alpha}{m(X)(\|X\|^2+a)^b}$$

for $0 < \alpha < 4\,b(p-2(b+1))\inf_{x\in\mathbb{R}^p}(m(x)(\|x\|^2+a)^{b-1})$.

Alternatively, if $m(x) \geq \frac{k}{(\|x\|^2+a)^c}$ for $0 < c < \frac{p-4}{2}$, δ_m is inadmissible and the above gives an explicit improvement upon substituting $c - 1$ for b. Note that the improved estimators shrink towards 0.

Suppose, for example, that $m(x) \equiv 1$. Then (8.31) is satisfied for $b \geq 1$. Here $\varphi_m(X) = X$ and $\delta_m(X) = p$. Choosing $b = 1$, an improved class of estimators is given by $p - \frac{\alpha}{\|X\|^2+a}$ for $0 < \alpha < 4\,(p-4)$. The case $a = 0$ is equivalent to Johnstone's result for this marginal.
(II) Suppose that $2\,(b+1) > p > 4$ and $a = 0$. Then,

$$\frac{\xi_b^2(x)}{|\Delta \xi_b(x)|} = \frac{1}{2\,b(2(b+1)-p)}\,\frac{1}{\|x\|^{2(b-1)}}.$$

A development similar to the above implies that, when $m(x)$ is greater than or equal to some multiple of $\|x\|^{2(1-b)}$, an improved estimator is

$$\delta_m(X) + \frac{\alpha}{m(X)\|X\|^{2b}}$$

for

$$0 < \alpha < 4\,b(2\,(b+1)-p)\inf_{x \in \mathbb{R}^p}\left(m(x)\|x\|^{2(b-1)}\right).$$

In this case, the correction term is positive and the estimators expands away from 0. Note also that this result only works for $a = 0$ and hence, applies to pseudo-marginals which are unbounded in a neighborhood of 0. Since all marginals corresponding to a generalized prior π are bounded, this result can never apply to generalized Bayes procedures, but only to pseudo-Bayes procedures.

Suppose, for example, that $m(x) = \|x\|^{2-p}$. Here $\varphi_m(X) = \left(1 - \frac{p-2}{\|X\|^2}\right)X$ is the James-Stein estimator and $\delta_m(X) = p - \frac{(p-2)^2}{\|X\|^2}$. In particular, the above applies for $b - 1 = \frac{p-2}{2}$, that is, for $b = \frac{p}{2} > \frac{p-2}{2}$. An improved estimator is given by $\delta_m(X) + \frac{\gamma}{\|X\|^2}$ for $0 < \gamma < 4p$. This again agrees with Johnstone's result for James-Stein estimators.

8.3 Quadratic Loss Estimation: Multivariate Normal with Unknown Variance

In Sect. 8.2 it was assumed that the covariance matrix was known and equal to the identity matrix I_p. Typically, this covariance is unknown and should be estimated. In the case where it is of the form $\sigma^2 I_p$ with σ^2 unknown, Wan and Zou (2004) showed that, for the invariant loss $||\varphi(X)-\theta||^2/\sigma^2$, Johnstone's (1988) result can be extended when estimating the loss of the James-Stein estimator. In fact, the general framework considered in Sect. 8.2 can be extended to the case where σ^2 is unknown, and we show that a condition parallel to Condition (8.17) can be found.

Suppose $X \sim \mathcal{N}_p(\theta, \sigma^2 I_p)$ and $S \sim \sigma^2 \chi_k^2$ is independent of X. Consider an estimator of θ of the form $\varphi(X, S) = X + S\, g(X, S)$ with $E_{\theta,\sigma^2}[S^2\, ||g(X, S)||^2] < \infty$, where E_{θ,σ^2} denotes the expectation with respect to the joint distribution of (X, S). Then, by Theorem 2.5, an unbiased estimator of the loss

$$\frac{||\varphi(X, S) - \theta||^2}{\sigma^2} \tag{8.32}$$

of

$$\varphi(X, S) = X + S\, g(X, S) \tag{8.33}$$

is

$$\delta_0(X, S) = p + S \left\{ (k+2)\, ||g(X, S)||^2 + 2\, \text{div}_X g(X, S) + 2\, S \frac{\partial}{\partial S} ||g(X, S)||^2 \right\}. \tag{8.34}$$

The following theorem provides an extension of results in Sect. 8.2 to the setting of an unknown variance. The necessary conditions to insure the finiteness of the risks are parallel to the case where the variance σ^2 is known. It should be noticed that the corresponding domination condition of $\delta(X, S)$ over $\delta_0(X, S)$, that is, for any $X \in R^p$ and any $S \in \mathbb{R}_+$, $(k+2)\, ||g(x, s)||^2 + 2\, \text{div}_x g(x, s) + 2\, s \frac{\partial}{\partial s} ||g(x, s)||^2 \leq 0$, entails that the two conditions $E_{\theta,\sigma^2}[(S\, \text{div} g(X, S))^2] < \infty$ and $E_{\theta,\sigma^2}\Big[\big(S^2 \frac{\partial}{\partial S} ||g(X, S)||\big)^2\Big]$ imply the condition $E_{\theta,\sigma^2}[S^2\, ||g(X, S)||^4] < \infty$. Also the derivation of the finiteness of $R(\theta, \sigma^2, \varphi)$ follows as in the known variance case.

Theorem 8.3 *Let $X \sim \mathcal{N}_p(\theta, \sigma^2 I_p)$ where θ and σ^2 are unknown and $p \geq 5$ and let S be a nonnegative random variable independent of X and such that $S \sim \sigma^2 \chi_k^2$. Let the estimator $\varphi(X, S)$ of θ (under loss (8.32)) be given by (8.33).*

For any twice weakly differentiable function $\gamma(X)$ such that $E_{\theta,\sigma^2}[\gamma^2(X)] < \infty$, the risk difference (under loss $(\delta - ||\varphi(X, S) - \theta||^2/\sigma^2)^2$)

$$\mathcal{D}(\theta, \sigma^2, \varphi, \delta) = \mathcal{R}(\theta, \sigma^2, \varphi, \delta) - \mathcal{R}(\theta, \sigma^2, \varphi, \delta_0)$$

between the estimators $\delta(X, S) = \delta_0(X, S) - S\, \gamma(X)$ and $\delta_0(X, S)$ is given by

$$E_{\theta,\sigma^2}\bigg[S^2 \Big\{\gamma^2(X) + \frac{2}{k+2}\, \Delta\gamma(X) + 4\, g(X, S)^{\mathrm{T}} \nabla\, \gamma(X) + 4\, \gamma(X)\, ||g(X, S)||^2\Big\}\bigg], \tag{8.35}$$

so that a sufficient condition for $\mathcal{D}(\theta, \sigma^2, \varphi, \delta)$ to be nonpositive, and for $\delta(X, S)$ to improve on $\delta_0(X, S)$, is

$$\gamma^2(x) + \frac{2}{k+2}\,\Delta\gamma(x) + 4\,g(x,s)^{\mathrm{T}}\,\nabla\gamma(x) + 4\,\gamma(x)\,||g(x,s)||^2 \le 0 \tag{8.36}$$

for any $x \in \mathbb{R}^p$ and any $s \in \mathbb{R}_+$, with strict inequality on a set of positive measure.

Note that, in Theorem 8.3, the estimation loss is invariant squared error loss $||\varphi(X,S) - \theta||^2/\sigma^2$ while the loss for estimating loss is squared error $(\delta - ||\varphi(X,S) - \theta||^2/\sigma^2)^2$.

Proof Consider the finiteness of the risk of the alternative loss estimator $\delta(X,S) = \delta_0(X,S) - S\,\gamma(X)$. It is easily seen that its difference in loss $d(\theta,\sigma^2,X,S)$ with $\delta_0(X,S)$ can be written as

$$\begin{aligned} &d(\theta,\sigma^2,X,S)\\ &= \left(\delta_0(X,S) - \frac{1}{\sigma^2}||\varphi(X)-\theta||^2 - S\,\gamma(X)\right)^2 - \left(\delta_0(X,S) - \frac{1}{\sigma^2}||\varphi(X)-\theta||^2\right)^2\\ &= S^2\,\gamma^2(X) - 2\,S\,\gamma(X)\,(\delta_0(X,S) - \frac{1}{\sigma^2}||\varphi(X)-\theta||^2)\,. \end{aligned} \tag{8.37}$$

Hence, since $E_{\theta,\sigma^2}[||\varphi(X,S) - \theta||^2/\sigma^2] < \infty$ the condition $E_{\theta,\sigma^2}[\gamma^2(X)] < \infty$ ensures that the expectation of the loss in (8.37), that is, the risk difference $\mathcal{D}(\theta,\sigma^2,\varphi,\delta)$ is finite. Then $\mathcal{R}(\theta,\sigma^2,\varphi,\delta) < \infty$ since $\mathcal{R}(\theta,\sigma^2,\varphi,\delta_0) < \infty$.

We now express the risk difference as $\mathcal{D}(\theta,\sigma^2,\varphi,\delta) = E_{\theta,\sigma^2}[d(\theta,\sigma^2,X,S)]$. Using (8.34) and expanding $||\varphi(X,S) - \theta||^2/\sigma^2$ we get that $d(\theta,\sigma^2,X,S)$ in (8.37) can be written as

$$d(\theta,\sigma^2,X,S) = A(X,S) + B(\theta,\sigma^2,X,S)$$

where

$$\begin{aligned} A(X,S) = S^2\,\gamma^2(X) &- 2\,p\,S\,\gamma(X) - 2\,(k+2)\,S^2\,\gamma(X)\,||g(X,S)||^2\\ &- 4\,S^2\,\gamma(X)\,\mathrm{div}_X g(X,S) - 4\,S^3\,\gamma(X)\,\frac{\partial}{\partial S}||g(X,S)||^2 \end{aligned} \tag{8.38}$$

and

$$\begin{aligned} B(\theta,\sigma^2,X,S) = 2\,\frac{S^3}{\sigma^2}\,\gamma(X)\,||g(X,S)||^2 &+ 2\,\frac{S}{\sigma^2}\,\gamma(X)\,||X-\theta||^2\\ &+ 4\,\frac{S^2}{\sigma^2}\,\gamma(X)\,(X-\theta)^{\mathrm{T}} g(X,S)\,. \end{aligned} \tag{8.39}$$

Through Lemma 2.3 (2) with $h(x,s) = 2\,\frac{s^3}{\sigma^2}\,\gamma(x)\,||g(x,s)||^2$, the expectation of the first term in the right hand side of (8.39) equals

$$E_{\theta,\sigma^2}\left[2\,\frac{S^3}{\sigma^2}\,\gamma(X)\,||g(X,S)||^2\right] = E_{\theta,\sigma^2}\Big[2\,(k+4)\,S^2\,\gamma(X)\,||g(X,S)||^2 + 4\,S^3\,\gamma(X)\,\frac{\partial}{\partial S}||g(X,S)||^2\Big]. \tag{8.40}$$

Also, a reiterated application of Lemma 2.3 (1) to the expectation of the second term in the right hand side of (8.39) allows us to write

$$\begin{aligned}E_{\theta,\sigma^2}\left[2\,\frac{S}{\sigma^2}\,\gamma(X)\,||X-\theta||^2\right] &= E_{\theta,\sigma^2}[2\,\frac{1}{\sigma^2}\,(X-\theta)^{\mathrm{T}}\,S\,\gamma(X)\,(X-\theta)]\\ &= E_{\theta,\sigma^2}[2\,\mathrm{div}_X\{S\,\gamma(X)\,(X-\theta)\}]\\ &= E_{\theta,\sigma^2}[2\,p\,S\,\gamma(X) + 2\,S\,(X-\theta)^{\mathrm{T}}\nabla\gamma(X)]\\ &= E_{\theta,\sigma^2}[2\,p\,S\,\gamma(X) + 2\,\sigma^2\,S\,\Delta\gamma(X)]\end{aligned}$$

which gives

$$E_{\theta,\sigma^2}\left[2\,\frac{S}{\sigma^2}\,\gamma(X)\,||X-\theta||^2\right] = E_{\theta,\sigma^2}\left[2\,p\,S\,\gamma(X) + 2\,\frac{S^2}{k+2}\,\Delta\gamma(X)\right]. \tag{8.41}$$

This follows since $S \sim \sigma^2\chi_k^2$ entails that $E[S^2/(k+2)] = \sigma^2 E[S]$ and since S is independent of X. As for the third term in the right hand side of (8.39), its expectation can also be expressed using Lemma 2.3 (1) as

$$\begin{aligned}E_{\theta,\sigma^2}\left[4\,\frac{S^2}{\sigma^2}\,\gamma(X)\,(X-\theta)^{\mathrm{T}}g(X,S)\right] &= E_{\theta,\sigma^2}[4\,S^2\,\mathrm{div}_X\{\gamma(X)\,g(X,S)\}]\\ &= E_{\theta,\sigma^2}[4\,S^2\,\gamma(X)\,\mathrm{div}_X\{g(X,S)\} + 4\,S^2\,g(X,S)^{\mathrm{T}}\,\nabla\gamma(X)]\,,\end{aligned} \tag{8.42}$$

by the property of the divergence.

Finally, gathering (8.40), (8.41), and (8.42) yields an expression of (8.39), which with (8.38) gives the integrand term of (8.35), the desired result. □

As an example, consider the James-Stein estimator $\varphi^{JS}(X,S) = X - \frac{p-2}{k+2}\frac{S}{||X||^2}X$ discussed in Sect. 2.4. Here the shrinkage factor $g(X,S)$ only depends on X and equals $g(X) = -\frac{p-2}{k+2}\frac{X}{||X||^2}$ so that, through routine calculation, the unbiased estimator of loss is $\delta_0(X,S) = p - \frac{(p-2)^2}{k+2}\frac{S}{||X||^2}$. For a correction of the form $\gamma(x) = -d/||x||^2$ with $d \geq 0$, it is easy to check that the expression in (8.36) equals

$$d^2 + 4\,\frac{p-4}{k+2}\,d - 8\,\frac{p-2}{k+2}\,d - 4\left(\frac{p-2}{k+2}\right)^2 d = d\left(d - \frac{4}{k+2}\left[p + \frac{(p-2)^2}{k+2}\right]\right)$$

which is negative for $0 < d < \frac{4}{k+2}\left[p + \frac{(p-2)^2}{k+2}\right]$ and gives domination of $p - \frac{(p-2)^2}{k+2}\frac{S}{||X||^2} + \frac{d}{||x||^2}$ over $p - \frac{(p-2)^2}{k+2}\frac{S}{||X||^2}$. This condition recovers the result of Wan and Zou (2004) who considered the case $d = \frac{2}{k+2}\left[p + \frac{(p-2)^2}{k+2}\right]$.

8.4 Extensions to the Spherical Case

8.4.1 Quadratic Loss Estimation: Spherically Symmetric Distributions with Known Scale

In the previous sections the loss estimation problem was considered for the normal distribution setting. In this section we consider loss estimation for the class of spherically symmetric distributions. As developed in Corollary 4.1 in Chap. 4 we will use the representation of a random variable from a spherically symmetric distribution, $X = (X_1, \ldots, X_p)^{\mathrm{T}}$, as $X \stackrel{d}{=} RU^{(p)} + \theta$, where $R = ||X - \theta||$ is a random radius, $U^{(p)}$ is a uniform random variable on the p-dimensional unit sphere, where R and $U^{(p)}$ are independent.

In Sect. 8.4.2 we extend these results to the case where the distribution of X is spherically symmetric and where a residual vector U is available (which allows an estimation of the variance σ^2).

Assume $X \sim SS_p(\theta)$ and suppose we wish to estimate $\theta \in \mathbb{R}^p$ by a decision rule $\varphi(X)$ using quadratic loss. Suppose that we also use the quadratic loss to assess the accuracy of the loss estimate $\varphi(X)$; then the risk of this loss estimate is given by (8.2). Fourdrinier and Wells (1995b) considers the problem of estimating the loss when $\varphi(X) = X$ is the estimate of the location parameter θ. The estimate φ is the least squares estimator and is minimax among the class of spherically symmetric distributions with bounded second moment. Furthermore, if one assumes the density of X exists and is unimodal, then φ is also the maximum likelihood estimator.

The unbiased constant estimate of the loss $||X - \theta||^2$ is $\delta_0 = E_\theta[R^2]$. Note that δ_0 is independent of θ, since $E_\theta[||X - \theta||^2] = E_0[||X||^2]$. Fourdrinier and Wells (1995b) showed that the unbiased estimator δ_0 can be dominated by $\delta_0 - \gamma$, where γ is a particular superharmonic function for the case where the sampling distribution is a scale mixture of normals and in a more general spherical case.

The development of the results depends on some interesting extensions of the classical Stein identities in (8.11) and (8.15) to the general spherical setting as in Sect. 5.2. Recall that the distribution of $X - \theta$ conditional on $R = ||X - \theta||$ is $\mathcal{U}_{R,\theta}$. Suppose γ is a weakly differentiable vector valued function, then by applying the divergence theorem for weakly differentiable functions.

$$E_\theta[(X-\theta)^{\mathrm{T}}\gamma(X) \mid ||X-\theta|| = R] = \int_{S_{R,\theta}} (x-\theta)^{\mathrm{T}}\,\gamma(x)\mathscr{U}_{R,\theta}\,(dx) \quad (8.43)$$
$$= \frac{R}{\sigma_{R,\theta}(S_{R,\theta})}\int_{B_{R,\theta}} \operatorname{div}\gamma(x)\,dx.$$

If γ is a real-valued function, then it follows from (8.43) and the product rule applied to the vector valued function $(x-\theta)\gamma(x)$ that

$$E_\theta[||X-\theta||^2\gamma(X) \mid ||X-\theta|| = R]$$
$$= \int_{S_{R,\theta}} (x-\theta)^{\mathrm{T}}\,(x-\theta)\,\gamma(x)\,\mathscr{U}_{R,\theta}\,(dx)$$
$$= \frac{R}{\sigma_{R,\theta}(S_{R,\theta})}\int_{B_{R,\theta}} [p\,\gamma(x) + (x-\theta)^{\mathrm{T}}\,\nabla\gamma(x)]\,dx. \quad (8.44)$$

Our first extension of Theorem 8.1 is to the class of spherically symmetric distributions that are scale mixtures of normal distributions as discussed in Lemma 5.1. Suppose

$$p_\theta(x|\theta) = \int_0^\infty \phi(x;\theta, I/t)G\,(dt) \quad (8.45)$$

so that $G(\cdot)$ is the mixing distribution on $\tau = \sigma^{-2}$.

In the scale mixture of normals setting the unbiased estimate, δ_0, of risk equals

$$\delta_0 = E[R^2] = E_\theta[||X-\theta||^2] = p\int_0^\infty t^{-1}G(dt). \quad (8.46)$$

It is easy to see that the risk of the unbiased estimator δ_0 is finite if and only if $E_\theta[||X-\theta||^4] < \infty$, which holds if

$$\int_0^\infty t^{-2}G(dt) < \infty. \quad (8.47)$$

The main theorem in Fourdrinier and Wells (1995b) is the following domination result of an improved estimator of loss over the unbiased loss estimator.

Theorem 8.4 *Let the distribution of X be a scale mixture of normal random variables as in* (8.45) *such that* (8.47) *is satisfied and*

$$\int_{\mathbb{R}_+} t^{p/2}\,G(dt) < \infty\,. \quad (8.48)$$

Estimating θ through X and estimating the loss $\|X-\theta\|^2$, consider the estimator of loss δ_0 in (8.46) (which is an unbiased estimate of risk of X). Let γ be a twice *weakly differentiable on* $\mathbb{R}^p$ *shrinkage function* γ *such that* $E_\theta[\gamma^2] < \infty$ *for every* $\theta \in \mathbb{R}^p$.

Then a sufficient condition for $\delta_0 - \gamma$ *to dominate* δ_0 *under loss* $(\delta - ||X - \theta||^2)^2$ *is that* γ *satisfies the differential inequality* $k\,\Delta\gamma + \gamma^2 < 0$ *with*

$$k = 2\,\frac{\int_{\mathbb{R}_+} t^{p/2}\,G(dt)}{\int_{\mathbb{R}_+} t^{p/2-2}\,G(dt)}\,. \tag{8.49}$$

As an example, let $\gamma(x) = c/||x||^2$ where c is a positive constant. Note that γ is only weakly differentiable (but not differentiable in the usual sense) and that its Laplacian exists as a locally integrable function only when $p > 4$ (see Appendices A.1, A.2, A.3, and A.4). Then it may be shown that $\Delta\gamma(x) = -2c(p-4)/||x||^4$. Hence, $k\Delta(x) + \gamma^2(x) = -2kc(p-4)/||x||^4 + c^2/||x||^4 < 0$ if $-2kc(p-4) + c^2 < 0$, that is, $0 < c < 2k(p-4)$. It is easy to see that the optimal value of c for which this inequality is the most negative equals $k(p-4)$, so an interesting estimate in this class of γ's is $\delta = \delta_0 - k(p-4)/||x||^2 (p > 4)$. This is precisely the estimate proposed by Johnstone (1988) in the normal distribution case $\mathcal{N}_p(\theta, I_p)$ where $k = 2$; recall, in that case, $\delta_0 = p$. In this example, we have assumed that the dimension p is greater than four. In general, we can have domination as long as the assumptions of the theorem are valid. Actually, Blanchard and Fourdrinier (1999) show explicitly (see Sect. 8.7) that, when $p \leq 4$, the only solution γ in $L^2_{\text{loc}}(\mathbb{R}^p)$ of the inequality $k\Delta\gamma + \gamma^2 \leq 0$ is $\gamma \equiv 0$ (a.e., with respect to the Lebesgue measure). Hence, in the normal case $\mathcal{N}_p(\theta, I_p/t)$, where $2t^{-2}\Delta\gamma + \gamma^2$ is an unbiased estimator of the risk difference (for dimensions four or less) it is impossible to find an estimator $\delta = \delta_0 - \gamma$ whose unbiased estimate of risk is always less that of δ_0.

In the case of scale mixtures of normal distributions, the conjecture of the admissibility of $\delta_0 - \gamma$ for lower dimensions (although it is probably true) remains open. Indeed, under the conditions of Theorem 8.4, $k\,\Delta\gamma + \gamma^2$ is no longer an unbiased estimator of the risk difference and $E_\theta[k\,\Delta\gamma + \gamma^2]$ is only its upper bound. The use of Blyth's method would need to specify the distribution of X (that is, the mixture distribution G). It is worth noting that dimension-cutoff also arises through the finiteness of $E_\theta[\gamma^2]$ when using the classical shrinkage function $c/||x||^2$.

In order to prove Theorem 8.4 we need some additional technical results. The first lemma gives some important properties of superharmonic functions and is found in du Plessis (1970) and the second lemma links the integral of the gradient on a ball with the integral of the Laplacian, see also Appendix A.8.

Lemma 8.3 *If* γ *is a real-valued superharmonic function then*

(1) $\int_{S_{R,\theta}} \gamma(x)\,\mathcal{U}_{R,\theta}(dx) \leq \int_{B_{R,\theta}} \gamma(x)\,\mathcal{V}_{R,\theta}(dx)$.
(2) *Both of the integrals in (1) are decreasing in R.*

Proof See Sections 1.3 and 2.5 in du Plessis (1970) and Appendix A.8. □

Lemma 8.4 *Suppose γ is a twice weakly differentiable function. Then*

$$\int_{B_{R,\theta}} (x-\theta)^{\mathrm{T}} \nabla\gamma(x)\, \mathscr{V}_{R,\theta}(dx) = \frac{p\,\Gamma(p/2)}{2\pi^{p/2}} \frac{1}{R^p} \int_0^R r \int_{B_{r,\theta}} \Delta\gamma(x)\, dx\, dr.$$

Proof Since the density of the distribution of the radius under $\mathscr{V}_{R,\theta}$ is $(p/R^p)r^{p-1}$, we have

$$\int_{B_{R,\theta}} (x-\theta)^{\mathrm{T}} \nabla\gamma(x)\, \mathscr{V}_{R,\theta}(dx) = \int_0^R \int_{S_{r,\theta}} (x-\theta)^{\mathrm{T}}\, \nabla\gamma(x)\, \mathscr{U}_{r,\theta}(dx)\, \frac{p}{R^p}\, r^{p-1}\, dr.$$

The result follows from applying (8.44) to the inner most integral of the right hand side of this equality and by recalling the fact that $\sigma_{r,\theta}(S_{r,\theta}) = (2\,\pi^{p/2}/\Gamma(p/2))\, r^{p-1}$. □

Proof of Theorem 8.4 The risk difference between δ_0 and $\delta_0-\gamma$ equals $\alpha(\theta)+\beta(\theta)$ where

$$\alpha(\theta) = 2\,p \int_{\mathbb{R}_+} \left(\frac{1}{t} - \frac{\delta_0}{p}\right) \int_{\mathbb{R}_+} \int_{S_{R,\theta}} \gamma(x)\, \mathscr{U}_{R,\theta}\,(dx)\, \rho_t(dt)\, G\,(dt)$$

and

$$\beta(\theta) = \int_{\mathbb{R}_+} \int_{\mathbb{R}^p} (2\,t^{-2}\, \Delta\gamma(x) + \gamma^2(x)) \left(\frac{t}{2\,\pi}\right)^{p/2} \exp\left(-\frac{t}{2}||x-\theta||^2\right) dx\; G(dt)\,.$$

We have from the definition of $\mathscr{V}_{R,\theta}$ and an application of Fubini's theorem

$$\begin{aligned}
&\int_{\mathbb{R}_+} R^2 \int_{B_{R,\theta}} \gamma(x)\, \mathscr{V}_{R,\theta}\,(dx)\, \rho(dR) \\
&\quad = p\, \frac{\Gamma(p/2)}{2\,\pi^{p/2}} \int_{\mathbb{R}_+} R^{2-p} \int_{B_{R,\theta}} \gamma(x)\, dx\, \rho(dR) \\
&\quad = p \frac{\Gamma(p/2)}{2\,\pi^{p/2}} \int_{\mathbb{R}^p} \gamma(x) \int_{||x-\theta||}^{+\infty} R^{2-p}\, \rho(dR)\, dx\,.
\end{aligned} \tag{8.50}$$

Now, for fixed $t \geq 0$, in the normal case $\mathscr{N}_p(\theta, I_p/t)$ the distribution ρ_t of the radius has the density f_t of the form $f_t(R) = t^{p/2}/\big(2^{p/2-1}\, \Gamma(p/2)\big)\, R^{p-1}\, \exp\{-t\, R^2/2\}$ and $\delta_0 = p/t$. Thus, expression (8.50) becomes

$$\begin{aligned}
&\int_{\mathbb{R}_+} R^2 \int_{B_{R,\theta}} \gamma(x) \mathscr{V}_{R,\theta}\,(dx)\, \rho(dR) \\
&\quad = \frac{p\, t^{p/2}}{(2\,\pi)^{p/2}} \int_{\mathbb{R}^p} \gamma(x) \int_{||x-\theta||}^{+\infty} R\, \exp\left\{-\frac{t\,R^2}{2}\right\} dR\, dx
\end{aligned}$$

$$= \frac{p\,t^{p/2-1}}{(2\pi)^{p/2}} \int_{\mathbb{R}^p} \gamma(x) \exp\left\{-\frac{t}{2}\,||x-\theta||^2\right\} dx$$
$$= \frac{p}{t} \int_{\mathbb{R}_+} \int_{S_{R,\theta}} \gamma(x)\, \mathcal{U}_{R,\theta}(dx)\, \rho_t(dR),$$

the last equality holding since ρ_t is the radial distribution. Using the mixture representation with mixing distribution G, the expression of $\alpha(\theta)$ is written as

$$\begin{aligned}
\alpha(\theta) &= 2p \int_{\mathbb{R}_+} \left(\frac{1}{t} - \frac{\delta_0}{p}\right) \int_{\mathbb{R}^p} \gamma(x) \left(\frac{t}{2\pi}\right)^{p/2} \exp\left(-\frac{t}{2}||x-\theta||^2\right) dx\, G(dt) \\
&= 2p\; Cov\left((\frac{1}{t} - \frac{\delta_0}{p}),\, E[\gamma(X) \mid \tau]\right) \\
&\le 0
\end{aligned}$$

since $E[\gamma(x) \mid \tau]$ is nondecreasing by Lemma 8.3. Note also, since $\delta_0 = \frac{p}{t}$, the expression for $\alpha(\theta)$ is a covariance with respect to G and is nonpositive by Lemma 8.3 and the covariance inequality.

We can now treat the integral of the expression $\beta(\theta)$ in the same manner. The function $x \to (x-\theta)^{\mathrm{T}}\nabla\gamma(x)$ and the function $x \to \Delta\gamma(x)$ taking successively the role of the function γ, we obtain

$$\begin{aligned}
&\int_{\mathbb{R}_+} \frac{R^2}{p} \int_{B_{R,\theta}} (x-\theta)^{\mathrm{T}}\nabla\gamma(x)\, \mathcal{V}_{R,\theta}(dx)\, \rho_t(dR) \\
&\quad = \frac{1}{t} \int_{\mathbb{R}_+} \int_{S_{R,\theta}} (x-\theta)^{\mathrm{T}}\nabla\gamma(x)\, \mathcal{U}_{R,\theta}(dx)\, \rho_t(dR) \\
&\quad = \frac{1}{t} \int_{R_+} \frac{R^2}{p} \int_{B_{R,\theta}} \Delta\gamma(x)\, dx\, \rho_t(dR) \\
&\quad = \frac{t^{p/2-2}}{(2\pi)^{p/2}} \int_{\mathbb{R}^p} \Delta\gamma(x) \exp\left\{-\frac{t}{2}\,||x-\theta||^2\right\} dx
\end{aligned}$$

applying (8.43) for the second equality and remembering that $\Delta\gamma = \mathrm{div}(\nabla\gamma)$. Therefore, by Fubini's Theorem, $\beta(\theta)$ can be reexpressed as

$$\beta(\theta) = \int_{\mathbb{R}^p} \left(2\,\Delta\gamma(x) \frac{\int_{\mathbb{R}_+} t^{p/2-2} \exp(-t||x-\theta||^2/2) G(dt)}{\int_{\mathbb{R}_+} t^{p/2} \exp(-t||x-\theta||^2/2) G(dt)} + \gamma^2(x)\right) \tag{8.51}$$
$$\times \int_{\mathbb{R}_+} \left(\frac{t}{2\pi}\right)^{p/2} \exp\left(-\frac{t}{2}||x-\theta||^2\right) G(dt)\, dx.$$

Note that the ratio of the integrals in (8.51) is bounded below by

$$\frac{\int_{\mathbb{R}_+} t^{p/2-2}\, G(dt)}{\int_{\mathbb{R}_+} t^{p/2}\, G(dt)} = \frac{k}{2},$$

for k in (8.49), since the family of distributions with densities proportional to $t^{p/2}\exp(-\alpha t)\, G(dt)$ has monotone decreasing liklihood in $\alpha(= ||x-\theta||^2/2)$ and since t^{-2} is decreasing. Now, by the superharmonicity condition on γ, assumption (8.49) gives

$$\beta(\theta) \le \int_{\mathbb{R}^p} (k\,\Delta\gamma(x) + \gamma^2(x)) \int_{\mathbb{R}_+} \left(\frac{t}{2\pi}\right)^{p/2} \exp\left(-\frac{t}{2}\,||x-\theta||^2\right) G(dt)\, dx$$
$$\le 0\,.$$

□

The improved loss estimator result in Theorem 8.4 for scale mixtures of normal distributions was extended to the more general family of spherically symmetric distributions in Fourdrinier and Wells (1995b). In this setting, the conditions for improvement rest on the generating function g of the spherical density p_θ. A sufficient condition for domination of δ_0 has the usual form $k\,\nabla\gamma + \gamma^2 \le 0$.

Theorem 8.5 (Fourdrinier and Wells 1995b) *Assume the spherical distribution of X with generating function g has a finite fourth moment. Estimating θ through X and estimating the loss $\|X-\theta\|^2$, consider the estimator of loss $\delta_0 = E_0[\|X\|^2]$ (which is an unbiased estimate of risk of X). Let γ be a twice weakly differentiable on $\mathbb{R}^p$ such that $E_\theta[\gamma^2] < \infty$ for every $\theta \in \mathbb{R}^p$.*

If, for every $s \ge 0$,

$$2\, g(s) \int_s^\infty g(z)\, dz \le p\,\delta_0 \tag{8.52}$$

and if there exists a constant k such that, for any $s \ge 0$,

$$0 < k < \frac{\int_s^\infty z\, g(z)\, dz - s\int_s^\infty g(z)\, dz}{2\, g(s)}, \tag{8.53}$$

then a sufficient condition for $\delta_0 - \gamma$ to dominate δ_0 under loss $(\delta - ||X-\theta||^2)^2$ is that γ satisfies the differential inequality: $k\,\Delta\gamma + \gamma^2 < 0$.

For $p \ge 5$, we have shown that one can dominate the unbiased constant estimator of loss (associated with the estimator of θ, $\varphi_0(X) = X$) by a shrinkage-type estimator. As in the normal case, one may wish to add the frequentist-validity constraint, $E_\theta[\delta(X)] \ge E_\theta[\delta_0(X)]$ for all θ, to the loss estimation problem. In fact, in the normal case, the only frequentist valid estimator with $\mathscr{D}(\theta, \varphi_0, \delta) \le 0$ is δ_0 itself. The proof of this result follows from a randomization of the origin technique as in Hsieh and Hwang (1993). It remains an open question whether this is true in the general setting of a spherically symmetric distribution.

8.4.2 *Quadratic Loss Estimation: Spherically Symmetric Distribution with a Residual Vector*

In this subsection, we extend the ideas of the previous sections to a spherically symmetric distribution with a residual vector. We largely follow the development of Fourdrinier and Wells (1995a) (see also Fourdrinier and Wells 2012). We first develop an unbiased estimator of the loss and then construct a dominating shrinkage-type estimator. An important feature of our results is that the proposed loss estimates dominate the unbiased estimates for the entire class of spherically symmetric distributions. That is, the domination results are robust with respect to spherical symmetry, just as the improved estimators of the mean developed in Chap. 6 are similarly robust.

Let $(X, U) \sim SS(\theta, 0)$ where $\dim X = \dim \theta = p$ and $\dim U = \dim 0 = k$ $(p + k = n)$. For convenience let (X, U) and $(\theta, 0)$ represent $n \times 1$ vectors. In this section we consider the usual quadratic loss in estimation of θ,

$$||\varphi(X) - \theta||^2 , \tag{8.54}$$

and not the scaled version $||\varphi(X) - \theta||^2/\sigma^2$.

In the spherical case in Sect. 8.4.2 with know scale, the risk of X was constant with respect to θ. Thus, this risk, $E[R^2]$, provides an unbiased estimator of the loss subject to the knowledge of $E[R^2]$. Its properties, as the properties of any improved estimator, may depend on the specific underlying distribution. An important feature of the results in this subsection is that there is an unbiased estimator δ_0 of the loss of X, which is available for every spherically symmetric distribution (with finite fourth moment), $\delta_0(x) = p||U||^2/k$. Thus, we do not need to know the specific distribution, and we get robustness with an estimator that is no longer constant. Notice δ_0 makes sense because $p < n$.

We will now consider the estimation of the loss of a class of shrinkage estimators considered in Sect. 6.1 (see Cellier and Fourdrinier 1995). That is, for location estimators of the form

$$\varphi_g = X - \frac{||U||^2}{k+2}\, g(X), \tag{8.55}$$

where g is a weakly differentiable function from $\mathbb{R}^p$ into $\mathbb{R}^p$. Recall Theorem 6.1 shows that, if $||g||^2 \leq 2 \operatorname{div} g/(k + 2)$, φ_g dominates X under quadratic loss for all spherically symmetric distributions with a finite second moment. A member of the class is $\varphi_{JS} = X - ||U||^2/(k+2)\ (p-2)\ X/||X||^2$, the James-Stein estimator used when the variance is unknown as in Sect. 8.3.

It follows from Theorem 6.1 and the above discussion that an unbiased estimator of the loss (8.54) of the shrinkage estimator φ_g is given by

$$\delta_0^g = \frac{p}{k}\,||U||^2 + \Big(||g(X)||^2 + 2\,\mathrm{div} g(X)\Big)\,\frac{||U||^4}{(k+2)^2}\,. \tag{8.56}$$

As shown in Theorem 8.6 below, the unbiased estimator of the loss can be improved by a shrinkage estimator of the loss of the form

$$\delta_\gamma^g = \delta_0^g - ||U||^4\,\gamma(X)\,, \tag{8.57}$$

where γ is a positive function provided $p \geq 5$. Note that (8.57) is a true shrinkage estimator, while Johnstone's (1988) optimal loss estimate for the normal case is an expanding estimator. This is not contradictory since we are using a different estimator than Johnstone and he is only dealing with the normal case with known σ^2.

Theorem 8.6 (Fourdrinier and Wells 1995a) *Assume that $p \geq 5$ and the distribution of (X, U) has a finite fourth moment. Estimate θ through φ_g in (8.55) and consider estimating the loss $\|\varphi_g - \theta\|^2$. Let γ be a twice weakly differentiable nonnegative function on $\mathbb{R}^p$.*

A sufficient condition under which the estimator δ_γ^g given in (8.57) dominates the unbiased estimator δ_0^g under loss $(\delta - ||\varphi_g - \theta||^2)^2$ is that γ satisfies the differential inequality

$$\gamma^2 + \frac{4}{(k+2)^2}\,\gamma\,\mathrm{div} g - \frac{4}{(k+2)(k+6)}\,\mathrm{div}(\gamma\,g) + \frac{2}{(k+4)(k+6)}\,\Delta\gamma \leq 0\,. \tag{8.58}$$

An immediate corollary for the estimator $\varphi_0(X) = X$ $(g \equiv 0)$ follows.

Corollary 8.1 *Let $\varphi_0(X) = X$, $\delta_0(X, ||U||^2) = p/k\ \ ||U||^2$, and $\delta(X) = \delta_0(X) - ||U||^4\ \gamma(X)$. Assume that $p \geq 5$, the distribution of (X, U) has a finite fourth moment, and the function γ is twice weakly differentiable on $\mathbb{R}^p$. A sufficient condition under which the estimator $\delta(X)$ dominates the unbiased estimator δ_0 is that γ satisfies the differential inequality*

$$\gamma^2 + \frac{2}{(k+4)(k+6)}\,\Delta\gamma \leq 0\,. \tag{8.59}$$

Example 8.1 (Loss estimator for $\varphi_0(X) = X$) The standard example is where $\gamma(t) = d/||t||^2$ for all $t \neq 0$ with $d > 0$ satisfying the conditions of the theorem. More precisely it is easy to deduce that $\Delta\gamma(t) = -2d\,(p-4)/\|||t||^4$ and thus Inequality (8.59) reduces to

$$d^2 - \frac{4\,(p-4)}{(k+4)\,(k+6)}\,d \leq 0 \tag{8.60}$$

so that the sufficient condition of Corollary 8.1 is written as $0 < d \le 4\,(p-4)/(k+4)(k+6)$, which only occurs when $p \geq 5$. Clearly the left-hand side of (8.60) is minimized for $d = 2\,(p-4)/\{(k+4)\,(k+6)\}$, which provides the greatest improvement over the unbiased estimator δ_0.

Before proving the theorem, we need some preliminary integration identities which generalize Lemma 6.1.

Lemma 8.5 *For every twice weakly differentiable function $g : \mathbb{R}^p \to \mathbb{R}^p$ and for every function $h : \mathbb{R}_+ \to \mathbb{R}$,*

$$E_{R,\theta}\left[h(||U||^2)(X-\theta)^{\mathrm{T}}\,g(X)\right] = E_{R,\theta}\left[\frac{H(||U||^2)}{(||U||^2)^{k/2-1}}\,\mathrm{div}\,g(X)\right] \tag{8.61}$$

where H is the indefinite integral, vanishing at 0, of the function $t \mapsto 1/2\,h(t)\,t^{k/2-1}$, and provided the expectations exist.

Proof As in the proof of Lemma 6.1, we have

$$\begin{aligned}
&E_{R,\theta}\left[h(||U||^2)(X-\theta)^{\mathrm{T}}\,g(X)\right] \\
&= C_R^{p,k}\int_{B_{R,\theta}} h(R^2-\|x-\theta\|^2)(x-\theta)^{\mathrm{T}}\,g(x)\left(R^2-\|x-\theta\|^2\right)^{\frac{k}{2}-1}dx \\
&= -\,C_R^{p,k}\int_{B_{R,\theta}} (\nabla H(R^2-\|x-\theta\|^2))^{\mathrm{T}}\,g(x)\,dx
\end{aligned}$$

since

$$\begin{aligned}
\nabla H(R^2-\|x-\theta\|^2) &= -2\,H'(R^2-\|x-\theta\|^2)(x-\theta) \\
&= -h(R^2-\|x-\theta\|^2)\left(R^2-\|x-\theta\|^2\right)^{k/2-1}(x-\theta)\,.
\end{aligned}$$

Then, by the divergence formula,

$$\begin{aligned}
E_{R,\theta}\left[h(||U||^2)(X-\theta)^{\mathrm{T}}g(X)\right] = &-\,C_R^{p,k}\int_{B_{R,\theta}}\mathrm{div}\left(H(R^2-\|x-\theta\|^2)\,g(x)\right)dx \\
&+C_R^{p,k}\int_{B_{R,\theta}} H(R^2-\|x-\theta\|^2)\,\mathrm{div}g(x)\,dx\,.
\end{aligned}$$

Now, if $\sigma_{R,\theta}$ denotes the area measure on the sphere $S_{R,\theta}$, the divergence theorem ensures that the first integral equals

$$\int_{S_{R,\theta}} (H(R^2-\|x-\theta\|^2)g(x))^{\mathrm{T}}\,\frac{x-\theta}{\|x-\theta\|}\,\sigma_{R,\theta}(dx)$$

and is null since, for $x \in S_{R,\theta}$, we have $R^2 - \|x-\theta\|^2 = 0$ and $H(0) = 0$. Hence, in terms of expectation, we have

$$\begin{aligned}
&E_{R,\theta}\left[h(||U||^2)\,(X-\theta)^{\mathrm{T}}g(X)\right] \\
&= C_R^{p,k}\int_{B_{R,\theta}} \frac{H(R^2-\|x-\theta\|^2)}{\left(R^2-\|x-\theta\|^2\right)^{k/2-1}}\,\mathrm{div}g(x)\left(R^2-\|x-\theta\|^2\right)^{k/2-1}dx \\
&= E_{R,\theta}\left[\frac{H(||U||^2)}{(||U||^2)^{k/2-1}}\,\mathrm{div}g(X)\right],
\end{aligned}$$

which is the desired result. □

Corollary 8.2 *For every twice weakly differentiable function $\gamma : \mathbb{R}^p \to \mathbb{R}_+$ and for every integer q,*

$$\begin{aligned}
E_{R,\theta}\left[||U||^q\,||X-\theta||^2\,\gamma(X)\right] &= \frac{p}{k+q}E_{R,\theta}\left[||U||^{q+2}\,\gamma(X)\right] \\
&\quad + \frac{1}{(k+q)(k+q+2)}E_{R,\theta}\left[||U||^{q+4}\Delta\gamma(X)\right].
\end{aligned}$$

provided the expectation exists.

Proof Take $h(t) = t^{q/2}$ and $g(x) = \gamma(x)\,(x-\theta)$ and apply Lemma 8.5 twice. □

Proof of Theorem 8.6 Since the distribution of (X, U) is spherically symmetric around θ, it suffices to obtain the result working conditionally on the radius. For $R > 0$ fixed, we can compute this using the uniform distribution $U_{R,\theta}$ on the sphere $S_{R,\theta}$. Hence, the risk of δ_γ^g equals

$$\begin{aligned}
E_{R,\theta}\left[(\delta_\gamma^g - ||\varphi-\theta||^2)^2\right] &= E_{R,\theta}\left[(\delta_0^g - ||\varphi-\theta||^2)^2\right] + E_{R,\theta}\left[||U||^8\,\gamma^2(X)\right] \\
&\quad - 2E_{R,\theta}\left[||U||^4\gamma(X)(\delta_0^g - ||\varphi-\theta||^2)\right].
\end{aligned}$$

Applying Lemma 8.5, it follows that

$$2\,E_{R,\theta}\left[||U||^6\,(X-\theta)^{\mathrm{T}}\gamma(X)\,g(X)\right] = \frac{2}{k+6}E_{R,\theta}\left[||U||^8\,\mathrm{div}\,(\gamma(X)\,g(X))\right].$$

Hence expanding the risk and Corollary 8.2, it follows that the risk of δ_γ^g equals

$$\begin{aligned}
&E_{R,\theta}\left[(\delta_0^g - ||\varphi-\theta||^2)^2\right] + E_{R,\theta}\left[||U||^8\gamma^2(X)\right] \\
&- \frac{8k}{(k)(k+4)}E_{R,\theta}\left[||U||^6\gamma(X)\right]
\end{aligned}$$

$$+ E_{R,\theta}\left[||U||^8\left\{\frac{4}{(k+2)^2}\,\gamma(X)\,\mathrm{div}\,g(X) + \frac{4}{(k+4)(k+6)}\,\mathrm{div}\,(\gamma(X)\,g(X))\right\}\right]$$

$$+ \frac{2}{(k+4)(k+6)} E_{R,\theta}\left[||U||^8 \Delta\gamma(X)\right].$$

Since the function γ is nonnegative, the third term on the right-hand side is negative; also the $||U||^8$ term is a factor in all the other expressions. Hence, the sufficient condition for domination is

$$\gamma^2 + \frac{4}{(k+2)^2}\gamma\,\mathrm{div}\,g + \frac{4}{(k+4)(k+6)}\,\mathrm{div}\,(\gamma\, g) + \frac{2}{(k+4)(k+6)}\Delta\gamma \le 0$$

in order that the inequality $R(\delta^g, \theta, \varphi) \le R(\delta_0^g, \theta, \varphi)$ holds. □

8.5 Applications to Model Selection

Loss estimation results discussed in the previous sections can be applied to the model selection problem. The loss estimation ideas in this chapter lay the theoretical foundation for the construction of model selection rules as well as give a decision theoretic analysis of their statistical properties. Fourdrinier and Wells (1994) and Boisbunon et al. (2014) show that improved loss estimators give more accurate model selection procedures. Bartlett et al. (2002) studied model selection strategies based on penalized empirical loss minimization and pointed out the equivalence between loss estimation and data-based complexity penalization. It is shown that any good loss estimate may be converted into a data-based penalty function and the performance of the estimate is governed by the quality of the loss estimate.

The principle of parsimony helps to avoid classical issues such as overfitting or computational error. At the same time, the model should capture sufficient information in order to comply with some objectives of good prediction, good estimation, or good selection and thus, it should not be too sparse. This principle has been elucidated by many statisticians as a trade-off between the goodness of fit to the data and the complexity of the model (see, for instance, Hastie et al. 2008). From the practitioner's point of view, model selection is often implemented through cross-validation (see Celisse and Arlot (2010) for a review on this topic) or the minimization of criteria whose theoretical justification relies on hypotheses made within a given framework.

In this section, we review the work in Boisbunon et al. (2014) and examine model selection measures, C_p and AIC, from a loss estimation point of view. We will focus on the linear regression model

$$Y = X\beta + \sigma\varepsilon \tag{8.62}$$

where Y is a random vector in $\mathbb{R}^n$, X is a fixed and known full rank design matrix containing p observed variables $\mathbf{x}^j$ in $\mathbb{R}^n$, β is the unknown vector in $\mathbb{R}^p$ of regression coefficients to be estimated, σ is the noise level and ε is a random vector in $\mathbb{R}^n$ representing the model noise with mean zero and covariance matrix $\sigma^2 I_n$. One subproblem of model selection is the problem of variable selection: only a subset of the independent variables X^j have nonredundant information on Y and we wish to recover this subset as well as correctly estimate the corresponding regression coefficients.

Early work treated the model selection problem from the hypothesis testing point of view. For instance, the Forward Selection and Backward Elimination procedures were stopped using appropriate critical values. This practice changed with Mallows' automated criterion known as C_p (Mallows 1973). Mallows' idea was to propose an unbiased estimator of the scaled expected prediction error $E_\beta[\|X\hat{\beta}_I - X\beta\|^2/\sigma^2]$ where $\hat{\beta}_I$ is an estimator of β based on the selected variable set $I \subset \{1, \ldots, p\}$, E_β denotes the expectation with respect to the sampling distribution in model (8.62) and $\|\cdot\|$ is the Euclidean norm on $\mathbb{R}^n$. Assuming Gaussian *i.i.d.* error terms, Mallows proposed the following criterion

$$C_p = \frac{\|Y - X\hat{\beta}_I\|^2}{\hat{\sigma}^2} + 2\widehat{df} - n \tag{8.63}$$

where $\hat{\sigma}^2$ is an estimator of the variance σ^2 based on the full linear model fitted with the least-squares estimator $\hat{\beta}^{LS}$, that is, $\hat{\sigma}^2 = \|Y - X\hat{\beta}^{LS}\|^2/(n-p)$, and $\widehat{df}$ is an estimator of the "degrees of freedom" (df), also called the effective dimension of the model (see Hastie and Tibshirani 1990). For the least squares estimator, df is the number k of variables in the selected subset I.

Mallows' C_p relies on the assumption that, if for some subset I of explanatory variables the expected prediction error is low, then those variables are relevant for predicting Y. In practice, the rule for selecting the "best" candidate is the minimization of C_p. However, Mallows argues that this rule should not be applied in all cases, and that it is better to look at the shape of the C_p-plot instead, especially when some explanatory variables are highly correlated.

In 1974, Akaike proposed different automatic criteria that would not need a subjective calibration of the significance level as in hypothesis testing based approaches. His proposal was more general with applications to many problems such as variable selection, factor analysis, analysis of variance, and order selection in autoregressive models (Akaike 1974). Also his motivation was different from Mallows. Akaike considered the problem of estimating the density $f(\cdot|\beta, \sigma)$ of an outcome variable Y where f is parameterized by $\beta \in \mathbb{R}^p$ and $\sigma \in \mathbb{R}_+$. Akaike's aim was to generalize the principle of maximum likelihood, enabling a selection between several maximum likelihood estimators $\hat{\beta}_I$ and $\hat{\sigma}_I^2$. Akaike showed that all the information for discriminating the estimator $f(\cdot|\hat{\beta}_I, \hat{\sigma}_I^2)$ from the true $f(\cdot|\beta, \sigma)$ could be summed up by the Kullback-Leibler divergence $D_{KL}(\hat{\beta}_I, \beta) = E[\log f(Y_{\text{new}}|\beta, \sigma)] - \mathbb{E}[\log f(Y_{\text{new}}|\hat{\beta}_I, \hat{\sigma}_I^2)]$ where the expectation is taken over new observations. By means of asymptotic analysis and by considering the expectation of D_{KL}, Akaike arrived at the following criterion

$$\text{AIC} = -2\sum_{i=1}^{n} \log f(y_i|\hat{\beta}_I, \hat{\sigma}_I^2) + 2|I|, \tag{8.64}$$

where $|I|$ is the number of parameters of $\hat{\beta}_I$. In the special case of a Gaussian distribution, AIC and C_p are equivalent up to a constant for model (8.62). Hence, Akaike described his AIC as a generalization of C_p to a more general class of models. Unlike Mallows, Akaike explicitly recommends the rule of minimization of AIC in order to identify the best model from data.

In the context of the model in (8.62), $AIC = -n\log\hat{\sigma}_I^2 - 2(|I|+1) - n - n\log(2\pi)$, where $\hat{\sigma}_I^2 = \|Y - X\hat{\beta}_I\|^2/n$ Thus the best model is determined by minimizing $n\log\hat{\sigma}_I^2 + 2|I|$ across all candidate models. Hurvich and Tsai (1989) showed that AIC leads to overfitting in small sample size and proposed a biased corrected version of AIC that selects the model that minimizes $\log\hat{\sigma}_I^2 + (n + |I|)/(n - |I| - 2)$ across all candidate models.

Ye (1998) extended AIC to more complex settings by replacing $|I|$ by the estimated degrees of freedom introduced by Efron (2004). For the model in (8.62) Ye's [E]xtended AIC is

$$\text{EAIC}(\hat{\beta}) = \frac{\|Y - X\hat{\beta}_I\|^2}{\hat{\sigma}^2} + 2\,\text{div}_Y(X\hat{\beta}_I). \tag{8.65}$$

where $\hat{\sigma}^2 = \|Y - X\hat{\beta}^{LS}\|^2/(n-p)$.

8.5.1 Model Selection in the Loss Estimation Framework

As seen in the previous sections of this chapter, the idea underlying the estimation of loss is closely related to Stein's Unbiased Risk Estimate (SURE). When considering the Gaussian model in (8.62), we have $\mu = X\beta$, we set $\hat{\mu} = X\hat{\beta}$ and $L(\hat{\beta}, \beta)$ is defined as the quadratic loss $\|X\hat{\beta} - X\beta\|^2$. Special focus will be given to the quadratic loss since it is the most commonly used and allows tractable calculations. In practice, it is a reasonable choice if we are interested in both good selection and good prediction at the same time. Moreover, quadratic loss allows us to link loss estimation with C_p and AIC.

In the following theorem, an unbiased estimator of the quadratic loss, under a Gaussian regression model, is developed using a result of Stein (1981).

Theorem 8.7 *Let $Y \sim \mathcal{N}_n(X\beta, \sigma^2 I_n)$ and $\hat{\beta} = \hat{\beta}(Y)$ be a function of the least squares estimator of β such that $X\hat{\beta}$ is weakly differentiable with respect to Y. Let $\hat{\sigma}^2 = \|Y - X\hat{\beta}^{LS}\|^2/(n-p)$.*

Then,

$$\delta_0(Y) = \|Y - X\hat{\beta}\|^2 + (2\,\mathrm{div}_Y(X\hat{\beta}) - n)\hat{\sigma}^2 \tag{8.66}$$

is the unbiased estimator of $\|X\hat{\beta} - X\beta\|^2$.

Proof The risk of $X\hat{\beta}$ at $X\beta$ is

$$\begin{aligned} E_\beta[\|X\hat{\beta} - X\beta\|^2] = E_\beta[\|X\hat{\beta} - Y\|^2 + \|Y - X\beta\|^2] \\ +\mathbb{E}_\beta[2(Y - X\beta)^{\mathrm{T}}(X\hat{\beta} - Y)]\,. \end{aligned} \tag{8.67}$$

Since $Y \sim \mathscr{N}_n(X\beta, \sigma^2 I_n)$, we have $E_\beta[\|Y - X\beta\|^2] = E_\beta[(Y - X\beta)^{\mathrm{T}} Y] = n\,\sigma^2$ it follows that

$$E_\beta[\|X\hat{\beta} - X\beta\|^2] = E_\beta[\|Y - X\hat{\beta}\|^2] - n\,\sigma^2 + 2\,\mathrm{tr}(\mathrm{cov}_\beta(X\hat{\beta}, Y - X\beta)).$$

Moreover, applying Stein's identity for the right-most part of the expectation in (8.67) with $g(Y) = X\hat{\beta}$ and assuming that $X\hat{\beta}$ is weakly differentiable with respect to Y, we can rewrite (8.67) as

$$E_\beta[\|X\hat{\beta} - X\beta\|^2] = E_\beta[\|Y - X\hat{\beta}\|^2] - n\,\sigma^2 + 2\,\sigma^2\, E_\beta\Big[\mathrm{div}_Y X\hat{\beta}\Big].$$

Because $\hat{\sigma}^2$ is an unbiased estimator of σ^2 and is independent of $\hat{\beta}^{LS}$ and therefore of $\hat{\beta}(Y)$, the right-hand side of this last equality is also equal to the expectation of $\delta_0(Y)$ given by Eq. (8.66). Hence the statistic $\delta_0(Y)$ is an unbiased estimator of $\|X\hat{\beta} - X\beta\|^2$. □

Efron (2004) introduced the concept of the generalized degrees of freedom of an estimator of the mean as $df = E_\beta\Big[\mathrm{div}_Y X\hat{\beta}\Big]$. Consequently $\widehat{df} = \mathrm{div}_Y X\hat{\beta}$ is an unbiased estimate of the degrees of freedom. Therefore, $\delta_0(Y)$ in Theorem 8.66 can also be expressed as $\|Y - X\hat{\beta}\|^2 + (2\,\widehat{df} - n)\hat{\sigma}^2$. For example in the case of a linear estimator of the mean $\hat{\mu} = SY$ it follows that $df = \sigma^2 \mathrm{Tr}(S)$. Specifically, if $\hat{\mu} = X\hat{\beta}^{LS} = X(X^{\mathrm{T}}X)^{-1}X^{\mathrm{T}}Y$ then $df = p\sigma^2$ and $\widehat{df} = p\hat{\sigma}^2$.

For invariant loss $\|X\beta - X\hat{\beta}\|^2/\sigma^2$ and $S = \|Y - X\hat{\beta}^{LS}\|^2$ the following result is a natural adaptation of Theorem 3.1 from Fourdrinier and Wells (2012).

Theorem 8.8 *Let* $Y \sim \mathscr{N}_n(X\beta, \sigma^2 I_n)$ *and* $n \geq 5$. *Let* $\hat{\beta} = \hat{\beta}(Y)$ *be an estimator of* β *weakly differentiable with respect to* Y *and independent of* $\|Y - X\hat{\beta}^{LS}\|^2$.

Then

$$\delta_0^{inv}(Y) = \frac{n - p - 2}{\|Y - X\hat{\beta}^{LS}\|^2}\|Y - X\,\hat{\beta}\|^2 + 2\,\mathrm{div}_Y(X\,\hat{\beta}) - n \tag{8.68}$$

is an unbiased estimator of the invariant loss $\|X\hat{\beta} - X\beta\|^2/\sigma^2$.

Note that, for more general estimators of the form $\hat{\beta}(Y, S)$, a correction term has to be added to (8.68). Thus, if $\hat{\beta}(Y, S) = \hat{\beta}^{LS}(Y) + g(\hat{\beta}^{LS}, S)$ for some function g, this correction is $4(X\hat{\beta}(Y, S) - Y)^{\mathrm{T}} X\, \partial g(\hat{\beta}^{LS}, S)/\partial S$. An analogous correction to (8.66) is needed in Theorem 8.7 if $\hat{\beta}$ is a function of S as well as of $\hat{\beta}^{LS}$. We omit the details.

In terms of predicting a future value $Y_0 \sim \mathcal{N}_n(X\beta, \sigma^2 I_n)$ it is easy to see with a calculation analogous to Theorem 8.7 that

$$\begin{aligned} E_\beta[\|Y_0 - X\hat{\beta}\|^2] &= E_\beta[\|(Y_0 - Y) + (Y - X\hat{\beta})\|^2] \\ &= E_\beta[\|(Y_0 - Y)\|^2 + \|Y - X\hat{\beta}\|^2 \\ &\qquad +2\,\langle Y_0 - X\beta, Y - X\hat{\beta}\rangle - 2\,\langle Y - X\beta, Y - X\hat{\beta}\rangle] \\ &= E_\beta[\|Y - X\hat{\beta}\|^2 + \widehat{df}]. \end{aligned}$$

Hence $\|Y - X\hat{\beta}\|^2 + \hat{\sigma}^2\widehat{df}$ is an unbiased estimator of the prediction error $[\|Y_0 - X\hat{\beta}\|^2]$. Recall the formulas for C_p, $EAIC$ and δ_0 in (8.63), (8.65), and (8.66), respectively, of the three criteria of interest under the Gaussian model . The links between different criteria for model selection are due to the fact that, under our working hypothesis (linear model, quadratic loss, normal distribution $Y \sim \mathcal{N}_n(X\beta, \sigma^2 I_n)$ for a fixed design matrix X), they can be seen as unbiased estimators of related quantities of interest. It can be easily seen that the three criteria differ from each other only up to a multiplicative and/or additive constant. Hence the models selected by the three criteria will be the same. There is also an equivalence with other model selection criteria, such as those investigated in Li (1985), Shao (1997) and Efron (2004).

The final objective is to select the "best" model among those at hand. This can be performed by minimizing either of the three proposed criteria, that is the unbiased estimator of loss δ_0, C_p, AIC or $EAIC$. The idea behind this heuristic is that the best model in terms of prediction is the one minimizing the estimated loss.

Note that C_p and $EAIC$ are developed first fixing σ^2 and then estimating it by $\hat{\sigma}^2$, while, for δ_0, the estimation of σ^2 is integrated into the construction process. It is then natural to gather the evaluation of β and σ^2 estimating the invariant loss

$$\frac{\|X\beta - X\hat{\beta}\|^2}{\sigma^2},$$

for which $\delta_0^{\text{inv}}(\hat{\beta})$ in an unbiased estimator. Note that $\delta_0^{\text{inv}}(\hat{\beta})$ involves the variance estimator $\|Y - X\hat{\beta}^{LS}\|^2/(n - p - 2)$ instead of $\|Y - X\hat{\beta}^{LS}\|^2/(n - p)$. This alternative variance estimator was also considered in the unknown variance setting for the construction of the modified C_p, which is actually equivalent to $\delta_0^{\text{inv}}(\hat{\beta})$, and the corrected AIC (see Davies et al. 2006 and Hurvich and Tsai 1989).

However, in practice, we might only have a vague intuition of the nature of the underlying distribution and we might not be able to give its specific form. Boisbunon et al. (2014) showed that δ_0, which is equivalent to the Gaussian AIC as we have just seen, can be also derived from a more general distribution context, that of spherically symmetric distributions, with no need to specify the precise form of the distribution. Consequently, δ_0 can be considered as a generalization of C_p for non-Gaussian distributions.

A number of new regularized regression methods have recently been developed, starting with ridge regression in Hoerl and Kennard (1970), followed by the LASSO in Tibshirani (1996), and the Elastic Net in Zou and Hastie (2005), and Efron et al. (2004). Each of these estimates is weakly differentiable and has the form of a general shrinkage estimate; thus the prediction error estimate in (8.68) may be applied to construct a model selection procedure. Zou et al. (2007), Tibshirani and Taylor (2012), and Narayanan and Wells (2015) used this idea to develop a model selection method for the Lasso. In some situations verifying the weak differentiability of φ may be complicated. See Boisbunon et al. (2014) for further discussion of these issues.

8.6 Confidence Set Assessment

In the previous sections of this chapter, the usual quadratic loss $L(\theta, \varphi(x)) = ||\varphi(x) - \theta||^2$ was considered to evaluate various estimators $\varphi(X)$ of θ. The squared norm $||x - \theta||^2$ was crucial in the derivation of the properties of the loss estimators in conjunction with its role in the normal density or, more generally, in a spherical density. One could imagine other losses, but, to deal with tractable calculations, it helps to keep the Euclidean norm as a component of the loss. Hence, a natural extension is to consider losses that are functions of $||\delta(x) - \theta||^2$, that is, of the form $c(||\delta(x) - \theta||^2)$ for a nonnegative function c defined on $\mathbb{R}_+$.

Brandwein and Strawderman (1980) (see Sects. 5.5 and 6.5) considered a nondecreasing and concave function c of $\|\delta(x) - \theta\|^2$ in order to compare various estimators $\delta(X)$ of θ. As in the case tackled by Johnstone (1988) and Fourdrinier and Wells (1995b), it is still of interest to assess the loss of $\delta(X) = X$, that is, to estimate $c\left(\|x - \theta\|^2\right)$.

Note that estimating $c(||x-\theta||^2)$ can be viewed as an evaluation of a quantity that is not necessarily a loss. Indeed, it includes the problem of estimating the confidence statement of the usual confidence set $\{\theta \in \mathbb{R}^p \mid \|x - \theta\|^2 \leq c_\alpha\}$ with the confidence coefficient $1 - \alpha$: $c(\cdot)$ is the indicator function $\mathbb{1}_{[0,c_\alpha]}$ (the confidence interval estimation example seen in Sect. 8.1 has illustrated the necessity of a confidence evaluation depending on the data).

The problem of estimating a function $c(\cdot)$ of $||x - \theta||^2$ was addressed by Fourdrinier and Lepelletier (2008) whose work we follow.

Let X be a random vector in $\mathbb{R}^p$ with a spherical density of the form $x \mapsto f(\|x - \theta\|^2)$ where θ is the unknown location parameter. For a given nonnegative function c on $\mathbb{R}_+$, we are interested in estimating the quantity $c(\|x - \theta\|^2)$ when x has been observed from X. In contrast to the previous sections, if only X is considered as an estimator of θ, the function $c(\cdot)$ intervenes in the quantity to estimate. A simple reference estimator is the unbiased estimator $\delta_0 = E_0[c(\|X\|^2)]$. Note that, in the confidence statement estimation problem, $\delta_0 = E_0[\,1\!\!1_{[0,c_\alpha]}] = 1 - \alpha$ while, in the loss estimation problem of $||x - \theta||^2$ considered in the previous sections, $\delta_0 = E_0[\|X\|^2]$ (that is, p in the normal case).

Since δ_0 is a constant estimator, it is natural to search for better estimators in terms of the risk (8.2), that is, estimators δ such that

$$\begin{aligned} R_c(\delta,\theta) &= E_\theta\Big[\big(\delta(X) - c\big(\|X-\theta\|^2\big)\big)^2\Big] \\ &\le E_\theta\Big[\big(\delta_0 - c\big(\|X-\theta\|^2\big)\big)^2\Big] \\ &= R_c(\delta_0,\theta) \\ &= R_c(\delta_0,0)\,. \end{aligned}$$

Improvement on δ_0 will be considered when its own risk is finite, that is, under the condition $E_0[c^2(\|X\|^2)] < \infty$, which also guarantees the existence of δ_0. We assume Condition (8.9) to assure the finiteness of the risk of $\delta(X) = \delta_0(X) + \gamma(X)$.

Due to the presence of the function $c(\cdot)$, repeated use of Stein's identity is not appropriate to deal with the risk difference

$$\begin{aligned} \mathscr{D}_c(\theta,\delta) &= \mathscr{R}_c(\theta,\delta) - \mathscr{R}_c(\theta,\delta_0) \\ &= E_\theta[2\,\{\delta_0 - c(\|X-\theta\|^2)\}\,\gamma(X) + \gamma^2(X)] \end{aligned} \tag{8.69}$$

between $\delta(X)$ and δ_0. As an alternative, the approach in Fourdrinier and Lepelletier (2008) consists of introducing the Laplacian of the correction function γ, say $\Delta(\gamma)$, under the expectation sign in the right hand side of (8.69) and in developing an upper bound of the risk difference in terms of the expectation of a differential expression of the form $k\,\Delta\gamma + \gamma^2$ where k is a constant different from 0. The underlying idea is based on two facts. We know that, in the normal setting, $\delta_0 = 1 - \alpha$ is admissible for estimating a confidence statement (see Brown and Hwang 1990) and $\delta_0 = p$ is admissible for estimating the loss $||x - \theta||^2$ (see Johnstone 1988) when $p \le 4$. For $p \ge 5$, improved estimators are available, mainly through simulations in Robert and Casella (1994) and formally thanks to the differential inequality $2\,\Delta(\gamma) + \gamma^2 \le 0$ in Johnstone (1988). In Sect. 8.7 it will be shown that inequalities of the form $k\,\Delta(\gamma) + \gamma^2 \le 0$ have no nontrivial solution γ when $p \le 4$. Therefore, it may be reasonable to think that, in (8.69), such operators of the form $k\,\Delta(\gamma) + \gamma^2$, the Laplacian of γ should play a role in obtaining improved estimators when $p \ge 5$.

Here we develop the principle that leads to the role of $\Delta(\gamma)$, assuming that suitable regularity conditions on the various functions in use are satisfied to make valid what it is stated; we will precise the appropriate conditions afterwards. First, it can be checked that, if K is the function depending on f and c defined, for any $t > 0$, by

$$K(t) = \frac{1}{p-2} \int_t^\infty \left[\left(\frac{y}{t} \right)^{p/2-1} - 1 \right] (\gamma_0 - c(y))\, f(y)\, dy.$$

then, for almost every $x \in \mathbb{R}^p$,

$$\Delta K(\|x-\theta\|^2) = 2(\gamma_0 - c(\|x-\theta\|^2))\, f(\|x-\theta\|^2).$$

Hence, the first part of the expectation in (8.69) can be written as

$$E_\theta[2\,(\gamma_0 - c(\|X-\theta\|^2))\,\gamma(X)] = \int_{\mathbb{R}^p} \Delta K(\|x-\theta\|^2)\,\gamma(x)\,dx\,. \tag{8.70}$$

Now, through an appropriate Green's formula, the Laplacian in (8.70) can be moved from the function K to the function γ, so that

$$\int_{\mathbb{R}^p} \Delta K(\|x-\theta\|^2)\,\gamma(x)\,dx = \int_{\mathbb{R}^p} K(\|x-\theta\|^2)\,\Delta\gamma(x)\,dx. \tag{8.71}$$

Hence, (8.70) can be written as

$$E_\theta[2\,(\gamma_0 - c(\|X-\theta\|^2))\,\gamma(X)] = E_\theta\left[\frac{K(\|X-\theta\|^2)}{f(\|X-\theta\|^2)}\,\Delta\gamma(X) \right]. \tag{8.72}$$

Therefore, it follows from (8.72) an expression of the risk difference in (8.69) involving $\Delta\gamma(X)$, that is,

$$\mathscr{D}_c(\theta, \Delta) = E_\theta\left[\frac{K(\|X-\theta\|^2)}{f(\|X-\theta\|^2)}\,\Delta\gamma(X) + \gamma^2(X) \right]. \tag{8.73}$$

In Fourdrinier and Lepelletier (2008), under the condition that $\delta_0 - c$ has only one sign change, two cases are considered for a domination result to be obtained: (1) when $\delta_0 - c$ is first negative and then positive, the Laplacian of γ is assumed subharmonic while, when $\delta_0 - c$ is first positive and then negative; and (2) the Laplacian of γ is assumed superharmonic. Then, relying on the fact that f is bounded from above by a constant M, it can be proved that

$$E_\theta\left[\frac{K(\|X-\theta\|^2)}{f(\|X-\theta\|^2)}\,\Delta s(X) \right] \leq E_\theta[k\,\Delta s(X)]$$

with

$$k = \frac{1}{M} E_0[K(\|X\|^2)]\,,$$

so that a sufficient condition for δ to dominate δ_0 is for γ to satisfy the partial differential inequality

$$k\,\Delta\gamma(x) + \gamma^2(x) \leq 0\,, \tag{8.74}$$

for any $x \in \mathbb{R}^p$.

Before commenting on this result, we specify the regularity conditions (that we will call Conditions ($\mathscr{C}$) in the following) on γ, f, and c under which the result holds. In addition to the usual requirements that $E_\theta[\gamma^2] < \infty$ and $\gamma \in W^{2,1}_{loc}(\mathbb{R}^p)$, it is assumed that there exists $r > 0$ such that $\gamma \in C^2_b(\mathbb{R}^p \setminus B_r)$ the space of the functions twice continuously differentiable and bounded on $\mathbb{R}^p \setminus B_r$. Also, it is supposed that the functions $f(\cdot)$ and $c(\cdot)$ are continuous on $\mathbb{R}^*_+ = \{x \mid x > 0\}$, except possibly on a finite set T, and that there exists $\epsilon > 0$ such that f and $f(\cdot)c$ belong to $S^{0,p/2+1+\epsilon}(\mathbb{R}^*_+\backslash T)$, the space of continuous functions v on $\mathbb{R}^*_+\backslash T$ such that

$$\sup_{x\in\mathbb{R}^*_+\backslash T;\beta\leq p/2+1+\epsilon} \|x\|^\beta\,|v(x)| < \infty\,.$$

Typical solutions of (8.74) are functions of the form $\gamma(x) = -\text{sgn}(k)\,d/||x||^2$ with $0 \leq d \leq |k|\,(p-4)$. Intuitively, estimating a loss as $||x-\theta||^2$ (i.e. $c(t) = t$) is different from estimating a confidence statement (i.e. $c(t) = 1\!\!1_{[0,c_\alpha]}(t)$) : we would like to deal with small losses and with large confidence statements. The two sign change conditions do report on these two situations. Thus, for the first problem, the function $\delta_0 - t = p - t$ is first positive and then negative; this is a case for which it can be shown that $k < 0$ (see Fourdrinier and Lepelletier 2008), so that a dominating loss estimator is $\delta(X) = \delta_0 - \gamma(X) = p + \text{sgn}(k)\,d/||x||^2 = p - d/||x||^2$ for $0 \leq d \leq -k\,(p-4)$. Now, for the second problem, the function $\delta_0 - 1\!\!1_{[0,c_\alpha]}(t) = 1 - \alpha - 1\!\!1_{[0,c_\alpha]}$ is first negative and then positive and it is shown in Fourdrinier and Lepelletier (2008) that $k > 0$, so that a dominating confidence set assessment estimator is $\delta(X) = \delta_0 - \gamma(X) = 1 - \alpha + \text{sgn}(k)\,d/||x||^2 = 1 - \alpha + d/||x||^2$ for $0 \leq d \leq k\,(p-4)$. Note that the correction to δ_0 is downward (upward) by $d/||x||^2$ for the first (second) problem.

The use of the property that the generating function f is bounded by M gives rise to a constant k, which may be small in absolute value and hence, may reduce the scope of the possible corrections γ leading to improved estimators δ. In Fourdrinier and Lepelletier (2008), an additional condition is given, relying on the monotonicity of the ratio K/f, which avoids the use of M. Here is their result.

Theorem 8.9 *Assume that Conditions ($\mathscr{C}$) are satisfied and that the function $\delta_0 - c(t)$ has only one sign change. In the case where $\delta_0 - c(t)$ is first negative and then positive (first positive and then negative), assume that the Laplacian of γ is subharmonic (superharmonic). Finally assume that the functions K and K/f have the same monotonicity (both nonincreasing or both nondecreasing).*

Then a sufficient condition for δ to dominate δ_0 is that γ satisfy the partial differential inequality

$$\forall x \in \mathbb{R}^p \quad \kappa\,\Delta\gamma(x) + \gamma^2(x) \leq 0 \tag{8.75}$$

with

$$\kappa = E_0\left[\frac{K(\|X\|^2)}{f(\|X\|^2)}\right].$$

Proof We consider the case where $\gamma_0 - c$ is first negative and then positive (the case where the function $\Delta\gamma$ is assumed to be subharmonic). The main point is to treat the left hand side of (8.72); it equals

$$\begin{aligned}
&E_\theta[K(\|X-\theta\|^2)\,\Delta\gamma(X)]\\
&=\int_{\mathbb{R}^p} K(\|x-\theta\|^2)\,\Delta\gamma(x)\,f(\|x-\theta\|^2)\,dx\\
&=\int_0^\infty\int_{S_{r,\theta}} \Delta\gamma(x)\,dU_{r,\theta}\,(x)\,K(r^2)\,\frac{2\pi^{p/2}}{\Gamma(p/2)}\,r^{p-1}\,f(r^2)\,dr \qquad (8.76)
\end{aligned}$$

where $U_{r,\theta}$ is the uniform distribution on the sphere $S_{r,\theta} = \{x \in \mathbb{R}^p \mid \|x-\theta\| = r\}$ of radius r and centered at θ. Note that the function $r \mapsto \frac{2\pi^{p/2}}{\Gamma(p/2)}\,r^{p-1}\,f(r^2)$ is the radial density, that is, the density of the radius $R = \|X-\theta\|$. Now the right hand side of (8.76) can be bounded above by

$$\begin{aligned}
&\int_0^\infty\int_{S_{r,\theta}} \Delta\gamma(x)\,d\mathcal{U}_{r,\theta}(x)\,\frac{2\pi^{p/2}}{\Gamma(p/2)}\\
&\qquad r^{p-1}\,f(r^2)\,dr \times \int_0^\infty \frac{K(r^2)}{f(r^2)}\,\frac{2\pi^{p/2}}{\Gamma(p/2)}\,r^{p-1}\,f(r^2)\,dr, \qquad (8.77)
\end{aligned}$$

by the covariance inequality, since K/f is nonincreasing and $r \mapsto \int_{S_{r,\theta}} \Delta\gamma(x)\,d\mathcal{U}_{r,\theta}(x)$ is nondecreasing by the subharmonicity of $\Delta\gamma$ (see e.g. Doob 1984). Therefore, we have obtained

$$E_\theta\left[\frac{K(\|X-\theta\|^2)}{f(\|X-\theta\|^2)}\,\Delta\gamma(X)\right] \le E_0\left[\frac{K(\|X\|^2)}{f(\|X\|^2)}\right]\,E_\theta[\Delta\gamma(X)] = \kappa\,E_\theta[\Delta\gamma(X)],$$

which, through (8.72), implies that the risk difference in (8.69) satisfies

$$\mathcal{D}_c(\theta,\delta) \le E_\theta[\kappa\,\Delta\gamma(X) + \gamma^2(X)]$$

and, finally, proves the theorem. □

8.7 Differential Operators and Dimension Cut-Off When Estimating a Loss

In the previous sections, we have seen that, in various distribution settings, unbiased estimators of loss can be improved when the dimension $p \geq 5$. In the normal case, Johnstone (1988) formally proved that when $p \leq 4$ the unbiased loss estimator $\Delta_0(X) \equiv p$ (based on the MLE) is admissible so that no (global) improvement over it cannot be expected. That situation parallels the dimension cut-off phenomenon discussed in Sect. 2.6, which occurs when estimating the mean θ: the MLE X is admissible when $p \leq 2$, but inadmissible when $p \geq 3$.

In this section, we give a result parallel to Theorem 2.8 in Sect. 2.6 which shows that when $p \leq 4$ there is no nontrivial solution to the relevant partial differential inequality

$$\mathscr{R}\gamma(x) = k\,\Delta\gamma(x) + \gamma^2(x) \leq 0\,, \tag{8.78}$$

for any constant k. We once again follow Blanchard and Fourdrinier (1999) who proved the nonexistence of nontrivial solutions for a general differential inequality. Their unified result covers both Theorems 2.8 and 8.10 below.

Theorem 8.10 *Let $k \in \mathbb{R}$ be fixed. When $p \leq 4$, the only twice weakly differentiable solution γ with $\gamma^2 \in L^1_{loc}(\mathbb{R}^p)$ of (8.78), is $\gamma = 0$ (a.e.) for any $x \in \mathbb{R}^p$.*

The proof is based on the same sequence of test functions $(\varphi_n)_{n\geq 1}$ used in the proof of Theorem 2.8 and defined in Eq. (2.43). Recall that, for any $n \geq 1$, the function φ_n has compact support B_{2n}, the closed ball of radius $2n$ and centered at 0 in $\mathbb{R}^p$. Also, since φ'' is bounded, a property analogous to (2.44) for the second derivative is that, for any $\beta \geq 2$ and for any $j = 1, \ldots, p$,

$$\left|\frac{\partial^2 \varphi_n^\beta}{\partial x_j^2}(x)\right| \leq \frac{K}{n^2}\,\varphi_n^{\beta-2}(x)\,. \tag{8.79}$$

Note that, as all the derivatives of φ vanish out of the compact $[1, 2]$ and φ is bounded by 1, (8.79) can be refined to

$$\left|\frac{\partial^2 \varphi_n^\beta}{\partial x_j^2}(x)\right| \leq \frac{K}{n^2}\,\mathbb{1}_{C_n}(x)\,. \tag{8.80}$$

where $\mathbb{1}_{C_n}$ is the indicator function of the annulus $C_n = \{x \in \mathbb{R}^p \mid n \leq ||x|| \leq 2n\}$.

Proof of Theorem 8.10 Let γ be a twice differentiable function with $\gamma^2 \in L^1_{loc}(\mathbb{R}^p)$ satisfying (8.78). Then, using the defining property of twice differentiable functions, we have, for any $n \geq 1$ and any $\beta > 2$,

$$\begin{aligned}\int_{\mathbb{R}^p} \gamma^2(x)\,\varphi_n^\beta(x)\,dx &\le -k\int_{\mathbb{R}^p} \Delta\gamma(x)\,\varphi_n^\beta(x)\,dx \\ &= -k\int_{\mathbb{R}^p} \gamma(x)\,\Delta\varphi_n^\beta(x)\,dx \\ &\le k\int_{\mathbb{R}^p} |\gamma(x)|\,|\Delta\varphi_n^\beta|\,dx. \end{aligned} \tag{8.81}$$

Then, using (8.79), it follows from (8.81) that there exists a constant $C > 0$ such that

$$\begin{aligned}&\int_{\mathbb{R}^p} \gamma^2(x)\,\varphi_n^\beta(x)\,dx \\ &\le \frac{C}{n^2}\int_{\mathbb{R}^p} |\gamma(x)|\,\varphi_n^{\beta-2}(x)\,dx \\ &\le \frac{C}{n^2}\left(\int_{\mathbb{R}^p} \varphi_n^{\beta-4}(x)\,dx\right)^{1/2}\left(\int_{\mathbb{R}^p} \gamma^2(x)\,\varphi_n^\beta(x)\,dx\right)^{1/2}, \end{aligned} \tag{8.82}$$

applying Schwarz's inequality with $\beta > 4$ and using

$$\gamma(x)\,\varphi_n^{\beta-2}(x) = \varphi_n^{\beta/2-2}(x)\,\gamma(x)\,\varphi_n^{\beta/2}(x)$$

since $\gamma^2 \in L^1_{loc}(\mathbb{R}^p)$.

Clearly, (8.82) is equivalent to

$$\int_{\mathbb{R}^p} \gamma^2(x)\,\varphi_n^\beta(x)\,dx \le \frac{C^2}{n^4}\int_{\mathbb{R}^p} \varphi_n^{\beta-4}(x)\,dx\,. \tag{8.83}$$

Thus, since $\varphi_n = 1$ on B_n and $\varphi_n \ge 0$,

$$\int_{B_n} \gamma^2(x)\,dx = \int_{B_n} \gamma^2(x)\,\varphi_n^\beta(x)\,dx \le \int_{\mathbb{R}^p} \gamma^2(x)\,\varphi_n^\beta(x)\,dx\,. \tag{8.84}$$

Then, since $\operatorname{supp}\varphi_n = B_{2n}$ and $0 \le \varphi_n \le 1$, using (8.83) gives

$$\int_{B_n} \gamma^2(x)\,dx \le \frac{C^2}{n^4}\int_{\mathbb{R}^p} \varphi_n^{\beta-4}(x)\,dx \le \frac{C^2}{n^4}\int_{B_{2n}} dx = A\,n^{p-4} \tag{8.85}$$

for some constant $A > 0$. Letting n go to infinity in (8.85) shows that, when $p < 4$, $\gamma = 0$ almost everywhere, which proves the theorem in that case. It also implies that γ is in $L^2(\mathbb{R}^p)$ when $p = 4$.

Consider now the case $p = 4$. The result will follow by applying (8.80). Indeed, it follows from (8.80) and the first inequality in (8.82) that, for some constant $C > 0$,

$$\int_{B_n} \gamma^2(x)\,dx \le \frac{C}{n^2}\int_{C_n} |\gamma(x)|\,dx$$
$$\le \frac{C}{n^2}\left(\int_{C_n} dx\right)^{1/2}\left(\int_{C_n} \gamma^2(x)\,dx\right)^{1/2} \tag{8.86}$$

by Schwarz's inequality. Now,

$$\int_{C_n} dx \le \int_{B_{2n}} dx \propto n^4 \tag{8.87}$$

since $p = 4$. Hence (8.86) and (8.87) imply that, for some constant $A > 0$,

$$\int_{B_n} \gamma^2(x)\,dx \le A\left(\int_{C_n} \gamma^2(x)\,dx\right)^{1/2} . \tag{8.88}$$

As $\gamma \in L^2(\mathbb{R}^p)$, we have

$$\lim_{n\to\infty} \int_{C_n} \gamma^2(x)\,dx = 0$$

and hence (8.88) gives rise to

$$0 = \lim_{n\to\infty} \int_{B_n} \gamma^2(x)\,dx = \int_{\mathbb{R}^p} \gamma^2(x)\,dx\,,$$

which implies that $\gamma = 0$ almost everywhere and gives the desired result for $p = 4$. □

8.8 Discussion

There are several areas of the theory of loss estimation that we have not discussed. Our primary focus has been on location parameters for the multivariate normal and spherical distributions. Loss estimation for exponential families is addressed in Lele (1992, 1993) and Rukhin (1988). Lele (1992, 1993) developed improved loss estimators for point estimators in the general setup of Hudson's (1978) subclass of continuous exponential families. Hudson's family essentially includes distributions for which the Stein-like identities hold; explicit calculations and loss estimators are given for the gamma distribution, as well as for improved scaled quadratic loss estimators in the Poisson setting for the Clevenson and Zidek (1975) estimator. Rukhin (1988) studied the posterior loss estimator for a Bayes estimate (under quadratic loss) of the canonical parameter of a linear exponential family.

As pointed out in the introduction, in the known variance normal setting, Johnstone (1988) used a version of Blyth's method to show that the constant loss estimate p, for estimating the loss of the estimator X, is admissible if $p \leq 4$. Lele (1993) gave some additional sufficient conditions for admissibility in the general exponential family and worked out the precise details for the Poisson model. Rukhin (1988) considers loss functions for the simultaneous estimation of θ and $L(\theta, \varphi(X))$ and deduced some interesting admissibility results.

Loss estimates have been used to derive nonparametric penalized empirical loss estimates in the context of function estimation, which adapt to the unknown smoothness of the function of interest. See Barron et al. (1999) and Donoho and Johnstone (1995) for more details.

A number of researchers have investigated improved estimators of a covariance matrix, Σ, under the Stein loss, $L_S(\hat{\Sigma}, \Sigma) = tr(\hat{\Sigma}\Sigma^{-1}) - \log|\hat{\Sigma}\Sigma^{-1}| - p$, using an unbiased estimation of risk technique. In the normal case, Dey and Srinivasan (1985), Haff (1979), Stein (1977a,b), and Takemura (1984) proposed improved estimators that dominate the sample covariance under $L_S(\hat{\Sigma}, \Sigma)$. In Kubokawa and Srivastava (1999), it is shown that the domination of these improved estimators over the sample covariance matrix are robust with respect to the family of elliptical distributions. To date, there has not been any work on improving the unbiased estimate of $L_S(\hat{\Sigma}, \Sigma)$.

Appendix

A.1 Weakly Differentiable Functions

For $\Omega \subset \mathbb{R}^p$ an open set and for $q \in \mathbb{R}$ such that $1 \leq q \leq \infty$, the space of functions f from Ω into $\mathbb{R}$ such that f^q is locally integrable is defined by

$$L^q_{loc}(\Omega) = \left\{ f : \Omega \to \mathbb{R} \mid \int_K |f(x)|^q \, dx < \infty \quad \forall K \subset \Omega \text{ with } K \text{ compact} \right\} .$$

A function $f \in L^1_{loc}(\Omega)$ is said to be weakly differentiable if there exist p functions $g_1, \ldots g_p$ in $L^1_{loc}(\Omega)$ such that, for any $i = 1, \ldots, p$,

$$\int_\Omega f(x) \, \frac{\partial \varphi}{\partial x_i}(x) \, dx = - \int_\Omega g_i(x) \, \varphi(x) \, dx \tag{A.1}$$

for any $\varphi \in \mathscr{C}^\infty_c(\Omega)$, where $\mathscr{C}^\infty_c(\Omega)$ is the space of infinitely differentiable functions from Ω into $\mathbb{R}$ with compact support (test functions).

The space of the functions f in $L^1_{loc}(\Omega)$ satisfying (A.1) is the Sobolev space $W^{1,1}_{loc}(\Omega)$. The functions g_i are the i-th weak partial derivatives of f and are denoted, as are the usual derivatives, by $g_i = \partial_i f$ or $g_i = \partial f / \partial x_i$. They are unique in the sense that any function $\tilde{g}_i$ which satisfies (A.1) is equal almost everywhere to g_i.[1] The vector $\nabla f = (\partial_i f, \ldots, \partial_p f) = (\partial f / \partial x_1, \ldots, \partial f / \partial x_p)$ is referred to as the weak gradient of f.

Note that, in (A.1), it is just required that the function f is locally integrable but not necessarily in $L^1(\Omega)$, as will be the case for many functions of interest

[1]This can be derived from the fact that, if $h \in L^1_{loc}(\Omega)$ is such that $\int_\Omega h(x) \, \varphi(x) \, dx = 0$ for any $\varphi \in \mathscr{C}^\infty_c(\Omega)$, then $h = 0$ a.e on Ω (see e.g. Chapter II of Schwartz 1961 for a detailed proof).

D. Fourdrinier et al., *Shrinkage Estimation*, Springer Series in Statistics,
https://doi.org/10.1007/978-3-030-02185-6

(see examples below). The subspace of $W^{1,1}_{loc}(\Omega)$ of the functions f in $L^1(\Omega)$ satisfying (A.1) is the Sobolev space $W^{1,1}(\Omega)$.

If f is continuously differentiable, then f is weakly differentiable, the usual derivative and the weak derivative of f coinciding. Thus (A.1) appears to be the usual integration by part formula, where the usual term in brackets vanishes since the function φ has compact support. In that sense, this notion extends the usual notion of differentiability, which remains a basis to determine the expressions of the weak derivatives as illustrated by examples in Sect. A.2.

A function $f = (f_1, \ldots, f_p)$ from Ω into $\mathbb{R}^p$ is said to be weakly differentiable if, for any $i = 1, \ldots, p$, the coordinate function f_i is weakly differentiable. Then $\operatorname{div} f = \sum_{i=1}^p \partial_i f_i = \sum_{i=1}^p \partial f_i / \partial x_i$ is referred to as the weak divergence.

A function $f \in L^1_{loc}(\Omega)$ is said to be twice weakly differentiable if there exist functions $\partial f_1 = \partial f/\partial x_1, \ldots, \partial f_p = \partial f/\partial x_p$ in $L^1_{loc}(\Omega)$ such that, for any $i = 1, \ldots, p$,

$$\int_\Omega f(x) \, \frac{\partial \varphi(x)}{\partial x_i} \, dx = - \int_\Omega \frac{\partial f(x)}{\partial x_i} \, \varphi(x) \, dx \, , \tag{A.2}$$

and also if there exist functions $\partial^2_{ij} f = \partial^2 f/\partial x_i \partial x_j$, for $1 \le i, j \le p$, in $L^1_{loc}(\Omega)$ such that

$$\int_\Omega f(x) \, \frac{\partial^2 \varphi(x)}{\partial x_j \partial x_i} \, dx = \int_\Omega \frac{\partial^2 f(x)}{\partial x_j \partial x_i} \, \varphi(x) \, dx \, , \tag{A.3}$$

for any $\varphi \in \mathscr{C}^\infty_c(\Omega)$. The space of the functions f in $L^1_{loc}(\Omega)$ having second weak partial derivatives in $L^1_{loc}(\Omega)$ is the Sobolev space $W^{2,1}_{loc}(\Omega)$. Note that the weak derivatives commute in the sense that, for $1 \le i, j \le p$, we have $\partial^2 f/\partial x_i \partial x_j = \partial^2 f/\partial x_j \partial x_i$. Also, if f has continuous second derivatives, then f is twice weakly differentiable and the usual derivatives and the weak derivatives of f coincide. Finally, $\Delta f = \sum_{i=1}^p \partial^2 f/\partial x_i^2$ is referred to as the weak Laplacian of f and satisfies

$$\int_\Omega f(x) \, \Delta\varphi(x) \, dx = \int_\Omega \Delta f(x) \, \varphi(x) \, dx \, , \tag{A.4}$$

for any $\varphi \in \mathscr{C}^\infty_c(\Omega)$.

There is a natural extension to higher order derivatives. Following the lines of the definition of the second weak derivative, to a p-dimensional multi-index, that is, a p-tuple $\alpha = (\alpha_1, \ldots, \alpha_p)$ of nonnegative integers with length $|\alpha| = \sum_{i=1}^p \alpha_i$, is associated the derivative $\partial^\alpha = \partial_1^{\alpha_1} \ldots \partial_p^{\alpha_p} = \partial^{|\alpha|}/\partial x_1^{\alpha_1} \ldots \partial x_p^{\alpha_p}$ of order $|\alpha|$. Extending (A.3), a function $f \in L^1_{loc}(\Omega)$ has α^{th} weak derivative $\partial^\alpha \in L^1_{loc}(\Omega)$ if

$$\int_\Omega f(x) \, \partial^\alpha \varphi(x) \, dx = (-1)^{|\alpha|} \int_\Omega \varphi(x) \, \partial^\alpha f(x) \, dx \, , \tag{A.5}$$

assuming jointly that the weak derivatives of order less than or equal to $|\alpha|$ exist. For k a nonnegative integer, this leads to the Sobolev space

$$W^{k,1}_{loc}(\Omega) = L^1_{loc}(\Omega) \cap \{f/\partial^\alpha f \in L^1_{loc}(\Omega),\, |\alpha| \leq k\}\,.$$

It is interesting to interpret a weak derivative as a distributional derivative in the sense of Schwartz (1961). For this, we need first the following notion of convergence: a sequence $(\varphi_{n\in\mathbb{N}})$ in $\mathscr{C}^\infty_c(\Omega)$ is said to converge to $\varphi \in \mathscr{C}^\infty_c(\Omega)$ if

(i) there exists an open set $\Omega' \in \mathbb{R}^p$ such that $\overline{\Omega'} \subset \Omega$ with $\text{supp}\varphi_n \subset \Omega'$ for every $n \in \mathbb{N}$;
(ii) $\partial^\alpha \varphi_n \to \partial^\alpha \varphi$ as $n \to \infty$ uniformly on Ω for every $\alpha = 0, \ldots, n$.

The convergence in (ii) is called the convergence in the sense of test functions and is denoted by $\varphi_n \to \varphi$ in $\mathscr{D}(\Omega)$. There exists a topology corresponding to that convergence (see Grubb 2009; Hunter 2014). Endowed with that topology, the set $\mathscr{C}^\infty_c(\Omega)$ is denoted by $\mathscr{D}(\Omega)$. A distribution on Ω is a continuous linear functional T from $\mathscr{D}(\Omega)$ into $\mathbb{R}$. For any φ, the value of T acting on φ is denoted by $\langle T, \varphi\rangle$. Linearity of T naturally means

$$\forall(\varphi, \psi) \in \mathscr{D}(\Omega)^2 \quad \forall(\alpha, \beta) \in \mathbb{R}^2 \qquad \langle T, \alpha\varphi + \beta\psi\rangle = \alpha\langle T, \varphi\rangle + \beta\langle T, \psi\rangle$$

and continuity of T is viewed as

$$\forall\varphi \in \mathscr{D}(\Omega) \quad \forall(\varphi_n)_{n\in\mathbb{N}} \in \mathscr{D}(\Omega)^{\mathbb{N}} \quad \langle T, \varphi_n\rangle \to \langle T, \varphi\rangle \quad \text{as} \quad \varphi_n \to \varphi \text{ in } \mathscr{D}(\Omega)\,.$$

The set of distributions on Ω is denoted by $\mathscr{D}'(\Omega)$.

A central example of distributions is the class of regular distributions defined as follows. Let $f \in L^1_{loc}(\Omega)$. It defines a distribution T_f through

$$\forall\varphi \in \mathscr{D}(\Omega) \quad \langle T_f, \varphi\rangle = \int_\Omega f(x)\,\varphi(x)\,dx\,. \tag{A.6}$$

Note that this integral is well defined since integration is made on the compact support of φ. Clearly, this functional is linear. Also, it is continuous since, if $\varphi_n \to \varphi$ in $\mathscr{D}(\Omega)$, for any open set Ω' such that $\overline{\Omega'} \subset \Omega$,

$$|\langle T_f, \varphi_n\rangle - \langle T_f, \varphi\rangle| \leq \int_{\Omega'} |f(x)|\,dx \sup_{x\in\Omega'} |\varphi_n(x) - \varphi(x)| \to 0\,,$$

the convergence to 0 following from the convergence in $\mathscr{D}(\Omega)$ of the test functions. Note also that $f \in L^1_{loc}(\Omega)$ and $g \in L^1_{loc}(\Omega)$ define the same distribution $T_f = T_g$ if and only if $f = g$ almost everywhere (see Schwartz 1961 for a detailed proof).

This fact leads to identify the locally integrable function f and its associated regular distribution T_f, so that we may write

$$\forall \varphi \in \mathscr{D}(\Omega) \quad \langle f, \varphi \rangle = \langle T_f, \varphi \rangle = \int_{\Omega} f(x)\, \varphi(x)\, dx\,. \tag{A.7}$$

Considering the regular distribution associated to a smooth function allows to define, in a natural way, the derivative of a distribution. Let f be a continuously differentiable function on $\mathbb{R}^p$ and let $1 \leq i \leq p$ be fixed. The ith partial derivative $\partial_i f$ is locally integrable and we have, for any $\varphi \in \mathscr{D}(\mathbb{R}^p)$,

$$\langle \partial_i f, \varphi \rangle = \int_{\mathbb{R}^p} \partial_i f(x)\, \varphi(x)\, dx = \int_{\mathbb{R}^{p-1}} \int_{-\infty}^{\infty} \partial_i f(x)\, \varphi(x)\, dx_i\, dx_{-i}\,, \tag{A.8}$$

by Fubini's theorem where $dx_{-i} = dx_1, \ldots, dx_{i-1}, dx_{i+1}, \ldots, dx_p$. Integrating by parts the most inner integral in (A.8) gives

$$\int_{-\infty}^{\infty} \partial_i f(x)\, \varphi(x)\, dx_i = -\int_{-\infty}^{\infty} f(x)\, \partial_i \varphi(x)\, dx_i$$

since φ is zero outside a compact set. Hence the right hand-side of the second equality in (A.8) is

$$-\int_{\mathbb{R}^{p-1}} \int_{-\infty}^{\infty} f(x)\, \partial_i \varphi(x)\, dx_i\, dx_{-i} = -\int_{\mathbb{R}^p} f(x)\, \partial_i\, \varphi(x)\, dx = -\langle f, \partial_i \varphi \rangle\,. \tag{A.9}$$

Finally (A.8) and (A.9) give

$$\langle \partial_i f, \varphi \rangle = -\langle f, \partial_i \varphi \rangle\,. \tag{A.10}$$

Then we are led to define the ith derivative $\partial_i T$ of any distribution T, for any $\varphi \in \mathscr{D}(\mathbb{R}^p)$, by

$$\langle \partial_i T, \varphi \rangle = -\langle T, \partial_i \varphi \rangle\,. \tag{A.11}$$

Equality (A.11) does define a distribution. Indeed, first, it is clearly a linear functional of φ. Secondly, by definition of the convergence in $\mathscr{D}(\mathbb{R}^p)$, if $\varphi_n \to \varphi$ in $\mathscr{D}(\Omega)$, then $\partial_i \varphi_n \to \partial_i \varphi$ in $\mathscr{D}(\Omega)$. Now, as T is a distribution, $\langle T, \partial_i \varphi_n \rangle$ converges to $\langle T, \partial_i \varphi \rangle$. Therefore, according to (A.11), $\langle \partial_i T, \varphi_n \rangle$ converges to $\langle \partial_i T, \varphi \rangle$, and hence, $\partial_i T$ is a distribution.

Returning to the regular distribution defined in (A.6), according to (A.11), its derivative satisfies

$$\langle \partial_i T, \varphi \rangle = -\langle T, \partial_i \varphi \rangle\,. \tag{A.12}$$

Now, if $\partial_i T$ is also a regular distribution associated to a certain function $g_i \in L^1_{loc}(\Omega)$, Equality (A.12) can be written as

$$\int_{\Omega} g_i(x)\, \varphi(x)\, dx = - \int_{\Omega} f(x)\, \partial_i \varphi(x)\, dx\,,$$

which is exactly Equality (A.1) defining the weak differentiability of f. Therefore as noticed by Hunter (2014), a locally integrable function is weakly differentiable if its distributional derivative is regular and its weak derivative is the locally integrable function corresponding to the distributional derivative.

Twice distributional derivatives can be defined following the above plan. For $i, j = 1, \ldots, p$, and for any $\varphi \in \mathscr{D}(\mathbb{R}^p)$,

$$\langle \partial_{ij} T, \varphi \rangle = -\langle \partial_j T, \partial_i \varphi \rangle = +\langle T, \partial_{ij} \varphi \rangle$$

and

$$\langle \partial_{ji} T, \varphi \rangle = -\langle \partial_i T, \partial_j \varphi \rangle = +\langle T, \partial_{ji} \varphi \rangle\,.$$

As φ is twice continuously differentiable, we have $\partial_{ij}\varphi = \partial_{ji}\varphi$. It follows that $\partial_{ij} T = \partial_{ji} T$. Setting $i = j$ and summing on j gives rise to the distributional Laplacian so that

$$\langle \Delta T, \varphi \rangle = \langle T, \Delta \varphi \rangle\,. \tag{A.13}$$

The above link between locally integrable functions and regular distributions can clearly be extended to twice differentiability. A locally integrable function is twice weakly differentiable if its first and second distributional derivatives are regular and its weak first and second derivatives are the locally integrable functions corresponding to the respective distributional derivatives. In particular, the weak Laplacian corresponds to the distributional Laplacian. Thus, identifying a twice weakly differentiable function f with its associated regular distribution T_f and identifying Δf with ΔT_f, Equality (A.13) is

$$\langle \Delta T_f, \varphi \rangle = \langle T_f, \Delta \varphi \rangle\,, \tag{A.14}$$

and can be viewed as

$$\langle \Delta f, \varphi \rangle = \langle f, \Delta \varphi \rangle\,, \tag{A.15}$$

which corresponds to (A.4).

Finally, in the same way, extension to higher order derivatives is done as follows. Let $T \in \mathscr{D}'(\Omega)$. For any p-dimensional multi-index $\alpha = (\alpha_1, \ldots, \alpha_p)$ with length

$|\alpha| = \sum_{i=1}^{p} \alpha_i$, the derivative of order $|\alpha|$ is $\partial^\alpha T \in \mathscr{D}'(\Omega)$ defined, for any $\varphi \in \mathscr{D}(\mathbb{R}^p)$, by

$$\langle \partial^\alpha T, \varphi \rangle = (-1)^{|\alpha|} \langle T, \partial^\alpha \varphi \rangle .$$

A.2 Examples of Weakly Differentiable Functions

For each function h in the following examples, weak differentiability is determined by the local integrability of h and its classical derivatives and its absolute continuity along almost all lines parallel to the axes (see Proposition 2.1).

Example A.1 (Weak differentiability of James-Stein type shrinkage factors) For $q \in \mathbb{R}$, define, for $x \in \mathbb{R}^p \backslash \{0\}$,

$$h(x) = \frac{x}{\|x\|^q} . \tag{A.16}$$

As each coordinate function h_j of h is continuously differentiable for all $x \neq 0$, the absolute continuity holds for every line not containing 0. Hence it suffices to check local integrability of h and its derivatives.

The function h is weakly differentiable if and only if $q < p$. This condition reflects the local integrability of the partial weak derivative of any of the components $x_j / \|x\|^q$:

$$\frac{\partial}{\partial x_i}\left(\frac{x_j}{\|x\|^q}\right) = \begin{cases} \dfrac{1}{\|x\|^q} + x_i \dfrac{\partial}{\partial x_i}\left(\dfrac{1}{\|x\|^q}\right) & \text{if } i = j \\ x_j \dfrac{\partial}{\partial x_i}\left(\dfrac{1}{\|x\|^q}\right) & \text{if } i \neq j \end{cases}$$
$$= \begin{cases} \dfrac{1}{\|x\|^q} - \dfrac{q\, x_i^2}{\|x\|^{q+2}} & \text{if } i = j \\ -\dfrac{q\, x_j\, x_i}{\|x\|^{q+2}} & \text{if } i \neq j \end{cases} . \tag{A.17}$$

Using Eq. (1.11), the local integrability of $\partial / \partial x_i (x_j / \|x\|^q)$ reduces to

$$\int_{B_R} \frac{1}{\|x\|^q}\, dx < \infty \Leftrightarrow \int_0^R r^{p-1-q}\, dr < \infty \Leftrightarrow p - 1 - q > -1 ,$$

for any ball B_R of radius R centered at 0, which is the announced condition. Note that, through similar arguments, the local integrability of the function h itself is $q < p + 1$ and hence is implied by the local integrability of its derivatives.

As a special case, when $q = 2$, the function h corresponds (up to a multiplicative constant) to the shrinkage factor of the James-Stein estimator which is, according to the above, weakly differentiable for $p \geq 3$. Also, for any $x \neq 0$, its weak divergence equals

$$\begin{aligned} \operatorname{div}\left(\frac{x}{\|x\|^2}\right) &= \sum_{i=1}^{p} \frac{\partial}{\partial x_i}\left(\frac{x_i}{\|x\|^2}\right) \\ &= \sum_{i=1}^{p}\left(\frac{\|x\|^2 - 2\,x_i^2}{\|x\|^4}\right) \\ &= \frac{p-2}{\|x\|^2}. \end{aligned} \tag{A.18}$$

As another example, it can be seen, following the above development, that the function $x \mapsto 1/||x||^q$ is weakly differentiable if and only if $q + 1 < p$.

Example A.2 (Weak differentiability of spherically symmetric estimators) Define, for $x \in R^p$,

$$h(x) = g(\|x\|^2)\, x$$

where g is a function from $\mathbb{R}_+$ into $\mathbb{R}$ such that $g(t)$ is absolutely continuous for $t > 0$. For any $j = 1, \ldots, p$, each coordinate function h_j is absolutely continuous on all lines not containing the origin. Then it suffices to check local integrability of this function and its partial derivatives. We have

$$\frac{\partial}{\partial x_i} h_j(x) = \begin{cases} g(\|x\|^2) + 2\,g'(\|x\|^2)\,x_i^2 & \text{if } i = j \\ 2\,g'(\|x\|^2)\,x_i\,x_j & \text{if } i \neq j \end{cases}. \tag{A.19}$$

Using Eq. (1.11), a sufficient condition for local integrability of $\partial/\partial x_i h_j(x)$ is

$$\int_0^R g(r^2)\, r^{p-1}\, dr < \infty \quad \text{and} \quad \int_0^R g'(r^2)\, r^{p+1}\, dr < \infty\,.$$

Similarly, a sufficent condition for local integrability of $h_j(x)$ is

$$\int_0^R g(r^2)\, r^p\, dr < \infty\,,$$

which is guaranteed by the first of the above two conditions. For example, if $g(\|x\|^2) = r(\|x\|^2)/\|x\|^q$ where r and r' are bounded, weak differentiability of h holds if $q < p$, which is the same condition as that for (A.16).

Example A.3 (Twice weakly differentiable functions) It can be seen, following Example A.1, that the function $x \mapsto 1/||x||^q$ is in $W^{2,1}_{loc}(\mathbb{R}^p)$ for $p > q + 2$, which is the integrability condition of the second derivatives. Thus, for $q = 2$, the function $x \mapsto 1/||x||^2$ is in $W^{2,1}_{loc}(\mathbb{R}^p)$ for $p \geq 5$; under that condition, its weak Laplacian equals $\Delta(1/||x||^2) = -2(p-4)/||x||^4$, which shows that this function is superharmonic. Also, taking $q = p - 2$ gives rise to the fundamental harmonic function $x \mapsto 1/||x||^{p-2}$ for $p \geq 3$. Note that, although it is an infinitely differentiable function in $\mathbb{R}^p \backslash \{0\}$ (and, in fact, it is an analytic function), it is not a twice weakly differentiable function on the entire space $\mathbb{R}^p$ (it does not belong to $W^{2,1}_{loc}(\mathbb{R}^p)$) since the above integrability condition is violated (with $q + 2 = p$).

Remark A.1 (Non-almost differentiability of the James-Stein shrinkage factor)
In Sect. 2.3, we mentioned that $h : x \mapsto x/\|x\|^2$ is weakly differentiable for $p \geq 3$ but is not almost differentiable in the sense of Stein for any p. Indeed we will show that, for any $x \neq 0$ in $\mathbb{R}^p$ and $z = -a\,x$ with $a > 1$, each coordinate $h_j(x) = x_j/\|x\|^2$ does not satisfy (2.5). First, we have

$$h_j(x - a\,x) - h_j(x) = \frac{a}{1-a}\,\frac{x_j}{\|x\|^2}\,. \tag{A.20}$$

Secondly, it can be checked that

$$\int_0^1 (-a\,x)^{\mathrm{T}}\, \nabla h_j\big(x + t\,(-a\,x)\big)\, dt = -a\,x^{\mathrm{T}}\, \nabla h_j(x) \int_0^1 \frac{1}{(1-a\,t)^2}\, dt \tag{A.21}$$

using (A.17) with $q = 2$. However, as $a > 1$, the last integral in (A.21) does not exist because of the singularity at $t = 1/a$. Hence (A.20) is not equal to (A.21) for $a > 1$. However, note that (A.20) is equal to (A.21) whenever $a < 1$ and, therefore, the only possible candidate for $\nabla h_j(x)$ is given in (A.17). Thus h is weakly differentiable but not almost differentiable.

A.3 Vanishing of the Bracketed Term in Stein's Identity

Proposition A.1 *For fixed $\theta \in \mathbb{R}$ and $\sigma^2 > 0$, let X be a random variable with normal distribution $N(\theta, \sigma^2)$. Denoting by E_θ the expectation with respect to that distribution, if g is an absolutely continuous function such that $E_\theta[|g'(X)|] < \infty$ then* $\lim_{x\to\pm\infty} g(x)\exp\{-(x-\theta)^2/2\sigma^2\} = 0$.

Proof Through the change of variable $t = (x - \theta)/\sigma$ one can see that it suffices to prove the result for $\theta = 0$ and $\sigma^2 = 1$. Denoting by ϕ the standard normal density, that is, $\phi(x) = (1/\sqrt{2\pi})\exp(-x^2/2)$ we will use the fact that its derivative satisfies $\phi'(x) = -x\,\phi(x)$ and hence $\phi(x) = \int_x^\infty y\,\phi(y)\,dy$.

As, by assumption, the expectation of g' exists, we have

$$\begin{aligned}\int_0^{\infty} g'(x)\,\phi(x)\,dx &= \int_0^{\infty} g'(x)\,dx \int_x^{\infty} y\,\phi(y)\,dy\,dx \\ &= \int_0^{\infty}\int_0^{y} g'(x)\,dx\; y\,\phi(y)\,dy \\ &= \int_0^{\infty} [g(y)-g(0)]\,y\,\phi(y)\,dy\,, \qquad \text{(A.22)}\end{aligned}$$

by the Fubini theorem for the second equality and expressing the absolute continuity of g in the third equality. Now, integrating by parts, we have

$$\begin{aligned}\int_0^{\infty} g'(x)\,\phi(x)\,dx &= [g(x)\,\phi(x)]_0^{\infty} - \int_0^{\infty} g(x)\,\phi'(x)\,dx \\ &= \lim_{x\to\infty} g(x)\,\phi(x) - g(0)\,\phi(0) + \int_0^{\infty} g(x)\,x\,\phi(x)\,dx. \qquad \text{(A.23)}\end{aligned}$$

Equating (A.22) and (A.23) gives

$$-g(0)\int_0^{\infty} y\,\phi(y)\,dy = \lim_{x\to\infty} g(x)\,\phi(x) - g(0)\,\phi(0)$$

and hence

$$\lim_{x\to\infty} g(x)\,\phi(x) = g(0)\left\{\int_0^{\infty} -y\,\phi(y)\,dy + \phi(0)\right\} = g(0)\left\{\int_0^{\infty} \phi'(y)\,dy + \phi(0)\right\} = 0$$

since

$$\int_0^{\infty} \phi'(y)\,dy = [\phi(y)]_0^{\infty} = -\phi(0)\,.$$

Finally, the fact that we also have $\lim_{x\to-\infty} g(x)\,\phi(x) = 0$ can be obtained in a similar way using the rewriting $\phi(x) = \int_{-\infty}^{x} -y\,\phi(y)\,dy$. □

A.4 Examples of Settings Where Stein's Identity Does Not Hold

Example A.4 (James-Stein shrinkage factor when $p = 1, 2$) In Example A.1, we showed that the James-Stein shrinkage factor $h : x \mapsto x/\|x\|^2$ is weakly differentiable if and only if $p \geq 3$ and that, in that case, its weak divergence equals $\operatorname{div} h(x) = (p-2)/\|x\|^2$. In all dimensions, this is also the classical divergence when $x \neq 0$ but Stein's identity fails to hold when $p \leq 2$. Indeed, when $p = 2$, the classical divergence is identically equal to 0 almost everywhere so that its expected value is identically equal to 0. On the other hand, when $\theta = 0$, the function $x \mapsto (x - \theta)^{\mathrm{T}} h(x)$ is identically equal to 1 so that its expected value equals 1, and hence Stein's identity fails to hold for the classical divergence when $\theta = 0$. Note that, when $\theta \neq 0$, both expected values fail to exist. If $p = 1$, the classical divergence is $\operatorname{div} h(x) = -1/x^2$ and $(x - \theta)^{\mathrm{T}} h(x) = 1 - \theta/x$. Both expected values fail to exist when $\theta \neq 0$ while only the second expected value exists when $\theta = 0$. Therefore Stein's identity fails to hold when $p = 1$.

Note that, when $p = 1$, the function h is not absolutely continuous while, when $p = 2$, its coordinate functions are absolutely continuous on every line parallel to the axes except for the axes themselves.

Example A.5 (The sign function) The sign function defined, for $x = (x_1, \ldots, x_p) \in \mathbb{R}^p$ such that $x_i \neq 0$ for all $i = 1, \ldots, p$, by $\operatorname{sgn}(x) = (\operatorname{sgn}(x_1), \ldots, \operatorname{sgn}(x_p)) = (x_1/|x_1|, \ldots, x_n/|x_p|)$ is not weakly differentiable (note that it is not necessary to define sgn everywhere). Indeed, noticing that it suffices to consider the case where $p = 1$, for $\varphi \in \mathscr{C}_c^\infty(\mathbb{R})$, we have

$$\int_{\mathbb{R}} \operatorname{sgn}(x)\, \varphi'(x)\, dx = \int_{-\infty}^{0} -\varphi'(x)\, dx + \int_{0}^{\infty} \varphi'(x)\, dx = -2\, \varphi(0)\,,$$

since φ has compact support in $\mathbb{R}$. On the other hand, since $\operatorname{sgn}(x)$ is constant and equal to -1 for $x < 0$ and equal to $+1$ for $x > 0$, the natural candidate for a weak derivative is the function identically equal to 0, for which we have

$$-\int_{\mathbb{R}} \operatorname{sgn}'(x)\, \varphi(x)\, dx = 0\,.$$

Hence these two last integrals cannot be equal for any choice of φ such that $\varphi(0) \neq 0$.

This non-weak differentiability is reflected in the fact that, when $X \sim \mathcal{N}(\theta, I_p)$, no unbiased estimator of $E_\theta[(X - \theta)^{\mathrm{T}} \operatorname{sgn}(X)]$ exists. First, it is easy to see that it suffices to consider $p = 1$ and it is straightforward to calculate the corresponding expectation $A_\theta = E_\theta[(X - \theta)\operatorname{sgn}(X)]$ since we have

$$\begin{aligned}
A_\theta &= \frac{1}{\sqrt{2\pi}} \int_{-\infty}^{\infty} \operatorname{sgn}(x)\, \frac{d}{dx}\left[-\exp\left(-\frac{(x-\theta)^2}{2}\right)\right] dx \\
&= \frac{1}{\sqrt{2\pi}} \left\{-\left[-\exp\left(-\frac{(x-\theta)^2}{2}\right)\right]_{-\infty}^{0} + \left[-\exp\left(-\frac{(x-\theta)^2}{2}\right)\right]_{0}^{\infty}\right\} \\
&= \frac{1}{\sqrt{2\pi}} \left\{\exp\left(-\frac{\theta^2}{2}\right) + \exp\left(-\frac{\theta^2}{2}\right)\right\} \\
&= \sqrt{\frac{2}{\pi}} \left\{\exp\left(-\frac{\theta^2}{2}\right)\right\}.
\end{aligned}$$

Now assume that there exists a function ψ such that

$$E_\theta[\psi(X)] = \sqrt{\frac{2}{\pi}} \left\{ \exp\left(-\frac{\theta^2}{2}\right) \right\}. \tag{A.24}$$

Then, deriving with respect to θ, it follows that

$$\frac{d}{d\theta} E_\theta[\psi(X)] = -\sqrt{\frac{2}{\pi}}\, \theta \exp\left(-\frac{\theta^2}{2}\right) = -\theta\, E_\theta[\psi(X)]$$

and hence, since we deal with an exponential family, deriving under the integral sign gives

$$\int_{-\infty}^{\infty} \psi(x) \frac{\partial}{\partial x} \left\{ \frac{1}{\sqrt{2\pi}} \exp\left(-\frac{(x-\theta)^2}{2}\right) \right\} dx = -\int_{-\infty}^{\infty} \psi(x)(x-\theta) \frac{1}{\sqrt{2\pi}} \exp\left(-\frac{(x-\theta)^2}{2}\right) dx,$$

that is, according to (A.24),

$$-\theta\, E_\theta[\psi(X)] = -E_\theta[X\, \psi(X)] - \theta\, E_\theta[\psi(X)],$$

which gives rise to

$$E_\theta[X\, \psi(X)] = 0. \tag{A.25}$$

Therefore, as (A.25) is satisfied for all θ, by completeness of the normal family, we have $x\, \psi(x) = 0$ almost everywhere and, consequently, $\psi(x) = 0$ almost everywhere. This contradicts (A.24) proving that ψ cannot be an unbiased estimator of $E_\theta[(X-\theta)\,\mathrm{sgn}(X)]$.

A.5 Stein's Lemma and Stokes' Theorem for Smooth Boundaries

In this section, we prove an extension of Stein's lemma (Theorem 2.1) for densities proportional to $\exp(-\varphi(x))$ where φ is a continuously differentiable function. Additionally, we give an extension of Theorem 2.7 when the sets of integration B_r with boundary S_r are replaced by $[\varphi \le r] = \{x \in \mathbb{R}^p : \varphi(x) \le r\}$ with boundary $[\varphi = r] = \{x \in \mathbb{R}^p : \varphi(x) = r\}$. We follow the development in Fourdrinier and Strawderman (2016). Here is an extension of Stein's lemma.

Lemma A.1 *Let φ be a continuously differentiable function from $\mathbb{R}^p$ into $\mathbb{R}_+$ such that $\phi : x \mapsto K \exp\big(-\varphi(x)\big)$ is a density, where K is a normalizing constant, and such that, for any $i = 1, \ldots, p$, $\lim_{|x_i| \to \infty} \varphi(x_1, \ldots, x_p) = \infty$. If*

$g = (g_1, \ldots, g_p)$ is a weakly differentiable function from $\mathbb{R}^p$ into $\mathbb{R}^p$ then, denoting by E the expectation with respect to ϕ, we have

$$E[\nabla\varphi(X)^{\mathrm{T}} g(X)] = E[\mathrm{div} g(X)] , \tag{A.26}$$

provided that either expectation exists.

Proof Let $x = (x_1, \ldots, x_p) \in \mathbb{R}^p$. For fixed $i = 1, \ldots, p$, set $x_{-i} = (x_1, \ldots, x_{i-1}, x_{i+1}, \ldots, x_p)$ and, with a slight abuse of notation, set $x = (x_i, x_{-i})$. Note that

$$\frac{\partial \phi(x)}{\partial x_i} = -\frac{\partial \varphi(x)}{\partial x_i}\, \phi(x)$$

so that $\phi(x)$ can be written as

$$\phi(x) = \int_{-\infty}^{x_i} -\frac{\partial \varphi(\tilde{x}_i, x_{-i})}{\partial \tilde{x}_i}\, \phi(\tilde{x}_i, x_{-i})\, d\tilde{x}_i = \int_{x_i}^{\infty} \frac{\partial \varphi(\tilde{x}_i, x_{-i})}{\partial \tilde{x}_i}\, \phi(\tilde{x}_i, x_{-i})\, dx_i ,$$

noticing that, by assumption, $\lim_{|x_i|\to\infty} \varphi(x_1, \ldots, x_p) = \infty$ implies

$$\lim_{|x_i|\to\infty} \phi(x_i, x_{-i}) = \lim_{|x_i|\to\infty} \exp\big(-\varphi(x_i, x_{-i})\big) = 0 . \tag{A.27}$$

Thanks to the existence of the expectations in (A.26), we can write, for almost every x_{-i},

$$\begin{aligned}
&\int_{-\infty}^{\infty} \frac{\partial g_i(x_i, x_{-i})}{\partial x_i}\, \phi(x_i, x_{-i})\, dx_i \\
&= \int_{-\infty}^{0} \frac{\partial g_i(x_i, x_{-i})}{\partial x_i} \int_{-\infty}^{x_i} -\frac{\partial \varphi(\tilde{x}_i, x_{-i})}{\partial \tilde{x}_i}\, \phi(\tilde{x}_i, x_{-i})\, d\tilde{x}_i\, dx_i \\
&\quad + \int_{0}^{\infty} \frac{\partial g_i(x_i, x_{-i})}{\partial x_i} \int_{x_i}^{\infty} \frac{\partial \varphi(\tilde{x}_i, x_{-i})}{\partial \tilde{x}_i}\, \phi(\tilde{x}_i, x_{-i})\, d\tilde{x}_i\, dx_i \\
&= \int_{-\infty}^{0} -\frac{\partial \varphi(\tilde{x}_i, x_{-i})}{\partial \tilde{x}_i}\, \phi(\tilde{x}_i, x_{-i}) \int_{\tilde{x}_i}^{0} \frac{\partial g_i(x_i, x_{-i})}{\partial x_i}\, dx_i\, d\tilde{x}_i \\
&\quad + \int_{0}^{\infty} \frac{\partial \varphi(\tilde{x}_i, x_{-i})}{\partial \tilde{x}_i}\, \phi(\tilde{x}_i, x_{-i}) \int_{0}^{\tilde{x}_i} \frac{\partial g_i(x_i, x_{-i})}{\partial x_i}\, dx_i\, d\tilde{x}_i \\
&= \int_{-\infty}^{\infty} \frac{\partial \varphi(\tilde{x}_i, x_{-i})}{\partial \tilde{x}_i}\, \phi(\tilde{x}_i, x_{-i})\, [g_i(\tilde{x}_i, x_{-i}) - g_i(0, x_{-i})]\, d\tilde{x}_i \\
&= \int_{-\infty}^{\infty} \frac{\partial \varphi(\tilde{x}_i, x_{-i})}{\partial \tilde{x}_i}\, \phi(\tilde{x}_i, x_{-i})\, g_i(\tilde{x}_i, x_{-i})\, d\tilde{x}_i ,
\end{aligned}$$

since, using again (A.27),

$$-\int_{-\infty}^{\infty} \frac{\partial \varphi(\tilde{x}_i, x_{-i})}{\partial \tilde{x}_i} \phi(\tilde{x}_i, x_{-i})\, d\tilde{x}_i = \int_{-\infty}^{\infty} \frac{\partial \phi(\tilde{x}_i, x_{-i})}{\partial \tilde{x}_i}\, d\tilde{x}_i = 0\,.$$

Then integrating with respect to x_{-i} gives

$$\begin{aligned} E\left[\frac{\partial g_i(X)}{\partial x_i}\right] &= \int_{\mathbb{R}^p} \frac{\partial g_i(x_i, x_{-i})}{\partial x_i} \phi(x_i, x_{-i})\, dx_i\, dx_{-i} \\ &= \int_{\mathbb{R}^p} \frac{\partial \varphi(x_i, x_{-i})}{\partial x_i} \phi(x_i, x_{-i})\, g_i(x_i, x_{-i})\, dx_i\, dx_{-i} \\ &= E\left[\frac{\partial \varphi(X)}{\partial x_i}\, g_i(X)\right], \end{aligned}$$

and hence, summing on i gives the desired result. □

Corollary A.1 *Let φ and g be as in Lemma* A.1. *For $\tau > 0$, let $\phi_\tau : x \mapsto K_\tau \exp(-\varphi(x)/\tau)$ be a density, where K_τ is a normalizing constant, and let E_τ be the expectation with respect to ϕ_τ. Then*

$$E_\tau[\nabla\varphi(X)^{\mathrm{T}} g(X)] = \tau\, E_\tau[\mathrm{div} g(X)]\,, \tag{A.28}$$

provided that either expectation exists.

Proof Then the result is immediate from Lemma A.1 since

$$\nabla\left(\frac{\varphi(x)}{\tau}\right) = \frac{1}{\tau}\, \nabla\varphi(x)\,.$$

□

In preparation for an extension of Theorem 2.7 the following integration by slice theorem used in Fourdrinier et al. (2003) is relevant and serves as a general replacement for spherical coordinates. This result can be derived from the co-area theorem stated by Federer (1969) (i.e. Theorem 3.2.12).

Lemma A.2 (Fourdrinier et al. 2003) *For any real number r, let $[\varphi = r]$ be the manifold in $\mathbb{R}^p$ associated with a given continuously differentiable function φ defined on $\mathbb{R}^p$ with nonnegative values whose gradient does not vanish at any point. Then, for any Lebesgue integrable function f, we have*

$$\int_{\mathbb{R}^p} f(x)dx = \int_{\{r\in\mathbb{R} \mid [\varphi=r]\neq\emptyset\}} \int_{[\varphi=r]} \frac{f(x)}{\|\nabla\varphi(x)\|} d\sigma_r(x)dr \tag{A.29}$$

where σ_r is the area measure defined on $[\varphi = r]$.

The following theorem is an extension of Theorem 2.7, i.e. an almost everywhere version of Stokes's theorem. As in typical statements of Stokes' theorem, we assume the set $[\varphi \le r]$ is bounded, and hence, $[\varphi \le r]$ is compact for every $r > 0$. As in Theorem 2.7, the proof relies on Lemma A.1 and completeness of a certain exponential family.

Theorem A.1 *Let φ be a continuously differentiable function from $\mathbb{R}^p$ into $\mathbb{R}_+$ whose gradient does not vanish at any point and is such that $\lim_{|x_i|\to\infty} \varphi(x_1, \ldots, x_p) = \infty$, for any $i = 1, \ldots, p$. Also assume that, for every $r > 0$, $[\varphi \le r]$ is compact and that φ determines a density ϕ as in Lemma* A.1. *Let g be a weakly differentiable function from $\mathbb{R}^p$ into $\mathbb{R}^p$. Then, for almost every $r > 0$,*

$$\int_{[\varphi=r]} \left(\frac{\nabla\varphi(x)}{\|\nabla\varphi(x)\|} \right)^{\mathrm{T}} g(x)\, d\sigma_r(x) = \int_{[\varphi\le r]} \operatorname{div} g(x)\, dx\,. \tag{A.30}$$

Further, for every r for which

$$\lim_{r'\to r^-} \int_{[\varphi=r']} \left(\frac{\nabla\varphi(x)}{\|\nabla\varphi(x)\|} \right)^{\mathrm{T}} g(x)\, d\sigma_{r'}(x) = \int_{[\varphi=r]} \left(\frac{\nabla\varphi(x)}{\|\nabla\varphi(x)\|} \right)^{\mathrm{T}} g(x)\, d\sigma_r(x) \tag{A.31}$$

the two integrals in (A.30) *are equal.*

Proof Let $X \sim \phi_\tau(x)$ with ϕ_τ as in Corollary A.1. We assume, without loss of generality, that $E_\tau[|g(X)|] < \infty$ since, as in the proof of Theorem 2.7, we may replace g by a sequence $(g_n)_{n\in\mathbb{N}}$ of functions with compact support.

By Lemma A.2, we have

$$\begin{aligned} E_\tau[\nabla\varphi(X)^{\mathrm{T}} g(X)] &= \int_{\mathbb{R}^p} \nabla\varphi(x)^{\mathrm{T}} g(x)\, K_\tau \exp\!\left(-\frac{\varphi(x)}{\tau}\right) dx \\ &= K_\tau\, \tau \int_0^\infty \int_{[\varphi=r]} \frac{\nabla\varphi(x)^{\mathrm{T}} g(x)}{\|\nabla\varphi(x)\|}\, d\sigma_r(x)\, \xi_\tau(r)\, dr \end{aligned} \tag{A.32}$$

with

$$\xi_\tau(r) = \frac{1}{\tau} \exp\left(-\frac{r}{\tau}\right). \tag{A.33}$$

We also have

$$\begin{aligned} \tau\, E_\tau[\operatorname{div} g(X)] &= \tau \int_{\mathbb{R}^p} \operatorname{div} g(x)\, K_\tau \exp\!\left(-\frac{\varphi(x)}{\tau}\right) dx \\ &= K_\tau\, \tau \int_{\mathbb{R}^p} \operatorname{div} g(x) \left[-\exp\left(-\frac{r}{\tau}\right)\right]_{\varphi(x)}^{\infty} dx \end{aligned}$$

$$= K_\tau\, \tau \int_{\mathbb{R}^p} \operatorname{div} g(x) \int_{\varphi(x)}^{\infty} \xi_\tau(r)\, dr\, dx$$

$$= K_\tau\, \tau \int_0^{\infty} \int_{[\varphi \le r]} \operatorname{div} g(x)\, dx\, \xi_\tau(r)\, dr\,, \tag{A.34}$$

by Fubini's theorem.

Therefore, it follows from (A.28) in Corollary A.1, (A.32) and (A.34) that, for all $\tau > 0$,

$$\int_0^{\infty} \int_{[\varphi = r]} \left(\frac{\nabla \varphi(x)}{\|\nabla \varphi(x)\|} \right)^{\mathrm{T}} g(x)\, d\sigma_r(x)\, \xi_\tau(r)\, dr = \int_0^{\infty} \int_{[\varphi \le r]} \operatorname{div} g(x)\, dx\, \xi_\tau(r)\, dr\,, \tag{A.35}$$

and hence, since the family $(\xi_\tau(r))_{\tau > 0}$ defined in (A.33) is complete as an exponential family, we have equality of the inner-most integrals in (A.35) for almost every $r > 0$. This gives the first result.

Finally, the right hand-side of (A.30) is absolutely continuous in r, since

$$\int_{[\varphi \le r]} \operatorname{div} g(x)\, dx = \int_0^r h(\rho)\, d\rho$$

where

$$h(\rho) = \int_{[\varphi = \rho]} \frac{\operatorname{div} g(x)}{\|\nabla \varphi(x)\|}\, d\sigma_{\rho,\theta}(x)\,,$$

and hence continuous, so that (A.31) implies the second result. □

A.6 Proof of Lemma 6.3

Denote by $\eta\big(X, \|U\|^2\big)$ the integrand of the second expectation, that is,

$$\eta\Big(X, \|U\|^2\Big) = \frac{1}{2} \frac{1}{\|U\|^{k-2}} \int_0^{\|U\|^2} \gamma(X, s)\, s^{k/2-1}\, ds\,.$$

Then conditionally on $X = x$, we have

$$E_{\theta,\sigma^2}\Big[\eta\Big(X, \|U\|^2\Big) \,|X = x\Big] = \frac{1}{K(\theta, \sigma^2, x)} \int_{\mathbb{R}^k} \frac{1}{2} \frac{1}{\|u\|^{k-2}} \int_0^{\|u\|^2} \gamma(x, s)\, s^{k/2-1}\, ds$$

$$\frac{1}{\sigma^{p+k}}\, f\left(\frac{\|x - \theta\|^2 + \|u\|^2}{\sigma^2} \right) du$$

where

$$K(\theta, \sigma^2, x) = \int_{\mathbb{R}^k} \frac{1}{\sigma^{p+k}} f\left(\frac{\|x-\theta\|^2 + \|u\|^2}{\sigma^2}\right) du\,.$$

Applying Fubini's theorem, we obtain

$$E_{\theta,\sigma^2}\left[\eta\left(X, \|U\|^2\right) | X = x\right] = \frac{1}{K(\theta, \sigma^2, x)} \int_0^\infty$$

$$\int_{\bar{B}(\sqrt{s})} \frac{1}{2} \frac{1}{\|u\|^{k-2}} \frac{1}{\sigma^{p+k}} f\left(\frac{\|x-\theta\|^2 + \|u\|^2}{\sigma^2}\right) du$$

$$\gamma(x, s)\, s^{k/2-1}\, ds$$

where $\bar{B}(\sqrt{s}) = \left\{u \in \mathbb{R}^k / \|u\| > \sqrt{s}\right\}$ is the complement of the ball of radius $\sqrt{s}$ centered at 0 in $\mathbb{R}^k$. As, in the inner most integral, the variable u intervenes through its norm $\|u\|$, we have, letting $\varsigma_k = 2\pi^{k/2}/\Gamma(k/2)$,

$$\begin{aligned}\int_{\bar{B}(\sqrt{s})} \frac{1}{\|u\|^{k-2}} f\left(\frac{\|x-\theta\|^2 + \|u\|^2}{\sigma^2}\right) du &= \varsigma_k \int_{\sqrt{s}}^\infty \frac{1}{r^{k-2}} f\left(\frac{\|x-\theta\|^2 + r^2}{\sigma^2}\right) r^{k-1}\, dr \\ &= \frac{\varsigma_k}{2} \int_s^\infty f\left(\frac{\|x-\theta\|^2 + t}{\sigma^2}\right) dt \\ &= \varsigma_k\, \sigma^2\, F\left(\frac{\|x-\theta\|^2 + s}{\sigma^2}\right).\end{aligned}$$

Hence

$$E_{\theta,\sigma^2}\left[\eta\left(X, \|U\|^2\right) | X = x\right] = \frac{\sigma^2}{2} \frac{\varsigma_k}{K(\theta, \sigma^2, x)} \int_0^\infty \frac{1}{\sigma^{p+k}}$$

$$F\left(\frac{\|x-\theta\|^2 + s}{\sigma^2}\right) \gamma(x, s)\, s^{k/2-1}\, ds$$

$$= \sigma^2 \int_0^\infty \frac{F\left(\frac{\|x-\theta\|^2+s}{\sigma^2}\right)}{f\left(\frac{\|x-\theta\|^2+s}{\sigma^2}\right)} \gamma(x, s) \frac{1}{2} \frac{\varsigma_k}{K(\theta, \sigma^2, x)} s^{k/2-1} \frac{1}{\sigma^{p+k}}$$

$$f\left(\frac{\|x-\theta\|^2 + s}{\sigma^2}\right) ds = \sigma^2 E_{\theta,\sigma^2}\left[\frac{F\left(\frac{\|x-\theta\|^2+\|U\|^2}{\sigma^2}\right)}{f\left(\frac{\|x-\theta\|^2+\|U\|^2}{\sigma^2}\right)} \gamma(x, \|U\|^2) \,|\, X = x\right]$$

using the radial density of $U|X = x$ as above. Consequently, unconditioning, we have

$$E_{\theta,\sigma^2}\left[\eta\left(X,\|U\|^2\right)\right] = \sigma^2\, E_{\theta,\sigma^2}\left[\frac{F\left(\frac{\|X-\theta\|^2+\|U\|^2}{\sigma^2}\right)}{f\left(\frac{\|X-\theta\|^2+\|U\|^2}{\sigma^2}\right)}\,\gamma(X,\|U\|^2)\right]$$
$$= \sigma^2\, c\, E^*_{\theta,\sigma^2}\left[\gamma\left(X,\|U\|^2\right)\right],$$

according to the definition of E^*_{θ,σ^2}, which is the desired result. □

A.7 An Expression of the Haff Operator

We follow Fourdrinier et al. (2016) to prove the expression of the Haff operator given in Proposition 6.5. To this end, we recall some known differential expressions.

Let U and T be $p \times p$ matrices, the elements of which being functions of $S = (S_{ij})$ and let $\tilde{\mathscr{D}}_S$ be a $p \times p$ matrix, the elements of which being linear combinations of $\partial/\partial S_{ij}$. Tsukuma and Konno (2006) recall the following result from Haff (1979) and (1982):

$$\tilde{\mathscr{D}}_S\, U\, T = \{\tilde{\mathscr{D}}_S\, U\}\, T + \left(U^{\mathrm{T}}\, \tilde{\mathscr{D}}_S^{\mathrm{T}}\right)^{\mathrm{T}} T\,. \tag{A.36}$$

In the particular case where $\tilde{\mathscr{D}}_S = \mathscr{D}_S$ with $(\mathscr{D}_S)_{ij} = 1/2\,(1+\delta_{ij})\,\partial/\partial S_{ij}$, we have $\mathscr{D}_S^{\mathrm{T}} = \mathscr{D}_S$ so that, if U is symmetric,

$$\mathscr{D}_S\, U\, T = \{\mathscr{D}_S\, U\}\, T + \left(U\, \mathscr{D}_S\right)^{\mathrm{T}} T\,. \tag{A.37}$$

Note that

$$\mathscr{D}_S\, S = \frac{p+1}{2}\, I_p \tag{A.38}$$

since

$$\begin{aligned}
\left(\mathscr{D}_S\, S\right)_{ik} &= \sum_{j=1}^{p} \frac{1}{2}\,(1+\delta_{ij})\,\frac{\partial S_{jk}}{\partial S_{ij}} \\
&= \frac{\partial S_{ik}}{\partial S_{ii}} + \frac{1}{2}\sum_{j\neq i}^{p} \frac{\partial S_{jk}}{\partial S_{ij}} \\
&= \delta_{ki} + \frac{1}{2}\sum_{j\neq i}^{p} \delta_{ki} \\
&= \delta_{ki} + \frac{p-1}{2}\,\delta_{ki} \\
&= \frac{p+1}{2}\,\delta_{ki}\,.
\end{aligned}$$

Proof of Proposition 6.5 First, we express $\mathrm{tr}\big(V\,\nabla_{V^{\mathrm{T}}}\,G^{\mathrm{T}}(X,S)\big)$ in term of S, we have

$$
\begin{aligned}
(V\,\nabla_{V^{\mathrm{T}}}\,G^{\mathrm{T}}(X,S))_{ii} &= \sum_{j=1}^{n-1}\sum_{k=1}^{p} V_{ij}\,(\nabla_{V^{\mathrm{T}}})_{jk}\,(G^{\mathrm{T}}(X,S))_{ki}\\
&= \sum_{j=1}^{n-1}\sum_{k=1}^{p} V_{ij}\,\frac{\partial (G^{\mathrm{T}}(X,S))_{ki}}{\partial V_{kj}}\\
&= \sum_{j=1}^{n-1}\sum_{k=1}^{p} V_{ij}\sum_{r\le l}^{p}\frac{\partial (G^{\mathrm{T}}(X,S))_{ki}}{\partial S_{rl}}\,\frac{\partial S_{rl}}{\partial V_{kj}}\\
&= \sum_{j=1}^{n-1}\sum_{k=1}^{p} V_{ij}\sum_{r\le l}^{p}\frac{\partial (G^{\mathrm{T}}(X,S))_{ki}}{\partial S_{rl}}\,(\delta_{rk}V_{lj}+\delta_{lk}V_{rj})
\end{aligned}
\tag{A.39}
$$

since $S = V\,V^{\mathrm{T}}$. From (A.39) we can write

$$
(V\,\nabla_{V^{\mathrm{T}}}\,G^{\mathrm{T}}(X,S))_{ii} = A + B + C \tag{A.40}
$$

where

$$
\begin{aligned}
A &= \sum_{j=1}^{n-1}\sum_{k=1}^{p} V_{ij}\sum_{r=l}^{p} 2\,\delta_{rk}V_{rj}\,\frac{\partial (G^{\mathrm{T}}(X,S))_{ki}}{\partial S_{rr}}\\
&= 2\sum_{j=1}^{n-1}\sum_{k=1}^{p} V_{ij}\,V_{kj}\,\frac{\partial (G^{\mathrm{T}}(X,S))_{ki}}{\partial S_{kk}}\\
&= 2\sum_{k=1}^{p} S_{ik}\,\frac{\partial (G^{\mathrm{T}}(X,S))_{ki}}{\partial S_{kk}},
\end{aligned}
\tag{A.41}
$$

$$
B = \sum_{j=1}^{n-1} V_{ij}\sum_{k<l}^{p} V_{lj}\,\frac{\partial (G^{\mathrm{T}}(X,S))_{ki}}{\partial S_{kl}},
$$

and

$$
C = \sum_{j=1}^{n-1} V_{ij}\sum_{k>r}^{p} V_{rj}\,\frac{\partial (G^{\mathrm{T}}(X,S))_{ki}}{\partial S_{rk}}.
$$

Now it is clear that

$$
\begin{aligned}
B + C &= \sum_{j=1}^{n-1} V_{ij} \sum_{k \neq l}^{p} V_{lj} \, \frac{\partial (G^{\mathrm{T}}(X,S))_{ki}}{\partial S_{kl}} \\
&= \sum_{k \neq l}^{p} S_{lj} \, \frac{\partial (G^{\mathrm{T}}(X,S))_{ki}}{\partial S_{kl}} . \qquad \text{(A.42)}
\end{aligned}
$$

Then, substituting (A.41) and (A.42) in (A.40), it follows that

$$
\begin{aligned}
\operatorname{tr}\big(V \, \nabla_{V^{\mathrm{T}}} \, G^{\mathrm{T}}(X,S)\big) &= 2 \sum_{i=1}^{p} \left\{ \sum_{k=1}^{p} S_{ik} \, \frac{\partial (G^{\mathrm{T}}(X,S))_{ki}}{\partial S_{kk}} + \frac{1}{2} \sum_{k \neq l}^{p} S_{il} \, \frac{\partial (G^{\mathrm{T}}(X,S))_{ki}}{\partial S_{kl}} \right\} \\
&= 2 \sum_{i=1}^{p} \left\{ \sum_{l=1}^{p} S_{il} \sum_{k=1}^{p} \frac{1}{2} \, (1 + \delta_{lk}) \, \frac{\partial (G^{\mathrm{T}}(X,S))_{ki}}{\partial S_{lk}} \right\} \\
&= 2 \sum_{i=1}^{p} \left\{ \sum_{l=1}^{p} S_{il} \sum_{k=1}^{p} (\mathscr{D}_S)_{lk}) \, \frac{\partial (G^{\mathrm{T}}(X,S))_{ki}}{\partial S_{lk}} \right\} \\
&= 2 \operatorname{tr}\big(S \, \mathscr{D}_S \, G^{\mathrm{T}}(X,S)\big) \\
&= 2 \operatorname{tr}\big((S \, \mathscr{D}_S)^{\mathrm{T}} \, G(X,S)\big) . \qquad \text{(A.43)}
\end{aligned}
$$

Secondly, by definition of $D^*_{1/2}$ recalled in (6.57), we have

$$
\begin{aligned}
2 \, D^*_{1/2}\big(S \, G(X,S)\big) &= 2 \operatorname{tr}\big(\mathscr{D}_S\{S \, G(X,S)\}\big) \\
&= 2 \operatorname{tr}\big(\{\mathscr{D}_S \, S\} \, G(X,S)\big) + 2 \operatorname{tr}\big(\{S \, \mathscr{D}_S\}^{\mathrm{T}} \, G(X,S)\big) , \qquad \text{(A.44)}
\end{aligned}
$$

according to (A.37) with $U = S$ and $T = G(X,S)$. Then, using (A.38) and (A.43) in the right hand-side of (A.44), we obtain

$$
2 \, D^*_{1/2}\big(S \, G(X,S)\big) = (p+1) \operatorname{tr}\big(G(X,S)\big) + \operatorname{tr}\big(V \, \nabla_{V^{\mathrm{T}}} \, G^{\mathrm{T}}(X,S)\big) ,
$$

which is the result given in (6.70). □

A.8 Harmonic, Superharmonic and Subharmonic Functions

A.8.1 Harmonic Functions

Let Ω be an open subset of $\mathbb{R}^p$. A measurable function f from Ω into $\mathbb{R}$ is said to be harmonic if it is locally integrable and possesses the sphere mean value property: for any $x \in \Omega$ and any ball $B_{r,x} \subset \Omega$ of radius r and centered at x, we have

$$f(x) = \int_{S_{r,x}} f(y)\, d\mathscr{U}_{r,x}(y)\,, \tag{A.45}$$

where $\mathscr{U}_{r,x}$ is the uniform distribution on the sphere $S_{r,x}$, the boundary of $B_{r,x}$. Such a harmonic function f is necessarily infinitely differentiable and satisfies the Laplace equation $\Delta f = 0$ on Ω (actually, $\Delta f = 0$ is equivalent to harmonicity of f). This may be seen as follows.

Let η be the function defined on $\mathbb{R}^p$ by

$$\forall x \in \mathbb{R}^p \quad \eta(x) = \begin{cases} C \exp\left[\frac{-1}{1-\|x\|^2}\right] & \text{if } \|x\| < 1 \\ 0 & \text{if } \|x\| \geq 1 \end{cases}$$

where the positive constant C is chosen such that

$$\int_{\mathbb{R}^p} \eta(x)\, dx = 1$$

and hence is equal to

$$C = \left\{ \sigma(S) \int_0^1 \exp\left[\frac{-1}{1-r^2}\right] r^{p-1}\, dr \right\}^{-1}, \tag{A.46}$$

using Lemma 1.4. We have $\eta \in \mathscr{C}_c^\infty(\mathbb{R}^p)$ (i.e. η is infinitely differentiable and has compact support $\overline{B}$). For any $\epsilon > 0$, consider the standard mollifier η^ϵ defined by

$$\forall x \in \mathbb{R}^p \quad \eta^\epsilon(x) = \frac{1}{\epsilon^p}\, \eta\left(\frac{x}{\epsilon}\right).$$

Then $\eta^\epsilon \in \mathscr{C}_c^\infty(\mathbb{R}^p)$ with $\operatorname{supp} \eta^\epsilon = \overline{B_\epsilon}$.

Now, if f is locally integrable in Ω, the convolution $f^\epsilon = \eta^\epsilon * f$ defined on the set

$$\Omega^\epsilon = \{x \in \Omega / \operatorname{dist}(x, \partial\Omega) > \epsilon\}$$

by

$$\forall x \in \Omega \quad f^\epsilon(x) = \int_\Omega \eta^\epsilon(x-y)\, f(y)\, dy$$

is a smooth approximation of f in the sense that $f^\epsilon \in \mathscr{C}^\infty(\Omega^\epsilon)$, which may be justified by use of the dominated convergence theorem. Then, integrating over the spheres of radius r and centered at x (as in Lemma 1.4), we have, for any $x \in \Omega^\epsilon$,

$$\begin{aligned} f^\epsilon(x) &= \int_{\mathbb{R}_+} \int_{Sr,x} \eta^\epsilon(x-y)\, f(y)\, \mathbb{1}_\Omega(y)\, d\sigma_{r,x}(y)\, dr \\ &= \int_{\mathbb{R}_+} \int_{Sr,x} \frac{C}{\epsilon^p} \exp\left[\frac{-1}{1-\|x-y\|^2/\epsilon^2}\right] \mathbb{1}_{B_{\epsilon,x}} f(y)\, \mathbb{1}_\Omega(y)\, d\sigma_{r,x}(y)\, dr \\ &= \frac{C}{\epsilon^p} \int_0^\epsilon \exp\left[\frac{-1}{1-r^2/\epsilon^2}\right] \sigma(S)\, r^{p-1} \int_{Sr,x} f(y)\, \mathbb{1}_\Omega(y)\, d\mathscr{U}_{r,x}(y)\, dr\,, \end{aligned}$$

since $B_{\epsilon,x} \subset \Omega$. Hence, if f is harmonic,

$$f^\epsilon(x) = \frac{C\,\sigma(S)}{\epsilon^p} \int_0^\epsilon \exp\left[\frac{-1}{1-r^2/\epsilon^2}\right] r^{p-1}\, dr\; f(x) = f(x)\,,$$

thanks to the change of variable $r = r'\,\epsilon$ and (A.46). Therefore $f \in \mathscr{C}^\infty(\Omega)$ as $f^\epsilon \in \mathscr{C}^\infty(\Omega^\epsilon)$ and $\Omega = \bigcup_{\epsilon>0} \Omega^\epsilon$.

For notational convenience, we will denote the above sphere mean by

$$\mathscr{S}_{r,x}(f) = \int_{S_{r,x}} f(y)\, d\mathscr{U}_{r,x}(y) \tag{A.47}$$

and the ball mean of f by

$$\mathscr{B}_{r,x}(f) = \int_{B_{r,x}} f(y)\, d\mathscr{V}_{r,x}(y) \tag{A.48}$$

where $\mathscr{V}_{r,x}$ is the uniform distribution on the ball $B_{r,x} = \{y \in \Omega \mid \|y-x\| \le r\}$. Note that, by definition of $\mathscr{V}_{r,x}$, provided $B_{r,x} \subset \Omega$,

$$\begin{aligned} \mathscr{B}_{r,x}(f) &= \frac{1}{\lambda(B)\, r^p} \int_{B_{r,x}} f(y)\, dy \\ &= \frac{p}{\sigma(S)\, r^p} \int_0^r \int_{S_{\rho,x}} f(y)\, d\sigma_{\rho,x}(y)\, d\rho \\ &= \frac{p}{r^p} \int_0^r \rho^{p-1}\, \mathscr{S}_{\rho,x}(f)\, d\rho\,, \end{aligned} \tag{A.49}$$

according to Lemma 1.4 and the relationship between the volume of the unit ball and the area measure of the unit sphere. Equality (A.49) shows that, if f is harmonic, then it satisfies the ball mean value property, that is, for any $x \in \Omega$ and any ball $B_{r,x}$ of radius r and centered at x such that $B_{r,x} \subset \Omega$, we have

$$f(x) = \mathscr{B}_{r,x}(f)\,. \tag{A.50}$$

The fact that an harmonic function f satisfies the Laplace equation can be derived from the following lemma which makes a link between the derivative of the sphere mean and the ball mean of its Laplacian.

Lemma A.3 *Let Ω be a domain of $\mathbb{R}^p$ and let f be a twice weakly differentiable function on Ω such that $\mathscr{S}_{r,x}(\nabla f)$ exists, for any $x \in \mathbb{R}^p$ and any $r \geq 0$ such that $\overline{B_{r,x}} \subset \Omega$. Then, for almost every such r,*

$$\frac{d}{dr}\mathscr{S}_{r,x}(f) = \frac{r}{p}\mathscr{B}_{r,x}(\Delta f)\,. \tag{A.51}$$

Proof According to (A.47), we have, through an obvious change of variable,

$$\begin{aligned} \frac{d}{dr}\mathscr{S}_{r,x}(f) &= \frac{d}{dr}\int_{S_{r,x}} f(y)\, d\mathscr{U}_{r,x}(y) \\ &= \frac{d}{dr}\int_S f(r\,z + x)\, d\mathscr{U}(z) \\ &= \int_S \frac{\partial}{\partial r} f(r\,z + x)\, d\mathscr{U}(z)\,, \end{aligned} \tag{A.52}$$

where the differentiation under the integral sign can be justified as follows. First, note that

$$\frac{\partial}{\partial r} f(r\,z + x) = \nabla f(r\,z + x) \cdot z \tag{A.53}$$

so that, for $z \in S$,

$$\left|\frac{\partial}{\partial r} f(r\,z + x)\right| \leq \|\nabla f(r\,z + x)\|\,\|z\| = \|\nabla f(r\,z + x)\|\,.$$

Hence

$$\int_S \left|\frac{\partial}{\partial r} f(r\,z + x)\right| dU(z) \leq \int_S \|\nabla f(r\,z + x)\|\, d\mathscr{U}(z) = \mathscr{S}_{r,x}(\|\nabla(f)\|) < \infty\,,$$

since, by assumption, $\mathscr{S}_{r,x}(\nabla f)$ exists. Therefore the last equality in (A.52) is valid by the Lebesgue dominated convergence theorem.

Then, according to (A.53), (A.52) can be rewritten as

$$\begin{aligned}\frac{d}{dr}\mathscr{S}_{r,x}(f) &= \int_S \nabla f(r\,z+x)\cdot z\,d\mathscr{U}(z)\\ &= \int_{S_{r,x}} \nabla f(y)\cdot\frac{y-x}{r}\,d\mathscr{U}_{r,x}(y)\\ &= \frac{1}{\sigma(S)\,r^{p-1}}\int_{S_{r,x}} \nabla f(y)\cdot\frac{y-x}{r}\,d\sigma_{r,x}(y)\end{aligned}$$

Hence, by Theorem 2.7 (Stokes theorem for weakly differentiable functions), for almost every r,

$$\begin{aligned}\frac{d}{dr}\mathscr{S}_{r,x}(f) &= \frac{1}{\sigma_{r,x}(S_{r,x})}\int_{B_{r,x}} \operatorname{div}(\nabla f(y))\,dy\\ &= \frac{1}{\sigma_{r,x}(S_{r,x})}\int_{B_{r,x}} \Delta f(y)\,dy\\ &= \frac{r}{p}\int_{B_{r,x}} \Delta f(y)\,d\mathscr{V}_{r,x}(y)\,,\end{aligned} \tag{A.54}$$

since $\sigma_{r,x}(S_{r,x}) = p\,\lambda(B_{r,x})/r$. This is the desired result. □

As announced above, if the function f in Lemma A.3 is harmonic, it satisfies the Laplace equation. Indeed the sphere means $\mathscr{S}_{r,x}(f)$ do not depend on the radius r so that, according to (A.51), for almost every r, the ball mean $\mathscr{B}_{r,x}(\Delta f)$ equals 0. In particular, according to (A.54), $\int_{B_{r,x}} \Delta f(y)\,dy = 0$, for almost every r, and hence, for all r since this integral is continuous in r; therefore $\Delta f(x) = 0$. Conversely, if f satisfies the Laplace equation, for $R > 0$ such that $\overline{B_{r,x}} \subset \Omega$, integrating (A.51) between 0 and R gives

$$0 = \int_0^R \frac{r}{p}\,\mathscr{B}_{r,x}(\Delta f)\,dr = \int_0^R \frac{d}{dr}\mathscr{S}_{r,x}(f)\,dr = \mathscr{S}_{R,x}(f) - f(x)\,,$$

so that f is harmonic in Ω.

A.8.2 Semicontinuous Functions

For superharmonic functions and subharmonic functions the equality in (A.45) is replaced by an inequality and their definitions require an additional semicontinuous property. This gives rise to functions which are less smooth than harmonic functions, and so, provides a flexible class of functions.

We recall first the notion of semicontinuity. Note that the functions we consider may have infinite values. To this end, we need the extended real field $\overline{\mathbb{R}} = \mathbb{R} \cup \{-\infty, \infty\} = [-\infty, \infty]$.

Definition A.1 Let Ω be an open subset of $\mathbb{R}^p$. A function f from Ω into $\overline{\mathbb{R}}$ is said to be lower semicontinuous (l.s.c.) on Ω

$$\text{(a)} \quad \forall x \in \Omega \ \liminf_{y \to x} f(y) \geq f(x)$$

and is said to be upper semicontinuous (u.s.c.) on Ω if

$$\text{(b)} \quad \forall x \in \Omega \ \limsup_{y \to x} f(y) \leq f(x) \,.$$

More explicitely, in (a) and in (b) above, the limits correspond to

$$\liminf_{y \to x, y \in \Omega} f(y) = \sup_{V \in \mathcal{N}_x} \inf_{y \in (V \cap \Omega) \setminus \{x\}} f(y) \quad \text{and}$$

$$\limsup_{y \to x, y \in \Omega} f(y) = \inf_{V \in \mathcal{N}_x} \sup_{y \in (V \cap \Omega) \setminus \{x\}} f(y) \,,$$

respectively, where $\mathcal{N}_x$ denotes the collection of all neighborhoods of x in Ω. Then we can express formally the lower semicontinuity of f at x as

$$\forall t < f(x) \, \exists V \in \mathcal{N}_x \quad y \in V \Rightarrow f(y) \geq t$$

and the upper semicontinuity of f at x as

$$\forall t > f(x) \, \exists V \in \mathcal{N}_x \quad y \in V \Rightarrow f(y) \leq t \,.$$

Then f is lower semicontinuous on Ω if and only if, for any $t \in \overline{\mathbb{R}}$, the subset $f^{-1}((t, \infty]) = \{x \in \Omega / t < f(x) \leq \infty\}$ is open (equivalently, if $f^{-1}([-\infty, t]) = \{x \in \Omega / -\infty \leq f(x) \leq t\}$ is closed). Indeed, if f is l.s.c. and if $x \in f^{-1}((t, \infty])$ then $f(x) > t$. Hence there exists $V \in \mathcal{N}_x$ such that $f(V) \subset (t, \infty]$. This means that $f^{-1}((t, \infty])$ is open. Assume now that, for any $t \in \overline{\mathbb{R}}$, the subset $f^{-1}((t, \infty])$ is open. Then, in particular, for any $t < f(x)$, the subset $f^{-1}((t, \infty])$ is open, which proves that this is a neighborhood of x where f is bounded from below by t. Therefore f is l.s.c. at x, and hence everywhere.

Similarly, f is upper semicontinuous on Ω if and only if, for any $t \in \overline{\mathbb{R}}$, the subset $f^{-1}([-\infty, t)) = \{x \in \Omega / -\infty \leq f(x) < t\}$ is open (equivalently, if $f^{-1}([t, \infty]) = \{x \in \Omega \mid t \leq f(x) \leq \infty\}$ is closed).

Clearly, it follows from the above that the indicator function of an open set is l.s.c. and that the indicator function of a closed set is u.s.c. Also, a function is continuous if and only if it is both l.s.c. and u.s.c. As a last simple example, the function f from $\mathbb{R}$ into $\mathbb{R}$ defined by $f(x) = \sin(1/x)$ if $x \neq 0$ and $f(0) = 1$ is u.s.c. at 0 but is not continuous (the function limits from the left or right at zero do not even exist).

A.8.3 Superharmonic and Subharmonic Functions

We now give the notion of superharmonicity and subharmonicity. For simplicity, we will assume that the open subset Ω is connex (Ω is connected or Ω is a domain), that is, Ω cannot be represented as the union of two or more disjoint nonempty open subsets. Actually, when Ω is not connex, in most of the statements below, it can be replaced by its connected components (the maximal connected subsets (ordered by inclusion) of Ω).

Definition A.2 Let Ω be a domain of $\mathbb{R}^p$. A function f from Ω into $(-\infty, \infty]$ is said to be superharmonic if

1. f is lower semicontinuous on Ω;
2. $f(x) \geq \mathscr{S}_{r,x}(f)$, for any $x \in \Omega$ and any $r \geq 0$ such that $\overline{B_{r,x}} \subset \Omega$ (superharmonic mean value property); and
3. $f \not\equiv \infty$ on Ω.

A function f from Ω into $[-\infty, \infty)$ is said to be subharmonic if $-f$ is superharmonic, that is, if

1. f is upper semicontinuous on Ω;
2. $f(x) \leq \mathscr{S}_{r,x}(f)$, for any $x \in \Omega$ and any $r \geq 0$ such that $\overline{B_{r,x}} \subset \Omega$ (subharmonic mean value property); and
3. $f \not\equiv -\infty$ on Ω.

It is easy to see that a function f is harmonic if and only if it is superharmonic and subharmonic. When $p \geq 3$, the function h defined by $h(x) = \|x\|^{2-p}$ is harmonic in $\mathbb{R}^p \setminus \{0\}$. This can be seen from the fact (proved below) that its sphere mean, for any $x \in \mathbb{R}^p$ and any $R > 0$, equals

$$\mathscr{S}_{R,x}(h) = \begin{cases} R^{2-p} & \text{if } \|x\| \leq R \\ \|x\|^{2-p} & \text{if } \|x\| > R\,. \end{cases} \tag{A.55}$$

Note that h is continuous on $\mathbb{R}^p \setminus \{0\}$ and that the harmonic mean value property is satisfied under the requirement $\overline{B_{R,x}} \subset \mathbb{R}^p \setminus \{0\}$ (second expression in (A.55)). Extending h at 0 with $h(0) = \infty$, we see that $\mathscr{S}_{R,0}(h) = R^{2-p} < h(0)$ which shows that the superharmonic mean value property is satisfied at 0. Also, as for any $t \in \overline{\mathbb{R}}$, the subset $h^{-1}((t, \infty])$ is the open ball $B_{1/t^{1/(p-2)}}$, then h is l.s.c. at 0, and hence, it is superharmonic.

The expressions in (A.55) can be found following Lemmas 3.21 and 3.22 of du Plessis (1970). First, we have

$$\mathscr{S}_{R,x}(h) = \int_{S_{R,x}} \|y\|^{2-p}\, d\mathscr{U}_{R,x}(y) = \int_{S_R} \|z + x\|^{2-p}\, d\mathscr{U}_R(z) = \int_{S_R} \|x - z\|^{2-p}\, d\mathscr{U}_R(z)\,,$$

using the orthogonal invariance of the uniform distribution U_R on the centered sphere S_R. Now, consider the spherical polar coordinates set in (1.9) under the following form: let $z = (z_1, \ldots, z_p)$ with

$$
\begin{aligned}
z_1 &= R \sin t_1 \sin t_2 \ldots \sin t_{p-2} \sin t_{p-1} \\
z_2 &= R \sin t_1 \sin t_2 \ldots \sin t_{p-2} \cos t_{p-1} \\
z_3 &= R \sin t_1 \sin t_2 \ldots \cos t_{p-2} \\
&\vdots \\
z_{p-1} &= R \sin t_1 \cos t_2 \\
z_p &= R \cos t_1 ,
\end{aligned}
$$

where $(t_1, \ldots, t_{p-1}) \in (0, \pi)^{p-2} \times (0, 2\pi)$. Choosing the angle between x and $z \in S_R$ as the first angle t_1 and expending

$$\|x - z\|^2 = R^2 - 2R\,\|x\| \cos t_1 + \|x\|^2 ,$$

we have that $\mathscr{S}_{R,x}(h)$ is proportional to

$$I_r = \int_0^{\pi} (R^2 - 2Rr \cos t_1 + r^2)^{(2-p)/2} \sin^{p-2} t_1 \, dt_1 ,$$

where $r = \|x\|$. Assuming $r < R$ and deriving with respect to r we have

$$
\begin{aligned}
\frac{d}{dr} I_r &= (2-p) \int_0^{\pi} (R^2 - 2Rr \cos t_1 + r^2)^{-p/2} (r - R \cos t_1) \sin^{p-2} t_1 \, dt_1 \\
&= (p-2) \int_0^{\pi} \sin^{p-2} u \cos u \, (R^2 \sin^2 u - r^2)^{-1/2} \, du , \qquad \text{(A.56)}
\end{aligned}
$$

setting $\cos u = (r - R \cos t_1)\,\rho^{-1}$, $\sin u = r \sin t_1 \,\rho^{-1}$ with $\rho = (R^2 - 2Rr \cos t_1 + r^2)^{1/2}$, which provides $\rho = \rho(u) = -R \cos u + (r^2 - R^2 \sin^2 u)^{1/2}$ and $dt_1 = \rho(u)\,(r^2 - R^2 \sin^2 u)^{-1/2}\,du$. Integrating between 0 and $\pi/2$ and between $\pi/2$ and π in (A.56), we can see, through the change of variable $u = \pi - v$, that this last integral is equal to 0 since $\sin u = \sin v$ and $\cos u = -\cos v$. Thus I_r does not depend on r and hence, letting $r \to 0$, equals R^{2-p}.

A direct alternative proof of the second expression in (A.55) can also be given as follows. Assume now that $r > R$. As above, we have that $r^{p-2}\,\mathscr{S}_{R,x}(h)$ is proportional to $r^{p-2} I_r$ whose derivative with respect to r, say $d/dr(r^{p-2} I_r) = (p-2)\,r^{p-3} I_r + r^{p-2}\, d/dr\, I_r$, can be seen to be equal to

$$(p-2)\,r^{p-3} R \int_0^{\pi} (R^2 - 2Rr \cos t_1 + r^2)^{-p/2} (R - r \cos t_1) \sin^{p-2} t_1 \, dt_1 \quad \text{(A.57)}$$

Noticing that the integral in (A.57) corresponds to the first integral in (A.56) with r and R interchanged, this derivative is equal to 0. Hence $r^{p-2}\,\mathscr{S}_{R,x}(h)$ is constant. Now $r^{p-2}\,\|x - z\|^{2-p} = (r/(R^2 - 2\,R\,r\,\cos t_1 + r^2)^{1/2})^{p-2}$ goes to 1 when $r \to \infty$ and is bounded above by $(r/(r - R))^{p-2}$ so that the Lebesgue dominated convergence theorem applies and implies that $r^{p-2}\,\mathscr{S}_{R,x}(h) = 1$. This is the second result in (A.55). We now give results implied by superharmonicity and subharmonicity.

Theorem A.2 *Let Ω be a domain of $\mathbb{R}^p$ and let f be a function from Ω into $(-\infty, \infty]$. If f is superharmonic (respectively subharmonic) then*

(i) *for any $x \in \Omega$* $\liminf_{y\to x} f(y) = f(x)$ *(respectively* $\limsup_{y\to x} f(y) = f(x)$*);*
(ii) *$f(x) \geq \mathscr{B}_{r,x}(f)$ (respectively $f(x) \leq \mathscr{B}_{r,x}(f)$) whenever $\overline{B_{r,x}} \subset \Omega$; and*
(iii) *either $f \equiv \infty$ (respectively $f \equiv -\infty$) or f is locally integrable on Ω.*

Proof We only prove the superharmonicity part, the subharmonicity part follows by using similar arguments.

(i) By lower continuity of f at any $x \in \Omega$, we have $\liminf_{y\to x} f(y) \geq f(x)$. If that inequality were strict, we would have $f(y) > f(x)$ for any $y \in B_{r,x} \setminus \{x\}$ for some $r > 0$, which would contradict the superharmonic mean value property of f in (ii) of Definition A.2.
(ii) According to (A.49) we have

$$\mathscr{B}_{r,x}(f) = \frac{p}{r^p}\int_0^r \rho^{p-1}\,\mathscr{S}_{\rho,x}(f)\,d\rho \leq \frac{p}{r^p}\int_0^r \rho^{p-1}\,f(x)\,d\rho = f(x)\,,$$

where the inequality expresses the superharmonic mean value property (ii) in Definition A.2.
(iii) Let

$$\Omega_0 = \{x \in \Omega \mid f \text{ is integrable over some neighborhood of } x\}\,.$$

By definition of Ω_0, if $x \in \Omega_0$, there exists $r > 0$ such that f is integrable over $B_{r,x}$. Then, for $y \in B_{r,x}$, as $B_{r/2,y} \subset B_{r,x}$, f is integrable over $B_{r/2,y}$. and hence, $y \in \Omega_0$. This shows that Ω_0 is open.

Now, for $x \in \Omega \setminus \Omega_0$, f is not integrable over any neighborhood of x. Furthermore, since f is l.s.c., it is bounded from below in any bounded neighborhood of x. Hence $\mathscr{B}_{r,x}(f) = \infty$ whenever $B_{r,x} \subset \Omega$ which implies, according to (ii), that $f(x) = \infty$.

Assume that $B_{r,x} \subset \Omega$. For $y \in B_{r/3,x}$, we have $B_{2r/3,y} \supset B_{r/3,x}$, f, and hence, f is not integrable over $B_{2r/3,y}$. Therefore $f(y) = \infty$ and so f is not integrable over any neighborhood of y, so that $y \in \Omega \setminus \Omega_0$. This shows that, if $x \in \Omega \setminus \Omega_0$, then $B_{r/3,x} \subset \Omega \setminus \Omega_0$, which means that $\Omega \setminus \Omega_0$ is open.

Finally, we have proved that Ω_0 and $\Omega \setminus \Omega_0$ are open which implies, as Ω is a domain, that either $\Omega_0 = \emptyset$ or $\Omega \setminus \Omega_0 = \emptyset$, which is (iii). □

The following theorem relates the superharmonicity and the subharmonicity of a function to the sign of its Laplacian.

Theorem A.3 *Let Ω be a domain of $\mathbb{R}^p$ and let f be a twice weakly differentiable function on Ω. Then f has the superharmonic mean value property (respectively the subharmonic mean value property) if and only if, for almost every $x \in \Omega$, we have $\Delta f(x) \leq 0$ (respectively $\Delta f(x) \geq 0$).*

Proof We only prove the superharmonicity part.

Assume that f has the superharmonic mean value property. For $R > 0$ such that $\overline{B_{R,x}} \subset \Omega$, integrating (A.51) between 0 and R gives

$$\int_0^R \frac{r}{p}\, \mathscr{B}_{r,x}(\Delta f)\, dr = \int_0^R \frac{d}{dr} \mathscr{S}_{r,x}(f)\, dr = \mathscr{S}_{R,x}(f) - f(x) \leq 0\,, \qquad \text{(A.58)}$$

by the superharmonic mean value property. As (A.58) is satisfied for any $R > 0$, the integrand of the first integral in (A.58) is nonpositive almost everywhere. Then, for almost every $0 < r < R$, we have $\mathscr{B}_{r,x}(\Delta f) \leq 0$, and hence, for almost every $x \in \Omega$, $\Delta f(x) \leq 0$.

Conversely, assume that, for almost every $x \in \Omega$, $\Delta f(x) \leq 0$. Then, according to (A.51), $\mathscr{S}_{r,x}(f)$ is nonincreasing in r. Also, when R goes to 0 in (A.58), the integral goes to 0 so that $\lim_{r \to 0} \mathscr{S}_{r,x}(f) = f(x)$. Hence $f(x) \geq \mathscr{S}_{r,x}(f)$ and so f has the superharmonic mean value property. □

The following corollary is immediate.

Corollary A.2 *Let Ω be a domain of $\mathbb{R}^p$ and let f be a function which is twice weakly differentiable and lower semicontinuous (respectively upper semicontinuous) on Ω. Then f is superharmonic (respectively subharmonic) if and only if, for almost every $x \in \Omega$, we have $\Delta f(x) \leq 0$ (respectively $\Delta f(x) \geq 0$).*

Corollary A.2 is usually stated for twice continuously differentiable functions f. du Plessis (1970) notices that, when f is not smooth, a notion of generalized Laplacian is needed and the Laplacian of f "can no longer be a point function", and must be a functional on the space of the test functions. In fact, the generalized Laplacian that du Plessis considers is the Laplacian of the regular distribution in (A.14) which has been seen to correspond to the weak Laplacian in (A.15).

The next theorem shows that, when a function is superharmonic or subharmonic, then its sphere mean and its ball mean are monotone functions of the radius.

Theorem A.4 *Let Ω be a domain of $\mathbb{R}^p$ and let f be a function which is twice weakly differentiable and superharmonic (respectively subharmonic) on Ω. Then, for $r \geq 0$ and $x \in \Omega$ such that $\overline{B_{r,x}} \subset \Omega$, the sphere mean $\mathscr{S}_{r,x}(f)$ and the ball mean $\mathscr{B}_{r,x}(f)$ are nonincreasing (respectively nondecreasing) functions of r.*

Proof The monotonicity of the sphere mean $\mathscr{S}_{r,x}(f)$ follows from (A.51) and Corollary A.2. Next the monotonicity of the ball mean $\mathscr{B}_{r,x}(f)$ follows from the fact that the right-hand side of the last equality in (A.49) can be written as $p \int_0^1 t^{p-1} \mathscr{S}_{r\,t,x}(f)\, dt$. □

A.9 Differentiation of Marginal Densities

When considering a prior density $\pi(\theta)$ for a normal model with density

$$\frac{1}{(2\pi)^{p/2}} \exp\left(-\frac{||x-\theta||^2}{2}\right),$$

differentiation of the marginal density

$$m(x) = \int_{\mathbb{R}^p} \frac{1}{(2\pi)^{p/2}} \exp\left(-\frac{||x-\theta||^2}{2}\right) \pi(\theta)\, d\theta$$

is easily tractable since we are dealing with an exponential family. Thus, deriving under the integral sign, we have that the gradient of m is expressed as

$$\begin{aligned}\nabla m(x) &= \int_{\mathbb{R}^p} \frac{1}{(2\pi)^{p/2}} \nabla \exp\left(-\frac{||x-\theta||^2}{2}\right) \pi(\theta)\, d\theta \\ &= \int_{\mathbb{R}^p} \frac{1}{(2\pi)^{p/2}} \exp\left(-\frac{||x-\theta||^2}{2}\right) (\theta - x)\, \pi(\theta)\, d\theta\,.\end{aligned}$$

The following lemma shows that, for a general spherically symmetric density $f(||x-\theta||^2)$ a similar formula is valid.

Lemma A.4 *Let $f(||x-\theta||^2)$ be a spherically symmetric density and $\pi(\theta)$ a prior density (possibly improper) such that, for any $x \in \mathbb{R}^p$, the marginal density*

$$m(x) = \int_{\mathbb{R}^p} f(||x-\theta||^2)\, \pi(\theta)\, d\theta$$

exists. If the generating function f is absolutely continuous then, for almost any $x \in \mathbb{R}^p$,

$$\begin{aligned}\nabla m(x) &= \int_{\mathbb{R}^p} \nabla f(||x-\theta||^2)\, \pi(\theta)\, d\theta \\ &= \int_{\mathbb{R}^p} 2\, f'(||x-\theta||^2)\, (x-\theta)\, \pi(\theta)\, d\theta\,.\end{aligned}$$

Proof Let $j \in \{1, \dots, p\}$ and $z \in \mathbb{R}$ fixed. For any $x = (x_1, \dots, x_{j-1}, x_j, x_{j+1}, \dots, x_p)$ in $\mathbb{R}^p$, denote by $x_{(z)} = (x_1, \dots, x_{j-1}, z, x_{j+1}, \dots, x_p)$ the vector obtained by replacing the j-th component x_j of x by z.

As f is absolutely continuous, the function $z \to f(||x_{(z)} - \theta||^2)$ is also absolutely continuous so that, for $a \in \mathbb{R}$, we have

$$f(||x_{(z)} - \theta||^2) - f(||x_{(a)} - \theta||^2) = \int_a^z \frac{\partial}{\partial x_j} f(||x - \theta||^2)\, dx_j \,.$$

Therefore

$$\begin{aligned} m(x_{(z)}) &= \int_{\mathbb{R}^p} \int_a^z \frac{\partial}{\partial x_j} f(||x - \theta||^2)\, dx_j\, \pi(\theta)\, d\theta + m(x_{(a)}) \\ &= \int_a^z \int_{\mathbb{R}^p} \frac{\partial}{\partial x_j} f(||x - \theta||^2)\, \pi(\theta)\, d\theta\, dx_j + m(x_{(a)}) \end{aligned} \tag{A.59}$$

by Fubini's theorem. Equation (A.59) means that the function $z \to m(x_{(z)})$ is absolutely continuous, and hence differentiable almost everywhere. More precisely, it entails that the partial derivative of $m(x)$ with respect to x_j equals

$$\frac{\partial}{\partial x_j} m(x) = \int_{\mathbb{R}^p} \frac{\partial}{\partial x_j} f(||x - \theta||^2)\, \pi(\theta)\, d\theta,$$

which is the desired result. □

We now prove Lemmas 5.5, 5.6 and 5.7 given in Sect. 5.4 (see Fourdrinier and Strawderman 2008a).

Proof of Lemma 5.5 It follows from (5.21) that the sign of $M'(t)$ is the same as the sign of $x \cdot \nabla M(x)$. As noted in the proof of Theorem 5.7 (after (5.25)), the function $H(u, t)$ defined in (5.22) is nonpositive. Hence, by Lemma 5.6, $x \cdot \nabla M(x) \le 0$ and $M'(t) \le 0$. □

Proof of Lemma 5.6 According to (5.11) and Lemma A.4, we have

$$\begin{aligned} x \cdot \nabla m(x) &= 2 \int_{\mathbb{R}^p} x \cdot (x - \theta)\, f'(\|x - \theta\|^2)\, \pi(\|\theta\|^2)\, d\theta \\ &= \int_0^\infty \int_{S_{R,x}} x \cdot (x - \theta)\, \pi(\|\theta\|^2)\, d\sigma_{R,x}(\theta)\, f'(R^2)\, dR \end{aligned}$$

where $\sigma_{R,x}$ is the uniform measure on the sphere $S_{R,x}$ of radius R centered at x.

Through $(\theta - x)/\|\theta - x\|$, the unit normal exterior vector at $\theta \in S_{R,x}$, we have

$$x \cdot \nabla m(x) = 2 \int_0^\infty \int_{S_{R,x}} -\pi(\|\theta\|^2)\, x \cdot \frac{\theta - x}{\|\theta - x\|}\, d\sigma_{R,x}(\theta)\, R\, f'(R^2)\, dR$$

$$= -2 \int_0^\infty \int_{B_{R,x}} \operatorname{div}_\theta\big(\pi(\|\theta\|^2)\, x\big)\, d\theta\, R\, f'(R^2)\, dR$$

by Stokes theorem and hence

$$\begin{aligned}
x \cdot \nabla m(x) &= -2 \int_0^\infty \int_{B_{R,x}} x \cdot \nabla_\theta\big(\pi(\|\theta\|^2)\big)\, d\theta\, R\, f'(R^2)\, dR \\
&= -4 \int_0^\infty \int_{B_{R,x}} x \cdot \theta \pi'(\|\theta\|^2)\, d\theta\, R f'(R^2)\, dR \\
&= -4\, \lambda(B) \int_0^\infty \int_{B_{R,x}} x \cdot \theta \pi'(\|\theta\|^2)\, \mathscr{V}_{R,x}(\theta)\, R^{p+1}\, f'(R^2)\, dR
\end{aligned}$$

according to the definition of $\mathscr{V}_{R,x}$. Then

$$\begin{aligned}
x \cdot \nabla m(x) &= \int_0^\infty H(R^2, \|x\|^2)\, R^{p+1}\, f'(R^2)\, dR \\
&= -2 \int_0^\infty H(u, \|x\|^2)\, u^{p/2}\, f(u)\, du
\end{aligned}$$

through the change of variable $u = R^2$.

This is the first result. The second result follows in the same way referring to (5.18). □

Proof of Lemma 5.7 The result will follow from the monotonicity in R of

$$\int_{S_{R,x}} x \cdot \theta\, \pi'(\|\theta\|^2)\, d\mathscr{U}_{R,x}(\theta) \tag{A.60}$$

since

$$\begin{aligned}
\int_{B_{R,x}} x \cdot \theta\, \pi'(\|\theta\|^2)\, d\mathscr{V}_{R,x}(\theta) &= \frac{R}{p} \int_0^R \tau^{p-1} \int_{S_{\tau,x}} x \cdot \theta\, \pi'(\|\theta\|^2)\, d\mathscr{U}_{\tau,x}(\theta)\, d\tau \\
&= p \int_0^1 u^{p-1} \int_{S_{Ru,x}} x \cdot \theta\, \pi'(\|\theta\|^2)\, d\mathscr{U}_{Ru,x}(\theta)\, du.
\end{aligned}$$

Now deriving (A.60) with respect to R gives through Lemma A.3

$$\frac{d}{dR} \int_{S_{R,x}} x \cdot \theta\, \pi'(\|\theta\|^2)\, d\mathscr{U}_{R,x}(\theta) = \frac{R}{p} \int_{B_{R,x}} \Delta\big(x \cdot \theta\, \pi'(\|\theta\|^2)\big)\, d\mathscr{U}_{R,x}(\theta)\,.$$

Since, as noticed above,

$$\Delta\big(x \cdot \theta\,\pi'(\|\theta\|^2)\big) = x \cdot \theta\,\varphi'(\|\theta\|^2)$$

where

$$\Delta\pi(\|\theta\|^2) = \varphi(\|\theta\|^2)\,,$$

we have

$$\frac{d}{dR}\int_{S_{R,x}} x \cdot \theta\,\pi'(\|\theta\|^2)\,d\mathscr{U}_{R,x}(\theta) = \frac{R}{p}\int_{B_{R,x}} x \cdot \theta\,\varphi'(\|\theta\|^2)\,d\mathscr{V}_{R,x}(\theta).$$

By assumption on the monotonicity of the Laplacian of $\pi(\|\theta\|^2)$, we have $\varphi'(\|\theta\|^2) \geq 0$ which, by Lemma 5.8, implies that the last integral is nonnegative since $\mathscr{V}_{R,x}$ is unimodal. □

Proof of Lemma 5.8 To express the integral

$$I(x) = \int_{\mathbb{R}^p} x \cdot \theta\,\psi(\theta)\,h(\|\theta - x\|^2)\,d\theta$$

we will use the orthogonal decomposition $\theta = \alpha + \beta$ with $\alpha \in \Delta_x$ and $\beta \in \Delta_x^\perp$ where Δ_x denotes the linear subspace of $\mathbb{R}^p$ spanned by x, $\Delta_x^\perp$ being its orthogonal. Then we have $I(x) = x \cdot A(x) + x \cdot B(x)$ with

$$A(x) = \int_{\Delta_x} \alpha \left[\int_{\Delta_x^\perp} \psi(\alpha + \beta)\,h\left(\|\alpha - x\|^2 + \|\beta\|^2\right)d\beta\right]d\alpha$$

and

$$B(x) = \int_{\Delta_x^\perp} \beta \left[\int_{\Delta_x} \psi(\alpha + \beta)\,h\left(\|\alpha - x\|^2 + \|\beta\|^2\right)d\alpha\beta\right]d\beta\,.$$

Note that $x \cdot B(x) = 0$, and hence $I(x) = x \cdot A(x)$, since $B(x) \in \Delta_x^\perp$ (actually, $B(x) = 0$ since, in the expression of $B(x)$, the most inner integral is a function of $\|\beta\|^2$ so that the most outer integral is the product of a real valued function of $\|\beta\|^2$ and β). Therefore

$$I(x) = \int_{\Delta_x} \alpha \left[\int_0^\infty \left(\int_{S_r} \psi(\alpha + \beta)\,h\left(\|\alpha - x\|^2 + r^2\right)\sigma_r(d\beta)\right)dr\right]d\alpha$$

where σ_r denotes the area measure on the sphere S_r in $\Delta_x^\perp$ of radius r and centered at 0. Then, through the change of variable $\alpha = z\,x$, we have

$$A(x) = \int_{-\infty}^{+\infty} z \left[\int_0^{+\infty} h\left((z-1)^2\|x\|^2 + r^2\right) \left(\int_{S_r} \psi(zx+\beta)\sigma_r(d\beta) \right) dr \right] dz$$
$$= x \int_0^{+\infty} z \left[\int_0^{+\infty} \left\{ h\left((z-1)^2\|x\|^2 + r^2\right) \int_{S_r} \psi(zx+\beta)\sigma_r(d\beta) \right.\right.$$
$$\left.\left. -h\left((z+1)^2\|x\|^2 + r^2\right) \int_{S_r} \psi(-zx+\beta)\sigma_r(d\beta) \right\} dr \right] dz.$$

Now, using invariance of σ_r by symmetry, we have

$$\int_{S_r} \psi(-zx+\beta)\,\sigma_r(d\beta) = \int_{S_r} \psi(-zx-\beta)\,\sigma_r(d\beta) = \int_{S_r} \psi(zx+\beta)\,\sigma_r(d\beta)$$

since ψ is symmetric. Therefore $A(x) = \gamma(x) \cdot x$ where

$$\gamma(x) = \int_0^{+\infty} z \left[\int_0^{+\infty} \left\{ h\left((z-1)^2\,\|x\|^2 + r^2\right) - h\left((z+1)^2\|\,x\|^2 + r^2\right) \right\} \right.$$
$$\left. \left(\int_{S_r} \psi(zx+\beta)\sigma_r(d\beta) \right) dr \right] dz\,.$$

Since h is nonincreasing and $\psi \geq 0$, we obtain that $\gamma(x) \geq 0$. Hence $I(x) = x \cdot A(x) = \gamma(x)\,\|x\|^2 \geq 0$, which is the desired result. □

A.10 Results on Expectations and Integrals

In the following, when X has a uniform distribution on a sphere of radius R, we consider expectations of $R^{2q}\,\|X\|^{-2q}$. We mention in our notations the dimension of the spaces in which spheres and balls lie. Thus $\mathcal{U}_{R,\theta}^{p+k}$ stands for the uniform distribution on the sphere $S_{R,\theta}^{p+k} = \{(x,u) \in \mathbb{R}^{p+k} \mid \|(x,u) - (\theta,0)\| = R\}$, in $\mathbb{R}^{p+k}$, of radius R and centered at $(\theta, 0) \in \mathbb{R}^{p+k}$, while $\mathcal{U}_{R,\theta}^{p}$ holds for the uniform distribution on the sphere $S_{R,\theta}^{p} = \{x \in \mathbb{R}^p \mid \|x-\theta\| = R\}$, in $\mathbb{R}^p$. We essentially extend a result given by Fourdrinier and Strawderman (2008b) who considered the case where $q = 1$. See also Fourdrinier et al. (2013).

Lemma A.5 *Let $q > 0$. Then, for any fixed $\theta \in \mathbb{R}^p$, the function*

$$f_\theta : R \longmapsto R^{2q} \int_{S_{R,\theta}^{p+k}} \frac{1}{\|x\|^{2q}}\, d\mathcal{U}_{R,\theta}^{p+k}(x,u) \tag{A.61}$$

is nondecreasing for $p \geq 2(q+1)$ and $k \geq 0$. Also, for any fixed R, this monotonicity is reversed in $\|\theta\|$.

Proof Note that, by invariance, f_θ depends on θ only through $\|\theta\|$. With the change of variable

$$(y, v) = \left(\frac{x-\theta}{R}, \frac{u}{R}\right),$$

we have

$$f_\theta(R) = \int_{S_{1,0}^{p+k}} \frac{1}{\|y + \frac{\theta}{R}\|^{2q}} \, d\mathscr{U}_{1,0}^{p+k}(y, v) = f_{\|\theta\|}^*(R)\,.$$

Hence, integrating with respect to the uniform distribution on $\{\theta \in \mathbb{R}^p \mid \|\theta\| = R_0\}$,

$$\begin{aligned} f_\theta(R) &= \int_{S_{R_0,0}^p} \int_{S_{1,0}^{p+k}} \frac{1}{\|y + \frac{\theta}{R}\|^{2q}} \, d\mathscr{U}_{1,0}^{p+k}(y, v)\, d\mathscr{U}_{R_0,0}^p(\theta) \\ &= \int_{S_{1,0}^{p+k}} \int_{S_{R_0,0}^p} \frac{1}{\|y + \frac{\theta}{R}\|^{2q}} \, d\mathscr{U}_{R_0,0}^p(\theta)\, d\mathscr{U}_{1,0}^{p+k}(y, v) \end{aligned}$$

by Fubini's theorem. In the inner integral, the change of variable $z = \theta/R + y$ leads to

$$f_\theta(R) = \int_{S_{1,0}^{p+k}} \int_{S_{R_0/R,y}^p} \frac{1}{\|z\|^{2q}} \, d\mathscr{U}_{R_0/R,y}^p(z)\, d\mathscr{U}_{1,0}^{p+k}(y, v)$$

As the function $1/\|z\|^{2q}$ is superharmonic for $p \geq 2\,(q+1)$, the inner integral is nonincreasing in R_0/R for each y, and hence, nondecreasing in R and, for any fixed R, nonincreasing in $R_0 = \|\theta\|$. □

The first part of Lemma A.5 can be extended thanks to an extension of Anderson's theorem (see Anderson 1955) given in Lemma 3 of Chou and Strawderman (1990) which we recall below.

Lemma A.6 (Chou and Strawderman 1990) *Let h and f be measurable functions from $\mathbb{R}^p$ into $\mathbb{R}_+$. Assume that h and f are symmetric about the origin, unimodal and such that $\int_{\mathbb{R}^p} f(x)\,dx < \infty$ and $\int_{\mathbb{R}^p} h(x)\,f(x)\,dx < \infty$. Then, for $y \in \mathbb{R}^p$ and $0 \leq k \leq 1$,*

$$\int_{\mathbb{R}^p} h(x)\, f(x + k\,y)\, dx \geq \int_{\mathbb{R}^p} h(x)\, f(x+y)\, dx \tag{A.62}$$

and hence $\phi(k) = \int_{\mathbb{R}^p} h(x)\, f(x + k\,y)\, dx$ is a nonincreasing function of k.

Proof See Chou and Strawderman (1990) for Inequality (A.62). As for the monotonicity part, for $k_1 < k_2$, it suffices to apply (A.62) with $k = k_1/k_2$ and $k_2\, y$ playing the role of y to obtain $\phi(k_1) \geq \phi(k_2)$. □

Lemma A.7 *Let $q > 0$ and let $r(t)$ be a nonnegative and nondecreasing function on $[0, \infty)$ such that $r(t)/t^q$ is nonincreasing. Then, for any fixed $\theta \in \mathbb{R}^p$, the function*

$$f_\theta : R \longmapsto R^{2q} \int_{S_{R,\theta}^{p+k}} \frac{r(\|x\|^2)}{\|x\|^{2q}} \, d\mathscr{U}_{R,\theta}^{p+k}(x, u) \tag{A.63}$$

is nondecreasing for $p \geq 1$ and $k \geq 2$.

Proof Under $\mathscr{U}_{R,\theta}^{p+k}$, it is well known that the marginal distribution of $(x, u) \mapsto x$ is absolutely continuous with unimodal density $\frac{1}{R^p}\psi\left(\frac{\|x-\theta\|^2}{R^2}\right)$ for all $k \geq 2$ where $\psi(t) \propto (1-t)^{k/2-1}\mathbf{1}_{[0,1]}(t)$ (see Theorem 4.10). Then f_θ can be written as

$$f_\theta(R) = \int_{B_{1,0}^p} \frac{r(R^2\,\|z + \frac{\theta}{R}\|^2)}{\|z + \frac{\theta}{R}\|^{2q}} \, \psi(\|z\|^2) \, dz \,.$$

For any $R_1 \leq R_2$, we have, by nondecreasing monotonicity of $r(t)$,

$$f_\theta(R_1) \leq \int_{B_{1,0}^p} \frac{r(R_2^2\|z + \frac{\theta}{R_1}\|^2)}{\|z + \frac{\theta}{R_1}\|^{2q}} \, \psi(\|z\|^2) \, dz \,.$$

Furthermore nonincreasing monotonicity of $r(t)/t$ in t implies that the function $r(R_2^2\,\|z+\theta/R_1\|^2)/\|z+\theta/R_1\|^{2q}$ is symmetric and unimodal in z about $-\frac{\theta}{R_1}$. Hence, by Lemma A.6,

$$\begin{aligned}
\int_{B_{1,0}^p} \frac{r(R_2^2\|z + \frac{\theta}{R_1}\|^2)}{\|z + \frac{\theta}{R_1}\|^{2q}} \, \psi(\|z\|^2) \, dz &\leq \int_{B_{1,0}^p} \frac{r(R_2^2\|z + \frac{\theta}{R_2}\|^2)}{\|z + \frac{\theta}{R_2}\|^{2q}} \, \psi(\|z\|^2) \, dz \\
&= f_\theta(R_2) \,.
\end{aligned}$$

□

As, if $(X, U) \sim \mathscr{U}_{R,\theta}^{p+2}$, then $X \sim \mathscr{V}_{R,\theta}^p$, the following corollary of Lemma A.7 is immediate.

Corollary A.3 *Under the conditions of Lemma A.7, the function*

$$R \longmapsto R^{2q} \int_{B_{R,\theta}^p} \frac{r(\|x\|^2)}{\|x\|^{2q}} \, d\mathscr{V}_{R,\theta}^p(x) \tag{A.64}$$

is nondecreasing for $p \geq 1$.

A.11 Modified Bessel Functions

We develop some results on ratio of Bessel functions $\rho_\nu(t) = I_{\nu+1}(t)/I_\nu(t)$ needed in Sect. 7.3. First, recall the *modified Bessel function of the first kind*, for $\nu > 0$,

$$I_\nu(x) = \sum_{k=0}^{\infty} \frac{x^{2k+\nu}}{2^{2k+\nu}\, k!\, \Gamma(k+\nu+1)},$$

is one of the solutions to the *modified Bessel differential equation* given by

$$x^2\, y'' + x\, y' - \left(\nu^2 + x^2\right)\, y = 0,$$

where $\Gamma(x)$ is the *gamma function.*

Lemma A.8

(a) (Watson 1983) *The function* $\rho_\nu(\cdot)$ *is increasing and concave on* $(0, \infty)$*, with* $\lim_{t\to 0^+} \rho_\nu(t) = 0$, $\lim_{t\to\infty} \rho_\nu(t) = 1$*; and* $\frac{\rho_\nu(t)}{t}$ *decreasing in* t *with* $\lim_{t\to 0^+} \frac{\rho_\nu(t)}{t} = \frac{1}{2(\nu+1)}$*. Also, we have the identity* $\frac{d}{dt}\rho_\nu(t) = 1-(1+2\nu)\,\frac{\rho_\nu(t)}{t} - \rho_\nu^2(t)$*, and the inequality* $\frac{d}{dt}\rho_\nu(t) \le \frac{\rho_\nu(t)}{t}$.
(b) (Amos 1974) *For all* $\nu \ge 0$ *and* $t > 0$*, we have*

$$L\left(\frac{2(\nu+1)}{t}, \frac{2(\nu+1)}{t}\right) \le \rho_\nu^2(t) \le L\left(\frac{2\nu}{t}, \frac{2(\nu+2)}{t}\right), \tag{A.65}$$

where $L(a, b) = \{a/2 + \sqrt{1+(b/2)^2}\}^{-2}$.

Lemma A.9

(a) *For all* $p \in \{3, 4, \ldots\}$ *and* $\alpha \ge 0$*, the function given by* $r\,\{1 - \rho_{p/2-1}(\alpha\, r)\}$ *is increasing in* r*;* $r \ge 0$*;*
(b) *For all* $p \in \{3, 4, \ldots\}$*, we have the inequality* $\rho_{p/2-1}(t) + t\,\rho'_{p/2-1}(t) \le 1$*, for all* $t > 0$*;*
(c) *For all* $p \in \{3, 4, \ldots\}$*, we have* $\lim_{t\to\infty} t\,\{1 - \rho_{p/2-1}(t)\} = (p-1)/2$.[2]

Proof

(c) The result follows from the fact that $t\rho_\nu'(t) \to 0$ as $t \to \infty$, which must be the case for part (b) to hold since $\rho_\nu(t) \to 1$ as $t \to \infty$, as well as the given expression for ρ_ν' given in Lemma A.8.

[2]Alternatively, the more general result $\lim_{t\to\infty} t\,(1 - \rho_\nu(t)) = \nu + 1/2$ holds for all $\nu > 0$ by the bounds (A.65) given by Amos (1974) for $\rho_\nu(t)$, $t > 0$. This is verified with the evaluation $\lim_{z\to\infty} z\,(1 - z/(\nu + 1/2 + \sqrt{z^2+b}) = \nu + 1/2$.

(b) Part (a) tells us (take $\alpha = 1$) that $t\,(1 - \rho_{p/2-1}(t))$ increases in t, in other terms: $\frac{\partial}{\partial t}\{t\,(1 - \rho_{p/2-1}(t))\} \geq 0$ which is equivalent to, and establishes, part (b).

(a) Begin with (7.10) which implies that $r\,\{1 - \rho_{p/2-1}(\alpha r)\} = E_\theta[W|\,R = r]$, with $W = \|X\| - \frac{\theta' X}{\|\theta\|}$, and $\theta \in S_\alpha$. It will hence suffice to show that a family of conditional distributions $\{W|\,R = r : r > 0\}$ satisfies (for $p \geq 3$) an increasing in W monotone likelihood ratio property, with parameter r. Observe also that the probability distribution of W remains unchanged with orthogonal transformations $X \to \Gamma X$ (and $\theta \to \Gamma\theta$), which permits us, since the actions are transitive on S_α, to set without loss of generality $\theta = \theta_0 = (\alpha, 0, \ldots, 0)'$. Pursue next with the joint density (for $\theta = \theta_0$ and $p > 1$) of $(Y_1 = X_1, Y_2 = X'X - X_1{}^2)$, given by:

$$f_{Y_1,Y_2}(y_1, y_2) \propto e^{-\frac{1}{2}[(y_1-\alpha)^2+y_2]}\; y_2^{\frac{p-1}{2}-1}\; \mathbb{1}_{(0,\infty)}(y_2)\,,$$

to derive the joint density of $(W = \sqrt{Y_1^2 + Y_2} - Y_1,\, R = \sqrt{Y_1^2 + Y_2})$,

$$f_{W,R}(w, r) \propto r\;\exp\{\,-\frac{r^2}{2}+\alpha(r-w)\,\}\;[w(2r-w)]^{\frac{p-3}{2}}\;\mathbb{1}_{(0,\,2r\,)}(w)\;1_{(0,\infty)}(r)\,,$$

and the conditional density[3]:

$$f_{W|R=r}(w) \propto \exp\{-\alpha w\}\;[w(2r - w)]^{\frac{p-3}{2}}\;\mathbb{1}_{(0,\,2r\,)}(w)\,;\, r > 0. \qquad \text{(A.66)}$$

To conclude, the result follows by checking that the ratio $\frac{f_{W|R=r_1}(w)}{f_{W|R=r_0}(w)}$ is nondecreasing in w for all $r_1 > r_0 > 0$. □

[3]Interestingly, for $p = 3$, the distribution $W|R = r$ is truncated exponential.

References

Aitchison J (1975) Goodness of prediction fit. Biometrika 62:547–554
Akaike H (1974) A new look at the statistical model identification. IEEE Trans Autom Control 19(6):716–723
Alam K (1973) A family of admissible minimax estimators of the mean of a multivariate normal distribution. Ann Stat 1:517–525
Amos DE (1974) Computation of modified Bessel functions and their ratios. Math Comput 28:239–251
Anderson TW (1955) The integral of a symmetric unimodal function over a symmetric convex set and some probability inequalities. Proc Am Math Soc 6:170–176
Anderson TW (1984) An introduction to multivariate statistical analysis. Wiley, New York
Baranchik A (1970) A family of minimax estimators of the mean of a multivariate normal distribution. Ann Math Stat 41:642–645
Barlow RE, Proschan F (1981) Statistical theory of reliability and life testing: probability models. TO BEGIN WITH, Silver Spring
Barron AR, Birgé L, Massart P (1999) Risk bound for model selection via penalization. Probab Theory Relat Fields 113:301–413
Bartlett P, Boucheron S, Lugosi G (2002) Model selection and error estimation. Mach Learn 48:85–113
Berger JO (1975) Minimax estimation of location vectors for a wide class of densities. Ann Stat 3(6):1318–1328
Berger JO (1976) Inadmissibility results for the best invariant estimator of two coordinates of a location vector. Ann Stat 4(6):1065–1076
Berger JO (1985a) Statistical decision theory and Bayesian analysis, 2nd edn. Springer, New York
Berger JO (1985b) In defense of the likelihood principle: axiomatics and coherency. In: Bernardo JM, Lindley DV, DeGroot MH, Smith AFM (eds) Bayesian statistics II. North Holland, Amsterdam, pp 33–65
Berger JO (1985c) Decision theory and Bayesian analysis. Springer, New York
Berger JO (1985d) The frequentist viewpoint and conditioning. In: Cam LL, Olshen R (eds) Proceedings of the Berkeley conference in honor of Jersy Neyman and Jack Kiefer, vol 1. Wadsworth, Monterey, pp 15–44
Berger JO, Srinivasan C (1978) Generalized Bayes estimators in multivariate problems. Ann Stat 6(4):783–801
Berger JO, Strawderman WE (1996) Choice of hierarchical priors. Admissibility in estimation of normal means. Ann Stat 24:931–951
Berry C (1990) Minimax estimation of a bounded normal mean vector. J Multivar Anal 35:130–139

D. Fourdrinier et al., *Shrinkage Estimation*, Springer Series in Statistics,
https://doi.org/10.1007/978-3-030-02185-6

Bickel PJ (1981) Minimax estimation of the mean of a normal distribution when the parameter space is restricted. Ann Stat 9(6):1301–1309
Bickel PJ, Doksum KA (2001) Mathematical statistics, vol I. Prentice Hall, Upper Saddle River
Blanchard D, Fourdrinier D (1999) Non trivial solutions of non-linear partial differential inequations and order cut-off. Rendiconti di Matematica 19:137–154
Blyth CR (1951) On minimax statistical procedures and their admissibility. Ann Math Stat 22:22–42
Bock ME (1985) Minimax estimators that shift towards a hypersphere for location vectors of spherically symmetric distributions. J Multivar Anal 17:127–147
Bock ME (1988) Shrinkage estimators: pseudo-Bayes rules for normal mean vectors. In: Gupta SS, Berger J (eds) Statistical decision theory and related topics 4, vol 1. Springer, New York, pp 281–298
Boisbunon A, Maruyama Y (2014) Inadmissibility of the best equivariant predictive density in the unknown variance case. Biometrika 101(3):733–740
Boisbunon A, Canu S, Fourdrinier D, Strawderman WE, Wells MT (2014) Akaike's information criterion, Cp and estimators of loss for elliptically symmetric distributions. Int Stat Rev 82(3):422–439
Bondar JV, Milnes P (1981) Amenability: a survey for statistical applications of Hunt-Stein theorem and related conditions on groups. Zeitschr Wahrsch Verw Geb 57:103–128
Brandwein AC, Strawderman WE (1978) Minimax estimation of location parameters for spherically symmetric unimodal distributions under quadratic loss. Ann Stat 6:377–416
Brandwein AC, Strawderman WE (1980) Minimax estimation of location parameters for spherically symmetric distributions with concave loss. Ann Stat 8:279–284
Brandwein AC, Strawderman WE (1991a) Generalizations of James-Stein estimators under spherical symmetry. Ann Stat 19:1639–1650
Brandwein AC, Strawderman WE (1991b) Improved estimates of location in the presence of unknown scale. J Multivar Anal 39:305–314
Brandwein AC, Ralescu S, Strawderman W (1993) Shrinkage estimators of the location parameter for certain spherically symmetric distributions. Ann Inst Stat Math 45(3):551–565
Bressan A (2012) Lecture notes on functional analysis with applications to linear partial differential equations, vol 143. American Mathematical Society, Providence
Brown LD (1966) On the admissibility of invariant estimators in one or more location parameter. Ann Math Stat 37:1087–1136
Brown LD (1968) Inadmissibility of the usual estimators of scale parameters in problems with unknown location and scale parameters. Ann Math Stat 39:24–48
Brown LD (1971) Admissible estimators, recurrent diffusions, and insoluble boundary value problems. Ann Math Stat 42:855–903
Brown LD (1979) A heuristic method for determining admissibility of estimators–with applications. Ann Stat 7(5):960–994. https://doi.org/10.1214/aos/1176344782
Brown LD (1986) Fundamentals of statistical exponential families with applications in statistical decision theory. Lecture notes-monograph series, vol 9. Institute of Mathematical Statistics, Hayward
Brown LD, Hwang JT (1982) A unified admissibility proof. In: Gupta SS, Berger J (eds) Statistical decision theory and related topics III, vol 1. Academic, New York, pp 205–230
Brown LD, Hwang JT (1990) Admissibility of confidence estimators. In: Proceedings of the 1990 Taipei symposium in statistics, Institute of Statistical Science, pp 1–10
Brown LD, Purves R (1973) Measurable selection of extrema. Ann Stat 1:902–912
Brown LD, Zhao LH (2012) A geometrical explanation of Stein shrinkage. Stat Sci 27:24–30
Brown LD, Johnstone I, MacGibbon KB (1981) Variation diminishing transformations: a direct approach to total positivity and its statistical applications. J Am Stat Assoc 76:824–832
Brown LD, George I, Xu X (2008) Admissible predictive density estimation. Ann Stat 36:1156–1170
Carvalho C, Polson N, Scott J (2010) The horseshoe estimator for sparse signals. Biometrika 97:465–480

Casella G, Berger RL (2001) Statistical inference, 2nd edn. Duxbury Press, Belmont
Casella G, Strawderman WE (1981) Estimating a bounded normal mean. Ann Stat 9:870–878
Celisse A, Arlot S (2010) A survey of cross-validation procedures for model selection. Stat Surv 4:40–79
Cellier D, Fourdrinier D (1990) Sur les lois à symétrie elliptique. In: Séminaire de probabilités (Strasbourg), vol 24. Springer, Berlin/Heidelberg/New York, pp 320–328
Cellier D, Fourdrinier D (1995) Shrinkage estimators under spherical symmetry for the general linear model. J Multivar Anal 52:338–351
Cellier D, Fourdrinier D, Robert C (1989) Robust shrinkage estimators of the location parameter for elliptically symmetric distributions. J Multivar Anal 29:39–52
Chou JP, Strawderman WE (1990) Minimax estimation of means of multivariate normal mixtures. J Multivar Anal 35(2):141–150
Clevenson M, Zidek J (1975) Simultaneous estimation of the mean of independent Poisson laws. J Am Stat Assoc 70:698–705
Davies SL, Neath AA, Cavanaugh JE (2006) Estimation optimality of corrected AIC and modified Cp in linear regression. Int Stat Rev 74(2):161–168
Dey DK, Srinivasan C (1985) Estimation of a covariance matrix under Stein's loss. Ann Stat 13:1581–1591
Diaconis P, Ylvisaker D (1979) Conjugate priors for exponential families. Ann Stat 7:269–281
Donoho DL, Johnstone IM (1995) Adapting to unknown smoothness via wavelet shrinkage. J Am Stat Assoc 90:1200–1244
Donoho DL, Liu RC, MacGibbon KB (1990) Minimax risk over hyperrectangles, and implications. Ann Stat 80(3):1416–1437
Doob JL (1984) Classical potential theory and its probabilistic counterpart. Springer, Berlin/Heidelberg/New York
du Plessis N (1970) An introduction to potential theory. Hafner, Darien
Eaton ML (1989) Group invariance applications in statistics. Institute of Mathematical Statistics, Hayward
Eaton ML et al (1992) A statistical diptych: admissible inferences–recurrence of symmetric Markov chains. Ann Stat 20(3):1147–1179
Efron B (2004) The estimation of prediction error: covariance penalties and cross-validation. J Am Stat Assoc 81:461–470
Efron B, Morris C (1976) Families of minimax estimators of the mean of a multivariate normal distribution. Ann Stat 4(1):11–21
Efron B, Morris C (1977) Stein's paradox in statistics. Sci Am 236(5):119–127
Efron B, Hastie T, Johnstone I, Tibshirani R (2004) Least angle regression. Ann Stat 32:407–499
Faith RE (1978) Minimax Bayes point estimators of a multivariate normal mean. J Multivar Anal 8:372–379
Fan J, Li R (2001) Variable selection via nonconcave penalized likelihood and its oracle properties. J Am Stat Assoc 96:1348–1360
Fang KT, Zhang YT (1990) Generalized multivariate analysis. Springer, Berlin/Heidelberg/New York/London/Paris/Tokyo/Hong Kong
Fang KT, Kotz KS, Ng KW (1990) Symmetric multivariate and related distributions. Chapman and Hall, New York
Federer H (1969) Geometric measure theory. Springer, Berlin
Feller W (1971) An introduction to probability theory and its application, Vol II (2nd ed), Wiley, New York
Fourdrinier D, Lepelletier P (2008) Estimating a general function of a quadratic function. Ann Inst Stat Math 60:85–119
Fourdrinier D, Marchand E (2010) On Bayes estimators with uniform priors on spheres and their comparative performance with maximum likelihood estimators for estimating bounded multivariate normal means. J Multivar Anal 101:1390–1399
Fourdrinier D, Ouassou I (2000) Estimation of the mean of a spherically symmetric distribution with constraints on the norm. Can J Stat 28:399–415

Fourdrinier D, Strawderman WE (1996) A paradox concerning shrinkage estimators: should a known scale parameter be replaced by an estimated value in the shrinkage factor? J Multivar Anal 59:109–140
Fourdrinier D, Strawderman WE (2003) On Bayes and unbiased estimators of loss. Ann Inst Stat Math 55:803–816
Fourdrinier D, Strawderman WE (2008a) Generalized Bayes minimax estimators of location vector for spherically symmetric distributions. J Multivar Anal 99(4):735–750
Fourdrinier D, Strawderman WE (2008b) A unified and generalized set of shrinkage bounds on minimax Stein estimates. J Multivar Anal 99(10):2221–2233
Fourdrinier D, Strawderman WE (2010) Robust generalized Bayes minimax estimators of location vectors for spherically symmetric distribution with unknown scale. IMS Collect Inst Math Stat Festschrift Lawrence D Brown 6:249–262
Fourdrinier D, Strawderman WE (2015) Robust minimax Stein estimation under invariant data-based loss for spherically and elliptically symmetric distributions. Metrika 78:461–484
Fourdrinier D, Strawderman WE (2016) Stokes' theorem, Stein's identity and completeness. Stat Probab Lett 109:224–231
Fourdrinier D, Wells MT (1994) Comparaisons de procédures de sélection d'un modèle de régression: Une approche décisionnelle. CR Acad Sci Paris Serie I 319:865–870
Fourdrinier D, Wells MT (1995a) Estimation of a loss function for spherically symmetric distributions in the general linear model. Ann Stat 23:571–592
Fourdrinier D, Wells MT (1995b) Loss estimation for spherically symmetric distributions. J Multivar Anal 53:311–331
Fourdrinier D, Wells MT (2012) On improved loss estimation for shrinkage estimators. Stat Sci 27:61–81
Fourdrinier D, Strawderman, Wells MT (1998) On the construction of Bayes minimax estimators. Ann Stat 26:660–671
Fourdrinier D, Strawderman WE, Wells MT (2003) Robust shrinkage estimation for elliptically symmetric distributions with unknown covariance matrix. J Multivar Anal 85:24–39
Fourdrinier D, Marchand E, Strawderman WE (2004) On the inevitability of a paradox in shrinkage estimation for scale mixtures of normals. J Stat Plan Inference 121:37–51
Fourdrinier D, Strawderman, Wells MT (2006) Estimation of a location parameter with restrictions or "vague information" for spherically symmetric distributions. Ann Inst Stat Math 58:73–92
Fourdrinier D, Kortbi O, Strawderman W (2008) Bayes minimax estimators of the mean of a scale mixture of multivariate normal distributions. J Multivar Anal 99(1):74–93
Fourdrinier D, Marchand E, Righi A, Strawderman WE (2011) On improved predictive density estimation with parametric constraints. Electron J Stat 5:172–191
Fourdrinier D, Mezoued F, Strawderman WE (2013) Bayes minimax estimation under power priors of location parameters for a wide class of spherically symmetric distributions. Electron J Stat 7:717–741
Fourdrinier D, Strawderman W, Wells MT (2014) On completeness of the general linear model with spherically symmetric errors. Stat Methodol 20:91–104
Fourdrinier D, Mezoued F, Wells MT (2016) Estimation of the inverse scatter matrix of an elliptically symmetric distribution. J Multivar Anal 143:32–55
George EI (1986a) A formal Bayes multiple shrinkage estimator. Commun Stat Theory Methods 15(7):2099–2114
George EI (1986b) Minimax multiple shrinkage estimation. Ann Stat 14(1):188–205
George EI, Xu X (2010) Bayesian predictive density estimation. In: Chen M-H, Müller P, Sun D, Ye K, Dey DK (eds) Frontiers of statistical decision making and Bayesian analysis in honor of James O Berger. Springer, New York, pp 83–95
George EI, Feng L, Xu X (2006) Improved minimax predictive densities under Kullback-Leibler loss. Ann Stat 34:78–91
Girshick MA, Savage LJ (1951) Bayes and minimax estimates for quadratic loss functions. In: Proceedings of second Berkeley symposium on mathematical statistics and probability, vol 1. University of California Press, Berkeley, pp 53–74

Gomez-Sanchez-Manzano E, Gomez-Villegas M, Marin J (2008) Multivariate exponential power distributions as mixtures of normal distributions with Bayesian applications. Commun Stat Theory Methods 37:972–985

Green EJ, Strawderman WE (1991) A James-Stein type estimator for combining unbiased and possibly biased estimators. J Am Stat Assoc 86:1001–1006

Griffin JE, Brown PJ (2010) Inference with normal-gamma prior distributions in regression problems. Bayesian Anal 5(1):171–188. https://doi.org/10.1214/10-BA507

Grubb G (2009) Distributions and operators. Springer, New York

Haff LR (1979) An identity for the Wishart distribution with applications. J Multivar Anal 9:531–544

Haff LR (1982) Identities for the inverse Wishart distribution with computational results in linear and quadratic discrimination. Sankhya, Ser B 44:245–258

Hartigan JA (2004) Uniform priors on convex sets improve risk. Stat Probab Lett 67:285–288

Hastie TJ, Tibshirani RJ (1990) Generalized additive models. Chapman and Hall, London

Hastie TJ, Tibshirani RJ, Friedman (2008) The elements of statistical learning: data mining, inference and prediction, vol 1, 2nd edn. Springer, New York

Hodges J Jr, Lehmann EL (1950) Some problems in minimax point estimation. Ann Math Stat 21:182–197

Hoerl AE, Kennard RW (1970) Ridge regression: biased estimation for nonorthogonal problems. Technometrics 12:55–67

Hoffmann K (1992) Improved estimation of distribution parameters: Stein-type estimators. Teubner-Verlag, Stuttgart/Leipzig

Hsieh F, Hwang JTG (1993) Admissibility under the frequentist's validity constraint in estimating the loss of the least-squares estimator. J Multivar Anal 44:279–285

Hudson HM (1978) A natural identity for exponential families with applications in multiparameter estimation. Ann Stat 6:473–484

Hunter J (2014) Notes on partial differential equations. Unpublished lecture notes. University of California, Davis

Hurvich CM, Tsai CL (1989) Regression and time series model selection in small samples. Biometrika 76:297–307

Hwang JT, Casella G (1982) Minimax confidence sets for the mean of a multivariate normal distribution. Ann Stat 10:868–881

Isawa M, Moritani Y (1997) A note on the admissibility of the maximum likelihood estimator for a bounded normal mean. Stat Probab Lett 32:99–105

James W, Stein C (1961) Estimation with quadratic loss. In: Proceedings of fourth Berkeley symposium on mathematical statistics and probability. University of California Press, pp 361–379

Johnson N, Kotz S (1972) Distributions in statistics: continuous multivariate distributions. Wiley, New York

Johnstone I (1988) On inadmissibility of some unbiased estimates of loss. In: Gupta SS, Berger J (eds) Statistical decision theory and related topics IV, vol 1. Springer, New York, pp 361–379

Johnstone IM, MacGibbon KB (1992) Minimax estimation of a constrained Poisson vector. Ann Stat 20:807–831

Johnstone IM, Silverman BW (2004) Needles and straw in haystacks: empirical Bayes estimates of possibly sparse sequences. Ann Stat 32(4):1594–1649

Kagan AM, Linnik UV, Rao CR (1973) Characterization problems in mathematical statistics. Wiley, New York

Katz MW (1961) Admissible and minimax estimates of parameters in truncated spaces. Ann Math Stat 32(1):136–142

Kavian O (1993) Introduction à la théorie des points critiques et applications aux problèmes elliptiques. Springer, Paris

Ki F, Tsui KW (1990) Multiple-shrinkage estimators of means in exponential families. Can J Stat/La Revue Canadienne de Statistique 18:31–46

Kiefer J (1957) Invariance, minimax sequential estimation, and continuous time processes. Ann Math Stat 28:573–601

Kiefer J (1975) Conditional confidence approach in multi-decision problems. In: Krishnaiah PR (ed) Multivariate analysis 4. Academic, New York, pp 143–158

Kiefer J (1976) Admissibility of conditional confidence procedures. Ann Stat 4:836–865

Kiefer J (1977) Conditional confidence statements and confidence estimators. J Am Stat Assoc 72:789–827

Komaki F (2001) A shrinkage predictive distribution for multivariate normal observables. Biometrika 88:859–864

Kubokawa T (1991) An approach to improving the James-Stein estimator. J Multivar Anal 36:121–126

Kubokawa T, Srivastava MS (1999) Robust improvement in estimation of a covariance matrix in an elliptically contoured distribution. Ann Stat 27:600–609

Kubokawa T, Srivastava MS (2001) Robust improvement in estimation of a mean matrix in an elliptically contoured distribution. J Multivar Anal 76:138–152

Kubokawa T, Strawderman WE (2007) On minimaxity and admissibility of hierarchical Bayes estimators. J Multivar Anal 98(4):829–851

Kubokawa T, Marchand E, Strawderman WE (2015) On predictive density estimation for location families under integrated squared error loss. J Multivar Anal 142:57–74

Kubokawa T, Marchand E, Strawderman WE (2017) On predictive density estimation for location families under integrated absolute error loss. Bernoulli 23(4B):3197–3212

Kucerovsky D, Marchand E, Payandeh AT, Strawderman WE (2009) On the bayesianity of maximum likelihood estimators of restricted location parameters under absolute value error loss. Stat Decis Int Math J Stoch Methods Model 27:145–168

Kullback S, Leibler RA (1951) On information and sufficiency. Ann Math Stat 22:79–86

Lehmann EL, Casella G (1998) Theory of point estimation, 2nd edn. Springer, New York

Lehmann EL, Sheffé H (1950) Completeness, similar regions and unbiased estimates. Sankhyā 17:305–340

Lele C (1992) Inadmissibility of loss estimators. Stat Decis 10:309–322

Lele C (1993) Admissibility results in loss estimation. Ann Stat 21:378–390

Lepelletier P (2004) Sur les régions de confiance: amélioration, estimation d'un degré de confiance conditionnel. Ph.D. thesis, Université de Rouen, France

Levit BY (1981) On asymptotic minimax estimates of the second order. Theory Probab Its Appl 25(3):552–568

Li KC (1985) From Stein's unbiased risk estimates to the method of generalized cross validation estimation. Ann Stat 13(4):135–1377

Liang F, Barron A (2004) Exact minimax strategies for predictive density estimation, data compression and model selection. IEEE Trans Inf Theory 50:2708–2726

Liese F, Miescke KJ (2008) Statistical decision theory. Springer, New York

Lindley DV (1962) Discussion on professor Stein's paper. J R Stat Soc 24:285–287

Lu KL, Berger JO (1989) Estimation of normal means: frequentist estimation of loss. Ann Stat 17:890–906

Mallows C (1973) Some comments on Cp. Technometrics 15:661–675

Marchand É, Perron F (2001) Improving on the MLE of a bounded normal mean. Ann Stat 29(4):1078–1093

Marchand É, Perron F (2002) On the minimax estimator of a bounded normal mean. Stat Probab Lett 58:327–333

Marchand É, Perron F (2005) Improving on the MLE of a bounded location parameter for spherical distributions. J Multivar Anal 92(2):227–238

Marchand É, Strawderman WE (2004) Estimation in restricted parameter spaces: a review. In: DasGupta A (ed) A Festschrift for Herman Rubin. Lecture notes–monograph series, vol 45. Institute of Mathematical Statistics, Beachwood, pp 21–44

Maruyama Y (1998) A unified and broadened class of admissible minimax estimators of a multivariate normal mean. J Multivar Anal 64:196–205

Maruyama Y (2003a) Admissible minimax estimators of a mean vector of scale mixtures of multivariate normal distributions. J Multivar Anal 84:274–283
Maruyama Y (2003b) A robust generalized Bayes estimator improving on the James-Stein estimator for spherically symmetric distributions. Stat Decis 21:69–77
Maruyama Y (2009) An admissibility proof using an adaptive sequence of smoother proper priors approaching the target improper prior. J Multivar Anal 100(8):1845–1853
Maruyama Y, Strawderman WE (2005) A new class of generalized Bayes minimax ridge regression estimators. Ann Stat 33:1753–1770
Maruyama Y, Strawderman WE (2009) An extended class of minimax generalized Bayes estimators of regression coefficients. J Multivar Anal 100:2155–2166
Maruyama Y, Strawderman WE (2012) Bayesian predictive densities for linear regression models under α-divergence loss: some results and open problems. In: Fourdrinier D, Marchand E, Rukhin AL (eds) Contemporary developments in Bayesian analysis and statistical decision theory: a Festschrift for William E. Strawderman, vol 8. Institute of Mathematical Statistics, Beachwood, pp 42–56
Maruyama Y, Takemura A (2008) Admissibility and minimaxity of generalized Bayes estimators for spherically symmetric family. J Multivar Anal 99:50–73
Muirhead RJ (1982) Aspects of multivariate statistics. Wiley, New York
Murray GD (1977) A note on the estimation of probability density functions. Biometrika 64:150–152
Nachbin L (1965) The Haar integral. D. Van Nostrand Company, Toronto/New York/London
Narayanan R, Wells MT (2015) Improved loss estimation for the LASSO: a variable selection tool. Sankhya B 77:45–74
Neyman J, Pearson E (1933) On the problem of the most efficient tests of statistical inference. Biometrika A 20:175–240
Ng VM (1980) On the estimation of parametric density functions. Biometrika 67:505–506
Peisakoff M (1950) Transformation parameters (unpublished). PhD thesis, Princeton University
Philoche JL (1977) Une condition de validité pour le test F. Statistique et Analyse des Données 1:37–59
Pitman E (1939) The estimation of the location and scale parameters of a continuous population of any given form. Biometrika 30(3/4):391–421
Rao CR (1973) Linear statistical inference and its applications. Wiley, New York
Robert CP (1994) The Bayesian choice: a decision theoretic motivation. Springer, New York
Robert C, Casella G (1994) Improved confidence estimators for the usual multivariate normal confidence set. In: Gupta SS, Berger JO (eds) Statistical decision theory and related topics 5. Springer, New York, pp 351–368
Robertson T, Wright FT, Dykstra RL (1988) Order restricted statistical inference. Wiley, New York
Rudin W (1966) Real and complex analysis. McGraw-Hill, New York
Rukhin AL (1988) Estimated loss and admissible loss estimators. In: Gupta SS, Berger J (eds) Statistical decision theory and related topics 4, vol 1. Springer, New York, pp 409–418
Sacks J (1963) Generalized Bayes solutions in estimation problems. Ann Math Stat 34:751–768
Sandved E (1968) Ancillary statistics and estimation of the loss in estimation problems. Ann Math Stat 39:1756–1758
Schervish M (1997) Theory of statistics, 2nd edn. Springer, New York
Schwartz L (1961) Méthodes Mathématiques pour la Physique. Hermann, Paris
Schwartz L (1973) Théorie des distributions. Hermann, Paris
Sengupta D, Sen PK (1991) Shrinkage estimation in a restricted parameter space. Sankhyā 53:389–411
Shao J (1997) An asymptotic theory for linear model selection. Statistica Sinica 7:221–242
Shao J (2003) Mathematical statistics. Springer, New York
Stein C (1956) Inadmissibility of the usual estimator for the mean of a multivariate normal distribution. In: Proceedings of the third Berkeley symposium on mathematical statistics and probability 1. University of California Press, Berkeley, pp 197–206

Stein C (1962) Confidence sets for the mean of a multivariate normal distribution. J R Stat Soc Ser B 24:265–296

Stein C (1964) Inadmissibility of the usual estimator for the variance of a normal distribution with unknown mean. Ann Inst Stat Math 16:155–160

Stein C (1973) Estimation of the mean of a multivariate normal distribution. In: Proceedings of Prague symposium asymptotic statistics, pp 345–381

Stein C (1977a) Estimating the covariance matrix, unpublished manuscript

Stein C (1977b) Lectures on the theory of estimation of many parameters. In: Ibragimov A, Nikulin MS (eds) Studies in the statistical theory of estimation, vol 74. Proceedings of scientific seminars of the Steklov Institute, Leningrad, pp 4–65

Stein C (1981) Estimation of the mean of multivariate normal distribution. Ann Stat 9:1135–1151

Stoer J, Witzgall C (1970) Convexity and optimization in finite dimensions I, Die Grundlehren der mathematischen Wissenschaften, vol 163. Springer, Berlin

Strawderman WE (1971) Proper Bayes minimax estimators of the multivariate normal mean. Ann Math Stat 42:385–388

Strawderman WE (1973) Proper Bayes minimax estimators of the multivariate normal mean vector for the case of common unknown variances. Ann Stat 1:1189–1194

Strawderman WE (1974a) Minimax estimation of location parameters for certain spherically symmetric distributions. J Multivar Anal 4(3):255–264

Strawderman WE (1974b) Minimax estimation of powers of the variance of a normal population under squared error loss. Ann Stat 2:190–198

Strawderman WE (1992) The James-Stein estimator as an empirical Bayes estimator for an arbitrary location family. In: Bayesian statistics, 4. Oxford University Press, New York, pp 821–824

Strawderman WE (2003) On minimax estimation of a normal mean vector for general quadratic loss. In: Moore M, Froda S, Leger C (eds) Mathematical statistics and applications: Festschrift for constance van Eeden. Lecture notes–monograph series, vol 42. Institute of Mathematical Statistics, Beachwood, pp 3–14

Strawderman RL, Wells MT (2012) On hierarchical prior specifications and penalized likelihood. In: Fourdrinier D, Marchand E, Rukhin AL (eds) Contemporary developments in Bayesian analysis and statistical decision theory: a Festschrift for William E. Strawderman, collections, vol 8. Institute of Mathematical Statistics, Beachwood, pp 154–180

Strawderman RL, Wells MT, Schifano ED (2013) Hierarchical Bayes, maximum a posteriori estimators, and minimax concave penalized likelihood estimation. Electron J Stat 7:973–990

Stroock DW (1990) A concise introduction to the theory of integration. World Scientific, Singapore

Takada Y (1979) Stein's positive part estimator and Bayes estimator. Ann Inst Stat Math 31:177–183

Takemura A (1984) An orthogonally invariant minimax estimator of the covariance matrix of a multivariate normal population. Tsukuba J Math 8:367–376

Tibshirani R (1996) Regression shrinkage and selection via the LASSO. J R Stat Soc Ser B 58(1):267–288

Tibshirani RJ, Taylor J (2012) Degrees of freedom in LASSO problems. Ann Stat 40:1198–1232

Tibshirani R, Saunders M, Rosset S, Zhu J, Knight K (2005) Sparsity and smoothness via the fused LASSO. J R Stat Soc Ser B 67(1):91–108

Titchmarsh EC (1932) Theory of functions. Oxford University Press, London

Tsukuma H, Konno Y (2006) On improved estimation of normal preicision and discriminant coefficients. J Multivar Anal 97:1477–1500

van Eeden C (2006) Restricted parameter space estimation problems: admissibility and minimaxity properties. Lecture notes in statistics, Springer, New York

Wald A (1939) Contributions to the theory of statistical estimation and testing hypotheses. Ann Math Stat 10(4):299–326

Wald A (1950) Statistical decision functions. Wiley, New York

Wan ATK, Zou G (2004) On unbiased and improved loss estimation for the mean of a multivariate normal distribution with unknown variance. J Stat Plann Inference 119:17–22

Watson G (1983) Statistics on spheres. Wiley, New York
Wells MT (1990) The relative efficiency of goodness-of-fit statistics in the simple and composite hypothesis-testing problem. J Am Stat Assoc 85(410):459–463
Wells MT, Zhou G (2008) Generalized Bayes minimax estimators of the mean of multivariate normal distribution with unknown variance. J Multivar Anal 99:2208–2220
Widder DV (1946) The Laplace transform. Princeton University Press, Princeton
Wither C (1991) A class of multiple shrinkage estimators. Ann Inst Stat Math 43:147–156
Ye J (1998) On meausuring and correcting the effects of data mining and model selection. J Am Stat Assoc 93:120–131
Yuan M, Lin Y (2006) Model selection and estimation in regression with grouped variables. J R Stat Soc Ser B 68(1):49–67
Zhang CH (2010) Nearly unbiased variable selection under minimax concave penalty. Ann Stat 38:894–942
Ziemer WP (1989) Weakly differentiable functions. Springer, New York
Zinodiny S, Strawderman WE, Parsian A (2011) Bayes minimax estimation of the multivariate normal mean vector for the of common unknown variance. J Multivar Anal 102(9):1256–1262
Zou H, Hastie T (2005) Regularization and variable selection via the elastic net. J R Stat Soc Ser B 67:301–320
Zou H, Hastie T, Tibshirani R (2007) On the degrees of freedom of the LASSO. Ann Stat 35:2173–2192

Author Index

D. Fourdrinier et al., *Shrinkage Estimation*, Springer Series in Statistics,
https://doi.org/10.1007/978-3-030-02185-6

Subject Index

D. Fourdrinier et al., *Shrinkage Estimation*, Springer Series in Statistics,
https://doi.org/10.1007/978-3-030-02185-6

FSC
www.fsc.org

MIX
Papier aus verantwortungsvollen Quellen
Paper from responsible sources
FSC® C105338

If you have any concerns about our products,
you can contact us on
ProductSafety@springernature.com

In case Publisher is established outside the EU,
the EU authorized representative is:
Springer Nature Customer Service Center GmbH
Europaplatz 3, 69115 Heidelberg, Germany

Printed by Libri Plureos GmbH
in Hamburg, Germany